Improved Agriculture and Allied Practices for Farmers

About the Editors

Dr. J.P.S. Dabas is working as Incharge of Centre for Agricultural Technology Assessment and Transfer (CATAT), aimed at assessment and refinement of technologies while developing different models and approaches. Till today many technologies developed by IARI has been efficiently and successfully implemented. He has a vast experience of more than 30 years in the service. He is well known for conducting strategic and action research as agricultural technologies assessment and gender empowerment through technology interventions, teaching post graduate students, facilitating national training programmes and workshops.

Dr. Ambrish Kumar Sharma, Principal Scientist 'Plant Breeding' in ICAR – Indian Agricultural Research Institute, New Delhi. During his experience of more than three decades, the collection of 57 thousand germplasm of Maize, Wheat, Barley, Triticale and Vegetables were characterized leading to identification of various promising genetic stocks. He has in his credit several research papers published in reputed journals, books and book chapters, technical bulletins and scientific and technical articles.

Dr. Nishi Sharma, Senior Scientist at Centre for Agricultural Technology Assessment and Transfer, Indian Agricultural Research Institute, New Delhi is has specialization in agricultural extension and is serving the agricultural profession for more than three decades. Dr. Sharma has vast experience in the field of extension research, transfer of agricultural technology and capacity building programmes of agricultural officials and progressive farmers. She has 26 research papers, 24 concept papers, four books, six book chapter, 65 training manuals, 35 folders/technical bulletins, several reports, and contributions (75) in national and International Seminars and conference to her credit.

Dr. B.K. Singh, Principal Scientist in Centre for Agricultural Technology Assessment and Transfer (CATAT) having experience of assessment and dissemination of agricultural technologies in peri-urban agriculture system and national extension programme. He has in his credit various researches as assessment of technologies in well established extension approached and system, guided post graduate students, conducted training programmes, seminars, conferences and workshops.

Dr. Pratibha Joshi, Scientist at Centre for Agricultural Technology Assessment and Transfer, ICAR-Indian Agricultural Research Institute, New Delhi with specialization of assessment of gender specific technologies available in NARES system. She is having experience of 8 years in this profession and 15 research papers, four books, 15 training manuals, 25 technical and popular articles besides contribution (20) in national and International Seminars and Conferences.

Improved Agriculture and Allied Practices for Farmers

J.P.S. Dabas
Ambrish K. Sharma
Nishi Sharma
B.K. Singh
Pratibha Joshi

NEW INDIA PUBLISHING AGENCY
New Delhi – 110 034

NEW INDIA PUBLISHING AGENCY
101, Vikas Surya Plaza, CU Block, LSC Market
Pitam Pura, New Delhi 110 034, India
Phone: +91 (11)27 34 17 17 Fax: +91(11) 27 34 16 16
Email: info@nipabooks.com
Web: www.nipabooks.com

Feedback at feedbacks@nipabooks.com

ISBN 978-93-86546-59-3

Composed, Designed & Printed in India

Preface

India has a wide diversity of climate and all variety of crops can be grown in one or the other part of the country. Although agriculture contributes only 21% of India's GDP, its importance in the country's economic, social, and political fabric goes well beyond this indicator. Current focus is on diversification, which is essential for nutritional security and higher return to the farmers. New technologies are needed to push yield frontiers further, utilize inputs more efficiently and diversify to more sustainable and higher value cropping patterns. At the same time there is urgency to exploit potential of rained and other less endowed areas in a better way if we are to meet targets of agricultural growth and poverty alleviation. Given the wide range of agro ecological setting, Indian agriculture is faced with a great diversity of needs, opportunities and prospects. Future growth needs to be targeted.

A lot can be achieved by continued emphasis on Human Resource Development to boost the technological skill of farmers and entrepreneurs for bringing transformation in growth of agriculture from homestead activity to the status of industry. Information and knowledge about the new technological breakthroughs in meeting the future demands in agriculture through expansion of existing area and production, improving productivity through Hi-tech interventions like adopting high yielding varieties and hybrids of cereals, vegetables, fruits and flowers high density planting, integrated nutrient and pest management strategies, protected cultivation, value-addition and processing, export promotion, improved methods of care of household animals, management of fodder and fuel, etc is the need of hour.

The book at hand, therefore attempts to highlight the dimensions of improved agricultural and allied technologies for farmers. The first part of the book covers the crop production technologies of Rice, Mustard, Lentil, Sugarcane besides IFS model and SRI techniques. The Segment of Horticultural Production Technology highlights the Protected Cultivation, Vegetable, Fruit, Floriculture and Diversification for nutritional security and prosperity. The Crop Protection unit includes information on Weed Management, Nematode Management and Integrated Pest Management. The Natural Resource Management components cover the topics on Soil Testing, Biofertilizer, Water Management and Biogas. The Post harvest and value addition highlights the grain and seed protection during storage, post harvest management of fruits and vegetables, value addition of agricultural produce and packing house operations of horticultural crops. Allied activities include information on apiculture, mushroom cultivation, and agricultural mechanization. The animal husbandry section compiles the information on dairy farming, fisheries based enterprises, scientific lac cultivation, pig farming and goat rearing. We hope this compilation of the book will be helpful to the users as first hand information for reference in the future.

Editors

Contents

Section 3: Crop Protection

Section 4: Natural Resource Management

Section 5: Post Harvest and Value Addition

Section 6: Agricultural and Allied Activities

Section 7: Animal Husbandry and Fisheries

Section 1: Crop Production Technology

Improved Agriculture and Allied Practices for Farmers, pp 1-11
New India Publishing Agency, New Delhi, India

1

Integrated Farming Systems for Enhancing Productivity

***Anil K. Choudhary*[1] *and A.K. Vyas*[2]**

[1]*Division of Agronomy, ICAR-Indian Agricultural Research Institute, New Delhi-110012*
[2]*Krishi Anusandhan Bhawan-II, ICAR, Pusa Campus, New Delhi-110012*

Introduction

Agriculture occupies a prominent place in Indian economy and rural livelihoods. But, it is faced with the challenge of producing food for a rapidly growing population whilst maintaining the nation's fragile resources.Efforts through green revolution brought out significant increase in food grain production by concentrating on the potential farms in terms of irrigation and other basic resources.Nationally and globally, it is necessary that the growth rate of food production must remain higher than the population growth rate to avoid hunger and overcome famines. But, food security one of the basic human rights is in jeopardy, today. Agriculture is crucial for ensuring the food and livelihood security of the country and hence it is important that this sector with diversified farming systems becomes resilient to increasing climatic variabilities and global economic changes. This concept encourages the farmers towards efficient utilization of natural resources and also proper channelization of by-products of one enterprise to the other enterprise in farming. The integrated farming systems, therefore, assume prime place for sound management of farm resources to enhance the farm productivity, reduce the environmental degradation, and improve the quality of life of resource poor farmers and to maintain the sustainability to meet the challenges of food security. In order to meet the soaring demand for food there is a need to go for integrated faming system.This approach facilitates in strengthening the socio-economic status of Indian farmers besides strengthening ecological balance. Integrated farming system is common to Indian farmers simply because of the preponderance of small farm systems which form the backbone of traditional agriculture. For small farmers, these systems enable a means of diversifying the use of meagre resources in the context of a rational means of reducing risk. Additionally, it also enables increased efficiency in the use of these same resources in a manner that there will be livelihood security and stability to farming systems. There is no scope for increasing the farm size. Yet steady increase in population with shrinkage of cultivated land as a result of industrialization and urbanization would pose serious hurdle for agricultural development. Only vertical expansion is possible by integrating appropriate farming components requiring lesser space and time ensuring continuous income to the farmers.

Need of Integrated Farming Systems

Rapidly increasing population, shrinking arable land resources for crop production and increasing concern for declining soil fertility and environmental degradation, spell out an urgent need for diversification of farming systems for enhancing and sustaining productivity of land through farm based integrated production systems. It is widely accepted as a means of achieving sustainable agriculture.Farming system approach is a multi-disciplinary whole-farm approach for solving the problems of small and marginal farmers besides management of natural and human resource at farm.The need of the hour is, therefore, to modern agriculture with integrated farming systems to enhance farm productivity and profitability. Globalization and climate change are posing serious negative impacts on agricultural systems. These systems and the natural resources that support them are already under severe strain from over-exploitation, the current issues further multiply other stresses. Integrated farming system is the integration or combination of different enterprises such as crop production, horticulture, agro-forestry, livestock (dairy, goatary, piggery, poultry, duckery etc.), bee-keeping, mushroom, aquaculture, sericulture and value addition with the main goal of increasing income and standard of living of small and marginal farmers besides poverty reduction, food security, competitiveness and sustainability. An integrated farming system represents multiple crops (cereals, legumes, tree crops, vegetables etc.) and multiple enterprises (animal farming, bee keeping, fish farming etc) in a single farm. Agriculture is the mainstay of rural masses in India where majority of rural population is primarily dependent on agriculture and allied activities for their livelihood and are threatened by the problems of unemployment and food security. Low land and labour productivity in agriculture are the real challenges resulting in widening gap between agriculture and non-agriculture sectors (Choudhary *et al.*, 2013).

Integrated Farming Systems: A Ray of Hope to Small and Marginal Farmers

This approach aims at enhancing farm income and employment from small-holdings by integrating various farm enterprises and recycling crop residues and by-products within the farm itself. Under shrinking arable land, it integrates farm enterprises within the agro-socio-economic environment of the farmers to bring profitability and better livelihoods to farmers while allocating land, labour, capital and management to different enterprises. Integrated Farming meets these potentially conflicting challenges at farm level in a manner that balances food production, profitability, safety, animal welfare, social responsibility and environmental care. Integrated farming seeks to reinforce the positive influences of agricultural production whilst reducing its negative impacts. The integrated farming system (IFS) approach introduces a change in the farming techniques for maximum production in the cropping patterns and takes care of optimal utilization of resources. The farm wastes are better recycled for productive purposes in the integrated system. A judicious mix of agricultural enterprises like dairy, poultry, piggery, fishery, sericulture etc. suited to the given agro-climatic conditions and socio-economic status of the farmers would bring prosperity in the farming. In IFS, livestock and crops are produced within a coordinated framework. The waste products of one component serve as a resource for the other. For example, manure is used to enhance crop production; crop residues and by-products feed the animals, supplementing often inadequate feed supplies, thus contributing to improved animal nutrition and productivity. The result of this cyclic combination is mixed farming system, which exists in many forms and represents the largest category of livestock systems in terms of animal numbers, productivity and number of people it serves. A typical integrated crop-livestock farming system is depicted in Fig. 1 (Choudhary and Thakur, 2011).

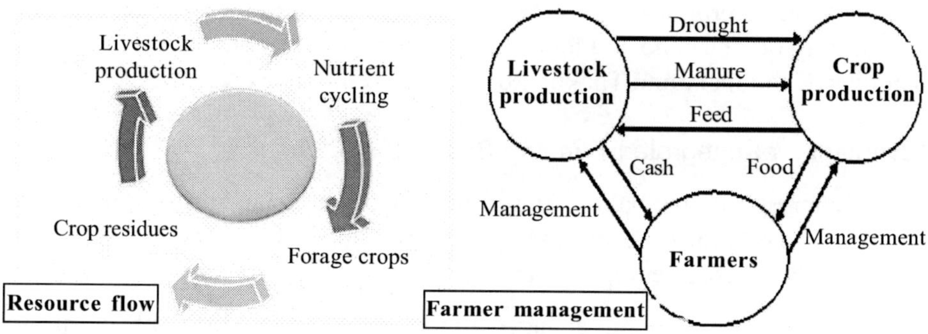

Fig. 1: Integrated crop-livestock farming system as integral part of farming systems

Integrated Farming System Approach: A Key to Household Food & Nutritional Security

The Indian economy is predominantly rural and agricultural, and the declining trend in size of land holding poses a serious challenge to the sustainability and profitability of farming. In view of the decline in per capita availability of land from 0.5 ha in 1950-51 to 0.15 ha by the turn of the century and a projected further decline to less than 0.1 ha by 2020, it is imperative to develop strategies and agricultural technologies that enable adequate employment and income generation, especially for small and marginal farmers who constitute more than 80% of the farming community. An integrated farming system represents multiple crops (cereals, legumes, tree crops, vegetables etc.) and multiple enterprises (animal farming, bee keeping, fish farming etc.) in a single farm. Integration of allied activities will result in the availability of nutritious food enriched with protein, carbohydrate, fat, minerals and vitamins.Integrated farming will help in environmental protection through effective recycling of waste from animal activities like piggery, poultry and pigeon rearing. Animals play key and multiple roles in the functioning of the farm, and not only because they provide livestock products (meat, milk, eggs, wool, and hides) or can be converted into prompt cash in times of need. It is widely accepted as a means of achieving sustainable agriculture. Farming system approach is a multi-disciplinary whole-farm approach for solving the problems of small and marginal farmers besides management of natural and human resource at farm. This approach aims at enhancing farm income and employment from small-holdings by integrating various farm enterprises and recycling crop residues and by-products within the farm itself. This approach aims at enhancing farm income and employment from small-holdings by integrating various farm enterprises within the farm itself. Thus, adoption of integrated farming system approach imbedded with combinations of crop, livestock, horticulture, fishery, sericulture and other feasible farm enterprises have great potential to augment the land and labour productivity on one hand and food and nutritional security to the rural poor on the other (Choudhary and Thakur, 2011).

Advantages of Integrated Farming Systems

- Better use of land, labour and capital: Better area land through adoption of crop rotations, steady employment of farm and family labour and more profitable use of equipment are obtained in diversified farming.
- The farmer and labour engaged all the year round in different activities. Less risk to crop failure and market price of the product.
- Regular and quicker return is obtained from various enterprises.
- Soil erosion can be checked as land kept under cultivated throughout the year.
- Diversified farming is less risky than specialized farming.

- Productivity, sustainability, recycling, balanced food, environmental safety.
- Profitability, income round the year, employment generation.
- Saving energy, meeting fodder crisis, agro-industries.
- Increasing input efficiency, increasing the standard of living of the farmers.

There are many components of integrated farming systems which need to be popularized to need based and location specific situations as discussed here under:

A. Cropping systems – As integral part of integrated farming system

In recent past, cropping system approach has gained importance in Indian agriculture to enhance farm productivity and profitability as a whole. Cropping system is an important component of a farming system and it represents the cropping pattern used on a farm and their interactions with farm resources other farm enterprises and available technology which determine their make-up. Cropping pattern means the proportion of area under various crops at a point of time in a unit area. If cropping pattern indicates the yearly sequence and spatial arrangement of crops and fallow on a given area, the crop rotation refers to recurrent succession of crops on the same piece of land either in a year or over a long period of time. Component crops are so chosen so that soil health is not impaired with an objective to get maximum profit from least investment without impairing the soil fertility. Thus, identification and popularization of diversified cropping systems is the need of the hour to increase agricultural production and ensure food security and rural livelihoods.In agriculture, management practices are usually formulated for individual crops. However, farmers are cultivating different crops in different seasons based on their adaptability to a particular season, domestic needs and farm profitability. In order to address above issues, NARS system in India comprising ICAR and SAUs' have developed many viable alternative options for diversification of crop based especially cereal based cropping system in varying socio-agro-economic regimes of India to enhance input-use-efficiency, productivity, farm profitability and sustainability under changing climate and production vulnerabilities in holistic manner. Some of the remunerative diversified crops and cropping system alternatives for major agro-ecological regions with inclusion of cereals and vegetable crops are presented in Table 1 & 2.

I. Crop based cropping systems

Cropping system approach has gained importance in Indian agriculture in recent past to enhance farm productivity and profitability as a whole. Crop based farming systems have been identified as profitable interventions in whole country. Better technology adoption and inclusion of HYV of crops like cereals, pulses and oilseeds etc. enhance the production of field crop based systems (Tables 1 & 2).

Table 1: Remunerative cereal based cropping systems for north Indian plains

Crops/Cropping System	Average yield (t/ha)	Sowing	Harvesting	Net returns (Rs/ha)
I. DOUBLE CROPPING SYSTEMS				
Rice-Wheat	5.0-6.0 4.5-5.0	June-July November	October April	90,000-1,10,000
Maize-Wheat	4.0-4.5 4.5-5.0	June November	October April	90,000-95,000
Pigeonpea-Wheat	1.8-2.0 4.0-4.5	June December	Nov./Dec. April	1,05,000-1,20,000
Rice-Chickpea	5.0-6.0 1.2-1.5	June-July October	October April	75,000-85,000
Hybrid Rice-Wheat	6.0-7.0 4.5-5.0	June November	October April	1,20,000-1,40,000
II. TRIPLE CROPPING SYSTEMS				
Maize Potato Wheat	4.0-4.5 18-20 3.5-4.0	Mid June September End Dec.	September December April	1,30,000-1,50,000
Maize Potato Moong	4.5-5.0 25-30 0.8-1.0	July October March	October February June	1,15,000-1,40,000
Rice Wheat Moong	5.0-6.0 4.5-5.0 0.8-1.0	June-July November April	October April June	1,20,000-1,40,000

(*Source:* IARI Publication, 2013)

Table 2: Vegetable and food grain based suitable cropping systems for north-western plains

Cropping System	Average yield (t/ha)	Sowing	Harvesting	Net return (Rs/ha)
Rice	5.5-6.0	June	September	1,45,000-1,75,000
Carrot	25-30	September	December	
Wheat	3.0-3.5	December	April	
Rice	5.5-6.0	June	September	1,37,000-1,70,000
Cauliflower	15-20	September	December	
Wheat	3.0-3.5	December	April	
Okra	13.0-14.0	June	September	1,35,000-1,62,000
Radish	15.0-20.0	September	November	
Wheat	4.0-4.5	November	April	

(*Source:* IARI Publication, 2013)

II. Vegetable based cropping systems

Vegetable based farming systems have been identified as profitable interventions in whole country. The technology involves growing vegetables alongwith other farm enterprises. Inclusion of HYV of vegetables like cabbage, cauliflower, chilli, tomato,French-bean, pumpkin, bottle-gourd, okra, brinzal, potato etc. enhance the production far above the other field crop based systems (Tables 3 & 4).

Table 3: Remunerative vegetable based suitable cropping systems for north Indian plains

Crops/Cropping System	Average yield (t/ha)	Sowing	Harvesting	Net returns (Rs/ha)
Brinzal	30-35	May-June	Sept.-Oct.	1,30,000-1,50,000
Fenugreek	6.0-7.0	October-Nov.	January	
Bottle-gourd	18-20	February-March	April-June	
Spinach	15-20	Sept.-Oct.	Dec.-January	1,15,000-1,25,000
Onion	15-18	Dec.-Jan.	May-June	
Okra	8.0-10	June-July	Aug.-Sept.	
Radish	20-22	Sept.-Oct.	Dec.-January	1,45,000-1,55,000
Tomato	30-32	Dec.-Feb.	May-June	
Bottle gourd	18-20	June-July	Aug.-Sept.	

(*Source:* IARI Publication, 2013)

Table 4: Vegetable based cropping systems for different agro-ecological situations

Zone	Cropping System	Average Yield (t/ha)	Sowing	Harvesting	Net Returns (Rs/ha)
North Indian Plains	Early Cauliflower	10-12	July	October	2,80,000-
	Peas	7-84	October	January	3,20,000
	Tomato	0-45	January	June	
	Okra	12-14	June	September	2,13,000-
	Carrot	25-30	October	December	2,57,000
	Cauliflower	25-30	December	March	
	Radish	15-18	April	May	
	Cucumber	12-13	July	September	2,60,000-
	Potato	16-18	October	December	3,00,000
	Onion	22.5-25	January	June	
North-Eastern plains	Cauliflower	10-11	June	August	3,00,000-
	Peas	8-10	September	November	3,50,000
	Radish	17-20	December	January	
	Capsicum	15-16	January	May	
	Green chilli	11-12	June	September	2,12,000-
	Broccoli	11-12	October	December	2,41,000
	Radish	15-16	January	February	
	Bottle-gourd	30-32.5	February	May	

(*Source:* IARI Publication, 2013)

B. Horticultural crops based farming systems

Different agro-ecological/phyto-geographical regions hold promises for cultivation of different horticultural crops. In view of their importance, they can be inducted in diversified horticultural crops based farming systems for sustainable food and nutritional security; to enhance input-use-efficiency, productivity, farm profitability and sustainability in holistic manner (Table 5).

Table 5 Fruit and food grain based cropping systems

Area	Cropping System	Average Yield (t/ha)	Net Returns (Rs./ha)
Northern Plain Area	Papaya (October transplanted)	37-45	1,33,000-1,67,000
	Maize	2.5-3.0	
	Wheat	1.8-2.5	
Eastern Plain Area	Banana +	22-26	1,55,000-2,00,000
	Elephant foot yam+	12-14.5	
	Turmeric	8.5-12	
North-East and South Mid Hilly areas	Banana +	12-15	96,000-1,21,000
	Pomegranate +	14-18.5	
	Ginger	7.5-10.5	

(*Source:* IARI Publication, 2013)

C. Livestock, fishery, duckery and piggery based farming systems

The livestock farming systems are generally based on traditional socio-economic considerations. The productivity of livestock in farming systems can be improved by increased fodder production as an intercrop with cereals, relay and alley cropping, forage production on bunds, improving the feeding value for enhanced utilization, improvement in productivity of grasses quantum and distribution decides the effective growing season and it becomes critical in selecting cropping systems for a given reason. Fish based farming system has been identified as profitable interventions in whole north-eastern region. This IFS module consists of poultry-fish-vegetable, duck-fish-vegetable, and pig-fish-vegetable cultivation. The technology involves growing vegetables around the fishery pond and rearing of poultry and pig and ruminants above or nearby the fishery pond. Poultry houses have been constructed just above the fish pond so that waste product of poultry would be fed to the fingerlings. Inclusion of HYV of vegetables enhance the production far above the previously grown local cultivars. Fingerlings of Rohu, Katla, Mrigal, Silver Carp, Grass carp etc. are best options for pisciculture among the stakeholders. Cross bred pig and poultry chick and duck are integrated with fish farming at homestead. Bio-wastes from piggery, poultry are recycled into the fishpond to substitute fertilizers and fish feed and also to use in the vegetables. These IFS modules consist of poultry-fish-vegetable, duck-fish-vegetable, and pig-fish-vegetable cultivation (Table 6). The technology involves growing vegetables around the fishery pond and rearing of poultry and pig and ruminants above or nearby the fishery pond. Poultry houses have been constructed just above the fish pond so that waste product of poultry would be fed to the fingerlings. Inclusion of vegetables enhances the farm production. Cross bred pig and poultry chick and duck are integrated with fish farming at homestead. Bio-wastes from piggery, poultry are recycled into the fishpond to substitute fertilizers and fish feed and also to use in the vegetables.

Table 6: Profitability of viable integrated farming system models for north-eastern Himalayas (1 ha size)

Integrated farming system models (1 ha size)	Total input (Rs.)	Net profit (Rs.)	B:C ratio (including labour component)	B:C ratio (excluding labour component)
Poultry-crop-fish-duck-horticulture-hedge-row	1,05,722	1,67,331	1.58	2.24
Crop-fish-poultry-MPTs	60,137	90,625	1.51	2.12
Crop-fish-goat-bamboo-MPTs	59,442	91880	1.55	2.40
Crop-fish-pig-hedge-row-MPTs-vermiculture	77,243	1,09,887	1.42	1.85
Crop-fish-dairy-mushroom-liquid manure-broom-horticulture-vermiculture	1,70,120	2,98,735	1.76	2.38
Crop and fishery (without integration)	31,773	34,894	1.09	1.50

(*Source:* http://www.kiran.nic.in)

D. Entrepreneurship based farming systems

In rural areas, livelihood and employment generation is a tedious challenge in current economic scenario. Thus, entrepreneurship based farming systems hold great promises besides fulfilling the objective of diversification of farming systems. Sharma *et al.*(2007) in a study on IFS in Jammu region observed that among component , cropping system as a whole recorded maximum net return, followed by dairy, poultry, and mushroom. Further, it was indicated that integration of all the components yielded higher return and employment opportunity than cropping alone (Table 7).

Table 7: Economics and employment generated in IFS

System	Gross return (Rs/ha)	Expenditure (Rs/ha)	Net income (Rs/ha)	Employment (M.D./ha/yr
Crop+ dairy+ Poultry+Mushroom	203044	84682	118362	154
Crop cultivation only	101742	26202	75540	64
Additional benefit	101302	58480	42822	90

Conclusion

With technological and infrastructural development, numerous new farming systems have emerged to boost the agrarian economy in India. The integration of various enterprises on land holdings of various sizes tend to be more profitable than arable farming alone. Therefore, there is a need to spread IFS models in different agro-ecological regions to promote and widespread adoption of appropriate diversified integrated farming systems for higher profitability. The resultant higher productivity and profitability under IFS models through various crops and farm enterprises to fetch better livelihoods strongly advocate to follow IFS principles and technology so as to transform less remunerative production systems into highly remunerative systems using available farm resources to generate better farm gains, climate resilience, livelihoods and employment on sustainable basis.

References

Choudhary, A.K.and Thakur, S.K. 2011. Development of integrated farming system model for marginal and small farmers of Mandi district of Himachal Pradesh: An innovative extension tool. In: *International conference on innovative approaches for agricultural knowledge management: Global extension experiences* held w.e.f. 9-12 Nov., 2011 at NAAS, New Delhi. *INSEE International Conference Abstracts*, p 554-555.

Choudhary, A.K., Thakur, S.K. and Suri, V.K. 2013. Technology transfer model on INM technology for sustainable crop production in high value cash crops and vegetables in NW Himalayas. *Communications in Soil Science and Plant Analysis* **44** (11): 1684-1699.

http://www.kiran.nic.in. *KIRAN (Knowledge Innovation Repository of Agriculture in the North-East)* Empowering Agricultural Knowledge and Innovation in North-East. An ICAR Initiative.

IARI Publication, 2013. Labhdayak phasal pranaliayan – Samasayaen avam samadhan. pp 28.

Sharma, A.K., Koushal, S. and Kachroo, D. 2007. Integrated farming systems for higher productivity.Extended summaries 3rd National Symposium on Integrated Farming System held at ARS, Durgapura (Jaipur), 26-28 October, 2007, Pp.49-50.

Improved Agriculture and Allied Practices for Farmers, pp 13-31
New India Publishing Agency, New Delhi, India

2

Scientific Cultivation of Rice

Dinesh Kumar and Shweta Mehrotra

Division of Agronomy, ICAR-Indian Agricultural Research Institute New Delhi-110012

Rice is the most important cereal crop of India occupying about 44.0 million hectares land. It provides food and employment to millions of people in the country. Despite development of high yielding varieties and technological interventions the yield of rice is still low (3.20 tonnes/ha) in India. The reasons for low yield of rice are:

- Lesser area under high yielding varieties
- Limited availability of quality seed
- Increased acidity/ sodicity in some parts of the country
- Declined soil fertility
- Unavailability of inputs, like fertilizers and seed, in time
- Inefficient nutrient management including imbalanced use of nutrients
- Improper crop rotations
- Reduced availability of irrigation water
- Inefficient management of weeds, diseases and insect-pests

Considering the significance of rice in Indian diet and food security, its productivity must be enhanced. To enhance productivity, proper and efficienct agronomic practices must be followed as discussed in the subsequent sections.

Rice is a staple food for about two-third population of the world. Milled rice usually contains 6-7% protein.The fat content in rice is low (2.0-2.5%) and much of fat is lost during milling. The by-products of rice milling are used for various purposes. Rice bran is used as cattle and poultry feed, and oil extraction. Rice bran (pericarp, seed coat, nucellus, and aleurone layer) accounts for about 8% of the paddy weight. It contains 15-20% oil (15% is recoverable). Of the total oil in the grain, about 75% is present in the bran itself. Global rice bran oil production varies from 1-1.4 million tonnes, of which India accounts for 0.7-0.9 m tonnes. Rice hulls can be utilized in manufacturing of insulation materials, cement, card board and as litter in poultry keeping. Rice straw can be used as cattle feed as well as litter during the winter season.

Rice ecosystems

A very broad based classification of rice environments or ecosystems used in India includcs following:

i) Irrigated rice ecosystem: Rice is grown under irrigated conditions in the states of Punjab, Haryana, Uttar Pradesh, Jammu & Kashmir, Andhra Pradesh, Tamil Nadu, Sikkim, Karnataka, Himachal Pradesh and Gujarat. Irrigated rice is grown in bunded (embanked) paddy fields. The area under irrigated rice accounts for approx 50% of the total area under rice cultivation in the country.

ii) Rainfed upland rice ecosystem: Upland rice areas lies in eastern zone comprising of Assam, Bihar, Eastern M.P., Orissa, Eastern U.P., West Bengal and North-Eastern Hill region. Upland rice fields are generally dry, unbunded, and directly seeded. Land utilized in upland rice production can be low lying, drought-prone, rolling, or steep sloping.

iii) Rainfed lowland rice ecosystem: Rainfed lowland farmers are typically challenged by poor soil quality, drought/flood conditions, and erratic yields. Production is variable because of the lack of technology used in rice production. The low land rice area accounts approx. 32% of the total area under rice cultivation in the country.

iv) Flood prone rice ecosystem: Flood-prone ecosystems are characterized by periods of extreme flooding and drought. Yields are low and variable. Flooding occurs during the wet season from June to November, and rice varieties are chosen for their level of tolerance to submersion.

Climatic requirements

Rice cultivation in India extends from 8° to 35° N latitude and from sea level to as high as 3,000 m elevation. Rice accommodates itself under an annual rainfall ranging from 1,000 to 1,500 mm or even more. High temperature, high humidity and high rainfall has considerable effect on growth and development of rice plant. At higher mean temperatures (25-32°C), early flowering is induced by reducing the growth duration; whereas the mean low temperature (<15°C) forces the crop to remain dormant with no flowering. Therefore, a day temperature of 25° to 33°C and night temperature of 15° to 20°C is preferable.

Depending on the temperature requirement, rice crop is grown during *kharif* season in north-western plains zone. However, in south and north-eastern parts of the country, it is grown in all the 3 seasons, as these areas do not have very cold weather during winter. Rice crop needs a hot and humid climate. It is essentially a short day C_3 plant.

Soil requirements

The wide range of agroclimatic conditions suggest an equally wide variety of soils. As regards texture, rice is grown on loamy sands in Punjab to heavy clay loams or clays in Andhra Pradesh and some other states. Whatever may be the soil type, one important characteristic of rice soils is that they remain submerged atleast for some time during the growth period of the crop. The submergence leads to gradual depletion of oxygen in the soil and this causes reduction of a number of ions such as nitrates, sulphates, iron, manganese etc. Soils having good water holding capacity with good amount of clay and organic matter are ideal for puddled rice cultivation.

Recommended varieties

For assured areas of irrigation and drainage, a semi-dwarf plant (90-110 cm) with sturdy stem, high and synchronous tillering ability, relatively erect dark green leaves, and high grain number seems to perform better. In rainfed low land situations, plants with a height of 120-140 cm with sturdy stem perform better. Quick growing, early maturing, relatively taller plant performs well under upland conditions.A good number of varieties/ hybrids combining resistance to biotic and abiotic stress and grain quality have been developed (Tables 1a, 1b, 1c, 1d).

Table 1(a): Rice Hybrids recommended for different states

State	Hybrid
Andhra Pradesh	APHR 1, APHR 2, DRRH 1, PHB 71, HRI 120, DRRH-3
Bihar	PHB 71, DRRH 1, PA 6201,
Haryana	Haryana Shankar Dhan-1 (HKRH-1), PHB 71, DRRH 1, PA 6201, Pusa RH 10,
Gujarat	DRRH-3
Karnataka	KRH1, KRH 2, HRI 120
Madhya Pradesh & Chattisgarh	JRH-6, JRH-5, DRRH-3
Maharashtra	Sahyadri-2, PHB 71, PA 6201, HRI 120
Orissa	CRHR-7 (Ajay), KRH 2, DRRH 1, PHB 71, HRI 120, Rajalaxmi

Punjab	Pusa RH 10, RH 204
Tamil Nadu	CORH 1, CORH 2, ADTRH 1, PHB 71, DRRH 1, DRH 2, HRI 120
Uttaranchal	Pant Sankar Dhan 1, PHB 71, HRI 120, Pant Sankar Dhan 6, Pant Sugandh Dhan-17
Uttar Pradesh	Pant Sankar Dhan 1, Pusa RH 10, Narendra Sankar Dhan 2 , PHB 71, PA 6201, HRI 120, UPRH 27
West Bengal	CNRH 3, PA 6201, KRH 2, PHB 71, 6201 (CPA 103)

Table 1(b) Basmati/scented varieties of rice recommended for cultivation in different states zones/states

Andhra Pradesh	Sumati, Kaki Rekhalu (HR 59), Sukhdas (HR 47)
Assam	Prasadbhog, Malbhog, Badshahbhog
Bihar & Jharkhand	Basmati 370, Badshahbhog, Mohanbhog, Ram Tulsi
Gujarat	Kamod 118, Pankhari 203, Zeerasal
Himachal Pradesh	Desi Basmati 23
Jammu & Kashmir	Ranbir Basmati, Muskh Budgi, Basmati 370
Karnataka	Jeerakasala, Gandhkasala, Sindigi local
Madhya Pradesh	Gopalbhog, Kali Kamod, Chattri
Maharashtra	Ambemoha 102, Ambemoha 147, Ambemoha 159, Krishnasal
Northern India (Haryana, Punjab, Uttar Pradesh)	Basmati-386, N 10 B, Basmati 370, Pusa Basmati 1, Hansraj, T 3, T 9, N 12, Taraori Basmati (Karnal local), Pusa Sugandh 2, Pusa Sugandh 3, Pusa Sugandh 4, Pusa Sugandh 5, Pusa Sugandh 6, Pusa Improved Basmati, PRH 10, Pant Dhan 15, Kalanamak, Punjab Basmati-1
Rajasthan	Mahasugandha, Kala Badal, Kamod, Nawabi Kolam
Tamil Nadu	Jeeraga Samba
Uttarakhand	Pant Dhan 15, Hansraj
West Bengal	Badshah Pasand, Govindabhog, Kataribhog, Sitabhog

Table 1(c) Rice varieties recommended for various states of India under irrigated conditions

State	Improved varieties
Andhra Pradesh	Amara, ASD-19, Cottondora Sannalu, Hamsa, Tella-Hamsa, Vashista, Mahsuri, Kasturi, Kakatiya, Prakash, Vikrant, Bhadra Kali, Jagtiala Mahsuri, Jagitiala Sannalu, Santhi, Apurva, Mahamaya, Nandyal Sannalu, Varalu, Bapatla Sannalu, Early Sambha, GK 5003
Bihar & Jharkhand	Pusa 33, Pusa 2-21, Pusa 169, Mehsuri, Saket 4, IR-8, IR-20, IR-24, BR 368, Jaya, Padma, Deepa, Sita, Bala, Kaveri, Ratna, Sabarmati, Janki, Rajendra Dhan 101, R. Dhan 202, Birsamati, Arvindha Amrit, JKRH-401
Gujarat	Gurjari, GR-6, Gau R 1, Gau R 10, Gau R 100, Dandi
Haryana	Pusa 33, Pusa 169, Ajay, Jaya, Bala, Ratna, Jhona 349, IR 8, PR 106, Kasturi, HKR-127

Himachal Pradesh	Bhrigu Dhan , Himalaya 2216, RP 2421, Himalaya-1, Himalaya-2, Himalaya-22, Himalaya-79
Jammu & Kashmir	SKAU 23, SKAU 27, Chena, Jhalum, Kohsaar
Karnataka	IR-30864, Prakash, Madhu, Vikram, Suma, Kusuma, Vani, Akash, Sharavati, Maha Maya, BR 2655-9-3-1, Pant Dhan 12, Sakuntala, Gautam, GK 5003
Kerala	KAUM 20-19-4, KAUM59-29-2-1-2, KAUM61-6-1-1-2, Anupurna, Aswathi, Bharti, Jyoti, Rohini, Triveni, Sabari, Kunju, Priya, Varsha, Kunju Varna, Gauri, Sweta
Madhya Pradesh & Chattisgarh	Prabhat, IR 8, IR 24, Jawahar 75, Mehsuri, Bala, Kaveri, Ratna, Jaya, Pragati, Jagrati, Kranti, Garima, Anupama, Akashi, Madhuri, Vani, Mahamaya
Maharashtra	Ratnagiri 24, Karjat 184, Karjat 2, Ratnagiri 3, PKV Makarand
Orissa	Rajeshwari, Hema, Sakti, Kumar, Bala, IR 20, Prakash, Vikram, Vani, JKRH-401
Punjab	Sahyadri-4, Jaya, Bala, Ajai, PR 105, PR 108, PR 109, IR 8, Palman 579, Mutant 95, Ratna, Sabarmati, Aravinda Amrit, PRH-122
Rajasthan	Jaya, Ratna, Pusa 33, Ajay, Sabarmati, Chambal PR 106, BK 29, BK 130, Vani
Tamil Nadu	COR-46, ADT (R) 45, ADT-42, ADT-43, ADT-44, ASD-20 Kari Kalan, Karuna, Bhawani, Vaghair, Bharti, Caveri, Krishna, Kanchi, PMK 2 and ADT 42
Uttar Pradesh	*Early duration varieties for plains:* Saket 4, Govind, Prasad, Manhar, Pusa 169, Pusa 33, Pusa 2-21, Ratna, Narendra 2 *Early duration varieties for Bundelkhand and Jhansi:* PNR 381, Saket 4, Prasad, Manhar, Pusa 169, Ratna, Ashwani *Early duration varieties for Hilly areas:* Ratna, VLK 35 *Medium duration varieties for plains:* Pant Dhan 4, Pant Dhan 10, Pant Dhan 11, IR 24, IR 36, Sarju 52, Jaya, NDR 80, Pant Dhan 12 *Medium duration varieties for Bundelkhand and Jhansi:* Pant Dhan 4, Jaya *Medium duration varieties for plains of Hilly areas:* VL 8, VL Dhan 16, VLK Dhan 39, P Dhan 6 *Long duration varieties for plains:* Mehsuri, Cross 116, T 100 *Long duration varieties for Hilly areas:* VL Dhan 221, VL 206
West Bengal and Assam	IR 20, Jayanthi, JKRH-401

Table 1(d) Promising rice varieties for rainfed/water scarce areas in different states

State	Varieties
Andhra Pradesh	Tulasi, Aditya, Prasanna, Rudramma, Krishna Hamsa, MTU 9993, Ravi, Varalu, Bapatla Sannalu, Vagad Dhan, MTU 17, Mettasanna, Tellavadlu, Anjali, Dhanrasi
Assam	Lum Pnah-1, Dumai, ARC 5929, Annada, Rasi, Tulasi, Anjali
Bihar & Jharkhand	Anjali, Tulasi, Birsa Dhan-106, Birsa Dhan-107, Birsa Dhan-108, Aditya, Rasi, PNR 381, Vandana, Birsa Dhan 101, BR 17,

	BR 16, Birsa Dhan-102, Birsa Dhan-103, Birsa Dhan-104, Birsa Dhan-105, Birsa Dhan-6, Birsa Dhan- 7, Birsa Dhan-201, Birsa Dhan-202, Birsa Dhan-221, Heera, Kanchan, Kalinga III, Vaidehi
Gujarat	GR 2, GR 3, GR 5, GR 6, Rasi, Prasanna, Tulasi, GR 8, Sathi 34-36, Early Kolam 161-172
Himachal Pradesh	Shankar Dhan 1, Lalna Kanda 41
Jammu & Kashmir	Ch 988, Ch 1039
Karnataka	Amrut, Mukti, IET 7564, Tulasi, Ravi, Krishna Hamsa, Avinash, Holomaldgiga, Mugad 161, Dhanrasi
Kerala	KAUA 4-4-2 ,Suvarna Modan, Onam, Harsha, PTB 10, PTB 28, PTB 29, PTB 30, Annapurna
Madhya Pradesh & Chattisgarh	Tulasi, Aditya, Heera, Purva, JR 75, Rathe, Annada, Jawahar Rice 3-45, Safed Dhan-3, Laloo 14
Maharashtra	Tulasi, Aditya, Heera, Annada, Tuljapur 4, Ratnagiri 73-1-41, Rasi, Ratnagiri 73-1, Kranti, Jalgaon 5, Nagpur 22, Ambemohar 159, Karjat 3, Dhanrasi
Meghalaya	Rasi, Heera, Ngoba, Krishna Hamsa, Tulasi, Lam Pnah-1, Bha Lum 1, Bha Lum 2, Shah Sarang 1
Orissa	Anjali, CR 683-123, Heera, Kalinga III, Annada, Parijat, Pathara, Kalyani II, Sattari, Neela, Rudra, Vanaprabha, Khandagiri, Nilgiri, Ghanteswari, Sneha, Badami, B 76, Kalakeri, BAM 12
Punjab	Lalnakanda
Tamil Nadu	ADT(R) – 48, IR 64, IR 50, TKM-9, Co 41, PMK 1, ASD 18, ADT 36, ADT 37, PKU 2, Tulasi, Krishna Hamsa, Aditya, ADT (R) 45, PMK(R)3, Dhanrasi, MDU-5
Tripura	Anjali, Tulasi, Heera, Rasi, Annada
Uttar Pradesh	Narendra Dhan 97, Narendra Dhan 18, Narendra Dhan 80, Narendra Dhan 118, Saket 14, Tulasi, Aditya, Jawahar Rice 3-45, Sudha, Bala, Kaveri, Pusa 2-21, Narmada, Saket 4, N-22, ND 2, Narendra Dhan 1, Govind, Jhona 349
Uttarakhand	NDR 84, NDR 118, VL Dhan 263, VL Dhan 221, VL Dhan 206, Shankar Dhan-1
West Bengal	Rasi, Khitish, Kiran, Bhupen, Khanika, Dular
For coastal and salt affected soils of West Bengal, Orissa, Kerala, Andhra Pradesh	Sumati, CSR 23, VTL-8, Narendra Usar Dhan-2008

Methods of cultivation/ sowing

Rice is generally grown under dry or wet cultivation methods, which are briefly described below:

Dry cultivation

Direct seeded rice: Dry system of rice cultivation is followed in uplands. Upland rice is grown on both level and slopy fields that are not bunded and are prepared and seeded dry, and the crop depends upon rainfall for its water requirement. Sowing of rice is usually done in May-June in case of the crop dependent on south-west monsoon and in September for the crop dependent on north-east monsoon.Three methods are commonly followed in sowing dry and semi-dry crop. These are broadcasting, drilling or sowing in furrows behind country plough, and dibbling in general, a seed rate of 30-50 kg ha^{-1} is required for drilling, while 60-100 kg ha^{-1} is required for broadcasting. A row spacing of 15-20 cm is optimum for upland rice.

Aerobic rice: "Aerobic rice" is a new concept of growing rice: it is high-yielding rice grown in non-puddled, aerobic soils under irrigation and high external inputs. To make aerobic rice successful, new varieties and management practices must be developed. Irrigation is applied when the soil becomes too dry, and the quantity of applied water is sufficient to bring the soil to field capacity. Aerobic rice should be responsive to high inputs (water, nutrients) to reach high yields under non-flooded conditions. The realization of water savings combined with high yields, however, depends on good water management.

Dry seeding of rice can be done by drilling the seed into a fine seedbed at a depth of 2–3 centimetres. Weed management is a critical factor in aerobic rice. Timely application of herbicides and two or three hand weedings may provide effective control. Seeding under dry situations is done in three different ways - viz. (i) drilling i.e. sowing in the furrow behind a plough, (ii) dibbling and (iii) broadcasting. But drilling method is ideal one. The light soils which generally come into conditions quickly, any method can be adopted. Seeding with drilling method has got a greater advantage over other methods, because of the uniformity of the stand and the control of the population of the plants per unit area. Heavy soils which do not come in conditions quickly, other methods except broadcasting are not feasible. It has been found that drilling or dibbling always gives considerably better yields than broadcasting system. It eliminates the use of seedlings and related operations such as seedling nursery preparation, care of seedlings, pulling, bundling, transporting, and transplanting.

Wet cultivation

Conventional rice cultivation: The conventional rice cultivation is a wet system of cultivation and rice is grown under wet season right from the start. In this system, the field is brought to a soil puddle by repeated ploughing with 5-7 cm standing water. After getting the requisite puddle, rice seedlings are tansplanted or sprouted seeds are direct seeded. The seedlings of rice are grown in nursery before transplanting. There are three methods of raising rice seedlings, namely, dry seed bed, wet seed bed and dapog method.

a) Dry Nursery: In this method the field is ploughed repeatedly to a fine tilth and raised beds of 4 to 5 cm height, 1 to 1.5 m width and of convenient length are prepared keeping 30 cm wide channel in between. A seed rate of 2.5-3 kg/100 m^2 nursery is appropriate.

b) Wet Nursery: It is the best method for obtaining well grown seedlings within a short period. For wet seed-bed nursery, land is first ploughed under dry or moist conditions and then puddled by ploughing. After this, beds are prepared as discussed for dry nursery and sprouted seeds are either broadcasted or sown in rows. Seed rate is the same as for dry nursery. For getting sprouted seeds, the seeds are soaked in ordinary water for 24 hours, incubated for 36 to 48 hours in a warm shady place and are never allowed to dry up by occasional moistening of seeds.

c) Dapog Method: This method of seedling raising is prevalent in Philippines and has been introduced in India. The essential feature of this method is to have a very thick nursery seeding without any contact with the soil. Raised beds are prepared similar to wet nursery. The beds are covered with banana leaves (mid-rib removed) or empty cement/fertilizer bags or polythene sheets to prevent the roots of seedlings from coming in contact with soil. A wall of 2 to 3 inches high is made of strips of banana sheets on the four sides of the bed to keep the seeds in place after they are sown. Pre-germinated seeds at the rate of 3 kg/m^2 are sown uniformly on the beds. Water is splashed on the germinating seeds and the seeds are pressed down slightly by hand or light wooden plank in the morning and afternoon for 3 to 6 days. This prevents exposure of roots to the air and helps germinated seeds to maintain touch with each other and with the ground. For 3 to 4 days the seedlings are kept moist by sprinkling water. Seedlings become ready for transplanting within 12 to 14 days, regardless of duration of the variety. Fertilizer is not needed since the seedlings are nourished by food in the seed.

Time of planting is the most important factor influencing the yield of the crop. In general, transplanting in the first fortnight of July gives best yield in *kharif.* Trials carried out in Assam indicated that planting of tall indica rice in the last week of December was ideal for high yield in *boro* crop. For timely transplanting in wet season, 20-25 days old seedlings are ideal. In general,t 20 cm x 10 cm, 20 cm x 15 cm or 15 cm x 15 cm spacings are ideal in transplanted rice. For realizing the yield potential of high yielding varieties of rice, 2-3 seedlings per hill is generally recommended. In rice hybrids where the tillering is more profuse, 1-2 seedlings per hill is sufficient. Depth of transplanting rice seedlings is 2-4 cm.

System of rice intensification (SRI)

The System of Rice Intensification synthesized in Madagascar by Fr. Henri de Laulanie′, S.J., has been gaining acceptance around the world because of the remarkable increases in factor productivity that its methods are making possible. Irrigated rice yields are being raised by 50–100%, and often more, without requiring the use of purchased inputs and without changing varieties. The higher yields with SRI practices are achieved with any and all rice varieties while using less water and with little use of mineral fertilizers.Because the resulting plant phenotypes are more resistant to damage by pests and diseases, there is usually little or no need for chemical crop protection. All this has been quite unexpected. While the mechanisms that make these benefits attainable are not yet fully understood, it appears that SRI effects are the result of greater root growth and of increases in soil biological activity that are induced by different cultural practices. Following are the components of SRI method of rice culture:

- Transplanting young seedlings, usually 8 to 12 days old and certainly less than 15 days old.
- Planting single seedlings rather than several seedlings together in a hill.
- Wide space planting (spacing of 25×25 cm or wider).
- Planting quickly and carefully to avoid trauma to seedling roots.
- Daily or intermittent irrigation before panicle initiation (PI) after which a thin layer of water (1-2 cm) is maintained. The fields are kept moist and not flooded.
- Hand or mechanical weeding, starting 10 days after transplanting and repeated 2-4 times at 10-12 d intervals until the canopy closes.
- Applying nutrients to soil preferably in organic form such as compost or mulch.

The comparison in production of rice by SRI and conventional rice culture (CRC) is given in Table 2.

Table 2: Differences in practices of SRI method and conventionally-grown rice

SRI method	Conventional Rice Culture (CRC)
Very young seedlings, 8–12 days old, are transplanted, at most 15 days old, i.e., before start of fourth phyllochron	*Mature seedlings,* transplanted at 3–4 weeks of age
Single seedlings are transplanted, very shallow (1–2 cm) into a muddy, unflooded field after uprooting them gently from a garden-like, unflooded nursery	*3–4 or more seedlings* per hill are transplanted in clumps by plunging them deeply into a flooded (hypoxic) soil environment
Sparse plant populations, widely spaced in a square pattern at least 25 x 25 cm, with a seeding rate of 5–10 kg ha^{-1}	*Large plant populations* are established in rows, with a seeding rate of 30–50 kg ha^{-1}
Soil aeration is maintained during the vegetative growth period, avoiding continuous soil saturation; after panicle initiation, shallow flooding (1–2 cm); alternate wetting and drying throughout cycle is preferable in some soils	*Soil saturation* with paddies kept flooded throughout the growing cycle
Weeds are controlled with a rotary hoe that aerates the soil as it eliminates weeds; weeding 3–4 times before canopy closes	*Weeds* are controlled by flooding and also by hand weeding and/or herbicides
Compost is recommended, as much as possible to build up soil organic matter, to support largerpopulations of soil biota	*Chemical fertilizer* is applied, providing up to 100–150 kg ha^{-1} N; soil organic matter is optional; little/ no attention to soil biota

Certainly there are some limitations on the use and adoption of SRI. The most objective one is the need of having good water control to get best results. Water savings can only be made- and soil can only be kept aerated enough to support the growth and functioning of aerobes in the soil - if farmers can apply limited amounts of water rather than keep their paddies continuously flooded. Many farmers do not have such water control as they operate in field-to-field (cascade) systems of distribution. Initially, SRI methods require more labour, as they need to be learned and mastered. In fact, this requirement is reduced in subsequent seasons, and SRI labour demand eventually becomes less than with conventional cultivation.

Balanced Fertilization and Integrated Nutrient Management

Balanced fertilization

It is well documented that unbalanced availability of nutrients can lead to mining of soil reserves for nutrients in short supply and to losses of plant nutrients supplied in excess. The long-term fertilizer experiments (LTFEs) conducted in

India have shown that soil productivity can be maintained by balanced nutrition of crop. Unbalanced plant nutrition has also resulted in micronutrient deficiencies in India. Zinc deficiency is the most widely occurring micronutrient deficiency problem limiting crop growth in many countries. Adequate supply of all the essential plant nutrients is a must for getting good yield of rice. This is particularly true for high yielding varieties and hybrids. In India, and for that matter in the entire Asia, fertilizer N, mostly urea has received more attention than phosphorus and potassium fertilizers but balanced fertilization is the key to sustain rice production.

Nitrogen

A dose 100-150 kg N/ha is generally recommended for high yielding varieties of rice. Hybrids where yield potential is 8-10 t/ha will need still higher doses of N (180-200 kg N/ha). Significant proportion of N is lost by denitrification, ammonia volatilization and leaching etc. in transplanted rice. Hence, there is a need to increase the nitrogen use efficiency (NUE) in rice. NUE can be increased by split application, deep placement and use of nitrification inhibitors and slow-release N fertilizers.Two split doses are recommended for short and medium duration varieties, while 3 split doses are recommended for long duration varieties. Nitrification inhibitors can also be used to reduce N losses. The most widely tested nitrification inhibitors are: N-Serve (2-chloro 6(trichloromethyl) pyridine), AM (2-amino-4-chloro-6 methyl pyrimidine), DCD (dicyandiamide) and ST (sulphathiazole). Similarly, neem oil coated urea (500-1000 ppm) enhances grain yield and nitrogen use efficiency in rice. This product is now available in the Indian market.

Phosphorus

A dose of 60 kg P_2O_5/ ha is generally recommended for rice. P application is generally recommended at final puddling before transplanting/final ploughing or at broadcasting in the case of direct seeded rice, however, in the case of non-availability of fertilizer in time, it can be top-dressed within 30 days of transplanting and at the most may be extended to 40 days on medium soils. Dipping of roots of rice seedlings in P-slurry has also been found reasonably good.

Potassium

A dose of 40 kg K_2O/ ha is generally recommended for rice. Response of kharif rice is generally higher than rabi rice. This could partly be due to the fact that disease incidence in rice is more during *kharif* season. Potassium application reduces the severity of diseases in rice.There are mainly two potassium fertilizers namely, muriate of potash (MOP) (KCl) containing 60% K_2O and sulphate of potash (SOP) (K_2SO_4) containing 48% K_2O. Due to high costs, mostly MOP is used for rice K fertilization. Potassium application is mostly recommended at the time of sowing/transplanting.

Sulphur

Substitution of high analysis fertilizers such as urea and DAP for ammonium sulphate and single superphosphate has been the main cause for wide spread S deficiency in Indian soils. In general, rice responds to 20-40 kg/ha. There are 2 kinds of sources of S. The first kind are the popular fertilizers containing S as the nutrient other than that for which fertilizer is applied. For example, single superphosphate (SSP) is marketed as phosphate fertilizer but contains 12% S. The other kind of sources of S are those materials which are marketed and used only as a source of S. These are referred to as S-fertilizers and include elemental S, gypsum (it is mostly used for reclaiming sodic soils), iron pyrites and phosphogypsum (a byproduct of phosphoric acid industry). S in pyrites is present as sulphide and must be oxidised before it becomes available to crop plants. Oxidation of pyrites is faster if it is applied on the surface rather than its incorporation in soil.

Micronutrients

Deficiency of 3 micronutrients (zinc, iron, boron) has been reported in rice-growing areas of India. Zinc sulphate (21% Zn) @ 20-25 kg/ha is generally recommended in zinc-deficient soils. Soil application of Zn is generally made at sowing/transplanting of rice. When Zn deficiency is noted in crop, one 0.5% Zn-sulphate spray is generally recommended. However, on Zn deficient soils, basal application is preferred to foliar application, because in most cases considerable damage is already done before a deficiency is detected. Dipping rice roots in 2-4% ZnO suspension has also been found to be very effective. Iron sulphate and FYM are the good sources of iron. FYM contains 1400-1500 ppm Fe, 115-220 ppm Mn, 8-10 ppm Cu, 42-247 ppm Zn and 17-40 ppm B. Fe chlorosis in rice is better controlled by foliar spray of 0.5% Fe-sulphate. To supply boron, borax and sodium tetraborate are used for soil application, while boric acid is used for foliar spray. Soil application of B should be made when deficiency of B has been established. Indiscriminate use may result in toxicity of plants. When foliar application is must, a 0.2% borax solution (with lime twice the weight of borax) is recommended to avoid injury to foliage.

Integrated nutrient management

The primary goal of integrated nutrient management (INM) is to combine old and new methods of nutrient management into ecologically sound and economically viable farming systems that utilize available organic and inorganic sources of nutrients in a judicious and efficient way. INM optimizes all aspects of nutrient cycling. It attempts to achieve tight nutrient cycling with synchrony between nutrient demand by the crop and nutrient release in the soil, while minimizing losses through leaching, runoff, volatilization and immobilization. A two-fold strategy may be useful in nutrient management. First, to continue to encourage the use of fertilizer for a faster growth in agriculture and second, to popularize the use of recycled crop residues, green manures, use of compost (both urban and rural) and biofertilizers as a supplementary source of nutrients, in conjunction with chemical plant nutrients. There is a need to add plant nutrients to the soil on the basis of soil fertility status. The application of nutrients according to soil-test-based-fertility-assessment ensures better efficiency and improved profitability.

Water Management

Rice is the largest consumer of water; about 3000- 5000 litres water is required to produce 1 kg rice. In general, water requirement from transplanting to harvest is between 800 and 1200 ha-mm, which amounts to a daily consumption of 6-10 ha-mm. In addition to the actual crop needs, a large amount of water is lost from rice fields. The principal water loss processes from paddy fields are: 1. Run-off, 2. percolation, 3. seepage and 4. evapotranspiration or consumptive water use. Lack of adequate moisture at these stages reduces the yield even upto the extent of 50 per cent. Continuous flooding at a static 2-5 cm depth provides the opportunity to produce potential yields. The submergence (2-5 cm) throughout the crop growth period is conducive to higher yields. Under shallow water table conditions, the practice of intermittent submergence i.e. submergence during the critical stages of initial tillering and flowering and maintenance of saturation or field capacity during the rest of the growth gave yields comparable to those obtained under continuous shallow submergence.

Weed management

Weeds compete with rice and take away a heavy toll of energy, water and plant nutrients. The following weed management practices are suitable for rice.

Manual weeding: Handweeding is the most widely used method for controlling weeds in rice. Normally, two weedings are done. Time of weeding slightly varies in different regions but weedings are generally done within 15 to 45 days after sowing.

Mechanical methoed: Weeding with some implements such as hand hoes, weeders and bullock drawn desi poughs can be done to control weeds in rice.

Chemical method: Use of herbicides is gradually increasing in rice culture. Herbicides are expensive to small farmers but not to the large farmers who face the problem of labour shortage. A list of herbicides to control the weeds is given in Table 3.

Table 3: List of herbicides suitable for application in rice under different situations

Rice	Herbicides	Dose (kg/ha)	Application time and remarks
Rice Nursery	Butachlor	1.0-1.5	Pre-emergence at 5-6 days after sowing (DAS); If moisture is less in soil, irrigation should follow immediately.
	Benthiocarb	"	— do —
	Pendimethalin	"	— do —
	Pretilachlor(S)	0.3-0.4	Pre-emergence at 3-5 DAS.
Direct-Seeded Upland Rice	Butachlor	1.0-1.5	To be applied before emergence of crop; One hand weeding at 30-35 DAS will supplement herbicide treatment.
	Benthiocarb	"	— do —
	Pendimethalin	"	— do —
	Pretilachlor(S)	0.3-0.4	Pre-emergence at 3-5 DAS.
	Oxadiazon	0.5-0.75	Pre- or early post-emergence for controlling annual grasses, sedges and some broad-leaved weeds.
	Metsulfuron-methyl	0.010-0.015	Post-emergence at 30-35 DAS; basically a broad-leaved weed killer and recommended as a substitute of 2,4-D.
	Bentazon	1.0-2.0	Post-emergence at 30-35 DAS; basically a broad-leaved weed killer, but controls *Cyperus esculentus*, a problematic sedge; could be a substitute of 2,4-D.
Direct-Seeded Puddled and Trans-planted Rice	Butachlor	1.0-1.5	Pre-emergence at 3-5 days after transplanting (DAT) on saturated soil; No irrigation or standing water impounded for at least 3 days after treatment.
	Benthiocarb	"	— do —
	Pendimethalin	"	— do —
	Pretilachlor(S)	0.3-0.4	Pre-emergence at 3-5 DAT

Anilofos	0.4	— do —
Metsulfuron-methyl	0.010-0.015	Post-emergence at 30-35 DAS; basically a broad-leaved weed killer and recommended as a substitute of 2,4-D.
Bentazon	1.0-2.0	At 30-35 DAT where broad-leaved weeds and sedges particularly *Cyperus esculentus* pose problem; could be a substitute of 2,4-D.
2,4-D	0.5	To be applied at 30-35 DAT
2,4-D	0.75-1.0	To be applied at 35-40 DAS/DAT, where broad-leaved weeds and sedges dominant. Field is to be drained before application & re-flooded after 2-3 days.
Oxadiazon	0.5-0.75	Pre- or early post-emergence; soil surface should not be disturbed after its application.

Disease management

A number of diseases infect rice crop and some time the effects could be devastating. The best way to overcome disease problem is to use disease free seed. For fungal and bacterial diseases, seed should be treated with bavistin @ 2 g kg^{-1} seed. For nematodes, the seed treatment involves hot water treatment at 55°C for 10-15 minutes. When a disease is noticed in the field, infected plants should be uprooted and burned. Fungicides can be sprayed to control a disease. The details of disease management are given in Table 4.

Table 4: Symptoms and integrated management of important diseases of rice in India

Disease name and (causal organism)	Symptoms	Integrated management
Leaf and neck blast (*Pyricularia grisea:* fungal disease)	Rice seedlings or plants at the tillering stage are often completely killed when leaves are severely infected. Initial symptoms on the leaves are white to grayish green circular lesions or spots with dark green borders. Later on, lesions on the leaves may enlarge and coalesce to kill the entire leaves. Lesions on the neck are grayish brown and causes the girdling of the neck and the panicle to	Cultural practices include early sowing of seeds after the onset of the rainy season and balanced use of fertilizers. Soils known to be low in plant-available silicon should be amended with calcium silicate slag. Planting resistant varieties against the rice blast is the most practical and economical way. Systemic fungicides, such as probenazole, tricyclazole, fthalide, pyroquilon and tricyclazole, are effective against the disease.

	fall over. If infection of the neck occurs before milk stage, no grain is formed, but if infection occurs later, grains of poor quality are formed.	
Bacterial Leaf Blight (*Xanthomonas oryzae* pv. *Oryzae)*	Symptoms include water-soaked to yellowish stripes on leaf blades or starting at leaf tips. Appearance of bacterial ooze that looks like a milky or opaque dewdrop on young lesions early in the morning. Severely infected leaves tend to dry quickly. Seedling wilt or kresek is observed 1-3 weeks after transplanting, in which green water-soaked layer along the cut portion or leaf tip develops, leaves wilt and roll up and become grayish green to yellow, finally entire plant wilt completely.	Preventive measures include field sanitation such as removing weed hosts, rice straws, ratoons, and volunteer seedlings.The use of resistant varieties is the most effective and the most common management practice. Seed treatment with bleaching powder (100μg/ml) and zinc sulfate (2%) reduce bacterial blight. Control of the disease with copper compounds, antibiotics and other chemicals has not proven highly effective.
Bakanae (*Gibberella fujikuroi* : fungal disease)	It is primarily a seedborne disease. Sowing ungerminated seeds in infested soil gives rise to infected seedlings. Infected plants are several inches taller than normal plants. Reduced tillering and drying leaves at late infection. Partially filled grains, sterile, or empty grains for surviving plant at maturity.	Clean seeds should be used. Seed treatment using fungicides such as thiram, thiophanate-methyl, or benomyl is effective. Triflumizole, propiconazole and prochloraz were found to be effective against strains that are resistant to benomyl and combination of thiram and benomyl.
False Smut (*Ustilaginoidea virens*: fungal disease)	Infection at flowering results in destruction of ovary, but the style, stigmas, and anther lobes remain intact and are buried in the spore mass. Infection may also occur when the grain is already mature. The spores accumulate in the glumes. The whole grain is replaced with a mass of spores (Mature spores are orange and turn yellowish green or greenish black).	Resistant or tolerant varieties should be grown. Destruction of straw and stubble from infected plants. Application of captan, captafol, fentin hydroxide, and mancozeb can inhibit conidial germination. At tillering and preflowering stages, spraying of carbendazim fungicide and copper base fungicide can effectively control the disease.

Bacterial Leaf Streak (*Xanthomonas oryzae* pv. *Oryzicola)*	Planting of infected seeds produce diseased seedlings. Initially, small, dark-green and water-soaked streaks on interveins from tillering to booting stage occur. Streaks are dark-green at first and later enlarge to become yellowish gray and translucent. Lesions turn brown to grayish-white then dry when disease is severe.	The disease can be controlled by proper application of fertilizers and proper plant spacing, the use of resistant varieties, and hot water treated seeds. Practicing field sanitation is important. Fallow field and allowing to dry thoroughly is also recommended
Sheath Blight (*Rhizoctonia solani*:fungal disease)	The disease is soilborne. The disease usually infects the plant at late tillering or early internode elongation stages. Initial lesions are water-soaked to greenish gray and later becomes grayish white with brown margin. Lesions may coalesce to form bigger lesions, and eventually death of the whole leaf.	Seeding rate or plant spacing should be optimized to avoid closer plant spacing. Need-based N application. Removal of weeds and infected crop residues from the field. Spraying of fungicides, such as benomyl and iprodione can be done.
Khaira disease (physiological disorder)	It is caused by the zinc deficiency. Zn deficiency can also increase spikelet sterility in rice. Midribs, particularly near the leaf base of younger leaves, become chlorotic and leaves lose turgor and turn brown as brown blotches and streaks appear on lower leaves, enlarge, and coalesce. A white line sometimes appears along the leaf midrib.	Soil application of zinc sulphate heptahydrate @ 20-25 kg/ha before sowing/ transplanting. Apply 0.5–1.5 kg Zn ha^{-1} as a foliar spray (e.g., a 0.5% $ZnSO_4$ solution at about 200 L water ha^{-1}) for emergency treatment of Zn deficiency in growing plants.

Insect- pests Management

Most of the rice pests are distributed throughout India. However, only a few pests are economically important in different regions. Agronomic practices such as summer ploughing, field sanitation, fertility management, water management, weed management and sowing and harvesting time with slight modifications can be exploited for reducing pest incidence without increasing any extra expenditure. Similarly, mechanical destruction of pests and infested plants or their parts should also be carried out. The preference should be given to the use of resistant cultivars wherever available. The use of selective pesticides including biopesticides should be stressed as they do not disrupt non-target organisms and environment. Pesticides and their doses recommended for control of a few pests are given in Table 5.

Table 5: Nature of damage and integrated management of major insect-pests of rice

Insect-pest	Nature of damage	Integrated management
Stem borer (yellow)	Symptoms of stem borer damage are deadhearts and whiteheads. Plants that have started bearing panicles when they were attacked usually produce whiteheads. Whiteheads are discolored panicles with empty or partially filled grains. Larvae feed on the tissues around the node. As a result, the infested stems are weakened and easily break. Occasionally, stem borers eat through the node and bore down the stem of stem borers that occur	Varietal resistance, cultural practices, insecticides, and biological control should be used compatibly. Adopt seedling root dip treatment in 0.05% Chlorpyriphos emulsion for one minute before transplanting in endemic areas. Apply Carbofuran 3G @ 20 kg ha^{-1} or Phorate 10G @ 12.5 kg ha^{-1} or spray Malathion 50EC @ 0.1% or Quinalphos 25EC @ 0.04% or Fenitrothion 50EC @ 0.1%. There are many natural enemies naturally in rice fields.
Brown plant hopper (BPH), and White backed plant hoppers (WBPH)	Plant hoppers are tiny brown-gray insects (0.1 - 0.4 cm long). High populations of plant hoppers cause leaves to initially turn orange-yellow before becoming brown and drying. This condition, called hopperburn, kills the plant. BPH can also transmit ragged stunt and grassy stunt diseases.	Outbreaks result when pesticides destroy natural enemies. Monitoring on a daily basis is necessary. Apply Carbofuran 3G @ 20 kg ha^{-1} or Phorate 10G @ 12.5 kg ha^{-1} or spray carbaryl 50WP 0.15% or monocrotophos 36WSC 0.04% or chlorpyriphos 20EC 0.05%. Direct the spray towards the base of the plants.
Green leaf hopper (GLH)	Green leaf hoppers (GLH) are the most common leaf hoppers in rice fields and are primarily important because they spread the viral disease tungro. Both nymphs and adults feed by extracting plant sap with their needle-shaped mouthparts. attack the GLH eggs, while	Use of GLH-resistant and tungro-resistant varieties. Nitrogen should be applied as needed. Good weed control in the field and on the bunds removes the preferred grassy hosts. Crop rotation with a non-rice crop. Various parasitoids predators and pathogens can also control GLH. Insecticides mentioned to control brown plant hoppers can be used.
Rice bug (Rice gundhi bug)	Adults are about 15 mm long and 3 mm wide; have long legs; and a brown-green body. An infested field can be recognized by the rice bugs' offensive odour. Bugs	Removal the weeds from fields and surrounding areas. Capturing rice bugs, in the early morning or late afternoon, by net can be effective at low rice bug

	damage rice by sucking out the contents of developing grains, from pre-flowering spikelets to soft dough stage. Immature (nymphs) and adult rice bugs both feed on rice grains.	densities. Biological control measures includes use of some wasps, grasshoppers and spiders that attack rice bugs or rice bug eggs. Direct control may be required if there population is high. Dust Carbaryl or Malathion @ 25-30 kg ha^{-1}or spray Malathion 50EC @ 0.1% or Fenitrothion 50EC @0.1% in early morning or late evening.
Armyworms or Earhead cutting caterpillar	Armyworm larvae feed on rice at any stage of the plant. The caterpillars feed on rice leaves and can also cut off young seedlings at the plant or panicle base. Panicles are cut off from the base.	Cultural measures removal of weeds outside and inside of the field. Avoid killing natural enemies (e.g., wasps and spiders) of armyworms. Flooding seedbeds is the best defense against armyworms. Insecticides should be the last resort for armyworm control. Use Carbaryl 5% dust @ 20 kg ha^{-1}
Gall midge	Infested tillers are transformed into tubular galls, resembling the leaf of onion (onion leaf). Infested tillers fail to produce panicles.	If there are 5% or more infested tillers per square metre, apply Carbofuran 3G @ 20 kg ha^{-1} or Phorate 10G @ 12.5 kg ha^{-1} or spray Malathion 50EC @ 0.1% or Quinalphos 25EC @ 0.04%

Harvesting and threshing

Different rice varieties mature in 100-150 days. The proper stage for harvesting is when about 80% spikelets in 80% panicles show ripening. The plant should be cut close to the ground and left in the field for a few days to dry. Later on, harvested produce should be collected in bundles and kept for threshing. Threshing can be accomplished by manual methods, pedal threshers or power driven stationary threshers. Combine machines can be employed for combined harvesting and threshing. The produce should be properly dried in the sun before storage. The optimum moisture content for storage of rice grains is 12%.

Yield

A well managed crop of mid-late duration (135-150 days) varieties like IR 8, IR 20, Jaya, etc. and hybrids yield about 6.0-7.0 tonnes grain/ha, whereas short duration cultivars yield about 4.5-5.5 tonnes grain/ha.

Improved Agriculture and Allied Practices for Farmers, pp 33-42
New India Publishing Agency, New Delhi, India

3

Improved Varieties of Rapeseed-Mustard Released for Different Agro-climatic Zones of India

Naveen Singh, Rajendra Singh, Yashpal, Sujata Vasudev Navinder Saini, V.V. Singh* and D.K. Yadava

Division of Genetics, ICAR-IARI, New Delhi-110012
**ICAR-DRMR, Bharatpur, Rajasthan 321303*

India ranks third after China and Canada in the world in production of edible oilseed crops. Average yield of rapeseed-mustard in the world is 2043 kg/ha. Average yield of European Union, Canada and China is higher than India. Higher yield in these countries is due to cultivation of long duration varieties suitable for their local climatic conditions. Duration of favorable climatic condition to grow rapeseed-mustard is shorter in India as compared to EU, Canada and China, which enforces cultivation of short duration varieties (130-145days). Rapeseed-mustard contribute one third in the production of edible oils in India. These crops ranks second in area and production after soybean. Total area and production under these crops is 5.80mh and 6.28mt respectively. It is necessary to mention that total water requirement of these crop is 80-240mm, which is very low as compared to other crops. These crops pay higher return by utilizing low inputs, thus, are highly remunerative for the farmers.

Rapeseed-mustard group of crops are mainly cultivated in irrigated areas where at least one irrigation facility is available and only 24.0% of area is under rain-fed situation. Five states of India namely, Rajasthan, Uttar Pradesh, Haryana, Madhya Pradesh and West Bengal contribute 82.00% and 85.67% in total area and production, respectively, under these crops. In Madhya Pradesh 53.2% of total area is irrigated with average yield of 1006 kg/ha, however, 76.6% of total area of country is irrigated even then average yield is only 1083 kg/ha. The lower average seed yield can be attributed to (i) cultivation of very short duration and low yielding *B. rapa* genotypes in Eastern and North-Eastern states; (ii) susceptibility of varieties to different pests; and (iii) high temperature stress at seedling and reproductive stage. Out of total area under these crops, Indian mustard (*Brassica juncea*) is grown over more than 80% area. Cultivation of

Indian mustard has also been taken up in non-traditional areas of southern states of India viz., Karnataka, Tamilnadu and Andhra Pradesh, recently. This has become possible due to its higher average yield, tolerance against different biotic and abiotic stresses, availability of varieties with 100-152 days maturity, and non-shattering behaviour.

TMOP, was started in 1985/86 and it has played a vital role in increasing area and production of rapeseed-mustard crops. Afterwards yield and production increased significantly by continuous efforts made by different research institutions and state agricultural universities. There is a huge diversity in the soil and climates in different states of India. Keeping this fact in view, rapeseed-mustard growing regions of the country have been divided into five zones. It is concluded from research experiments that average yield of rapeseed-mustard can be increased significantly by adopting suitable varieties and their matching production and protection technologies. Improved varieties of Rapeseed-mustard are being developed and released according to the soil-climatic conditions of different zones, but optimum yield per unit area from improved varieties could not be achieved due to following reasons.

1. Cultivation of old varieties despite availability of new improved varieties.
2. Cultivation of varieties other than notified for that region
3. Sowing of improved varieties either before or after the recommended time
4. Cultivation of varieties not suitable for the cropping system
5. Non availability of improved quality seed.
6. Weeds, diseases and insect-pests are not adequately managed.
7. Non-availability of good quality irrigation water.

Most of above listed factors can be controlled only by selection of varieties and following matching/recommended package of practices. To realize the potential yield, the varieties should be cultivated only in/at the recommended area, time, cropping system and other conditions. Varieties of Indian mustard can be divided in three groups according to sowing time, specific adaptation and cropping pattern. Details of varieties released and notified by AICRPR&M since 2004-05 are listed below:

1. Early sown varieties

These varieties are suitable for cultivation in the month of September. These varieties takes about 100-120 days to mature and can be harvested from the last week of December to second week of January. Farmers can grow late sown wheat, onion, garlic, early crop of sugarcane and summer vegetables after harvesting of the crop.

Variety	Developed by	Maturity duration (days)	Oil content (%)	Average yield (kg/ha)	Recommended area	Major characteristics
Pusa Mahak (JD-6)	IARI, N. Delhi	81-114	39-44	1049	Orissa, W. Bengal, Bihar, Jharkhand, Chhattisgarh and Assam	For early sowing in rainfed areas
Pusa Tarak (EJ 9912-13)	IARI, N. Delhi	100-120	38-42	1852-1996	Delhi & NCR	Early sowing in September
Pusa Mustard-25 (NPJ-112)	IARI, N. Delhi	94-120	36-41	1324-1654	Delhi, Haryana, J.& K., Punjab, Rajasthan &W. UP	Suitable for early sowing in irrigated areas, tolerant to high temperature at seedling stage
Pusa Mustard-27 (EJ-17)	IARI, N. Delhi	108-135	40-45	1437-1659	MP, Uttrakhand and Rajasthan	Suitable for early sowing in irrigated areas,lerant to high temperature at seedling stage
Pusa Mustard-28 (NPJ-124)	IARI, N. Delhi	97-131	40-42	1912-2098	Delhi, Haryana, Punjab, Jammu and parts of Rajasthan	Suitable for early sowing in irrigated areas, tolerant to high temperature, powdery mildew and at seedling stage salinity upto12dS/m
Rajvijay Mustard 1	RVRSAU, Morena	98-121	40-43	1704	UP. MP and parts of Rajasthan	Suitable for early sowing in September
GM 3 (Gujrat Mustard 3)	Sardar Krushinagar Dantiwada, KVV, Sardar Krushinagar	107	38	2175	Gujarat	Suitable for early sowing
GDM 4 (Gujrat Dantiwada Mustard 4)	Sardar Krushinagar Dantiwada, KVV, Sardar Krushinagar	112	39	2417	Gujarat	

2. Normal sown varieties

These varieties take about 130-150 days to mature and can be sown in mid of October. The varieties recommended for these conditions are possessing conventional as well as improved oil and seed meal quality traits.

2(A). Conventional varieties

Variety	Developed by	Maturity duration (days)	Oil content (%)	Average yield (kg/ha)	Recommended area	Major characteristics
RCC -4	HPKVV, Kangda	155-160	39-40	1000-1200	Himachal Pradesh	
Laxmi (RH 8812)	HAU, Hisar	145	40	2200	Haryana	Bold seeds
PBR -97	PAU, RS, Bathinda	136	41	1532-2269	Haryana, HP, Punjab, Jammu and Rajasthan	Rain-fed areas
Pusa Jagannath	IARI, N. Delhi	112-138	37-43	1605-1975	MP, UP and Rajasthan	Normal sown condition in irrigated areas
Pusa Vijay (NPJ -93)	IARI, N. Delhi	135-154	35-41	1870-2715	Delhi and NCR	
RGN-48	SKKVV, Sriganga nagar	138-157	39-41	1505-2158	Delhi, Haryana, Punjab, Rajasthan and J&K	Suitable for rain-fed condition
RGN -73	SKKVV, Sriganga nagar	127-136	38-42	1771-2226	MP, UP, Uttarakhand and some parts of Rajasthan	Suitable for irrigated conditions
GDM 5 (Gujrat Dantiwada Mustard 5)	Sardar Krushinagar Dantiwada, KVV, Sardar Krushinagar	144	40	2213	-	Suitable for normal sown, rain-fed condition
NRCDR-02	DRMR, Bharatpur	131-156	36-42	1951-2626	Haryana, Punjab, Jammu, some parts of Rajasthan and Delhi	Suitable for irrigated conditions
Giriraj (DRMRIJ -31)	DRMR, Bharatpur	137-153	42	2246-2757	Haryana, Punjab, Jammu, some parts of Rajasthan and Delhi	Suitable for irrigated conditions

PR-20 (PRKS-28)	GBPUA&T, Pantnagar	116-131	40	1966	Plains of uttrakhand	Suitable for irrigated and rain-fed conditions
Pant Rai-21 (PRB-2008-5)	GBPUA&T, Pantnagar	122-127	39-41	1192-1856	Plains of Uttrakhand	Suitable for irrigated conditions
PBR -357	PAU, Bathinda	145	39	2664	Haryana, Punjab, Jammu, some parts of Rajasthan and Delhi	Suitable for timely sown irrigated conditions
PBR -378	PAU, Bathinda	147	40	2449	Haryana, Punjab and some partsof Rajasthan	Suitable for rain-fed conditions
RH 0406	HAU, Hisar	142-145	39	2200-2300	Haryana, Punjab, Jammu, some parts of Rajasthan and Delhi	Bold seeds, suitable for normal sown irrigated, less lodging conditions
RH 0749	HAU, Hisar	146-148	39	2600 -2800	Haryana, Punjab, some parts of Rajasthan and Delhi	Suitable for normal sown irrigate conditions, less lodging
Bayer Mustard 5450	Bayer Bioscience Ltd..	130-135	41	1800-2000	UP, Uttrakhand and eastern Rajasthan	
JK Samaridhi Gold (JKMS 2)	JK Agri-genetics Ltd..	125	40-42	2500	UP, Haryana, Rajasthan, Bihar and MP	
Divya 33	MR Seeds Pvt. Ltd., Sriganganagar	148	39	2494	Haryana, Jammu, Punjab and Rajasthan	

2(B). White rust resistant varieties

Variety	Developed by	Maturity duration (days)	Oil content (%)	Average yield (kg/ha)	Recommended area	Major characteristics
JM-2	JNKVV Centre, Morena	135-138	38-41	1632-2588	MP and Chhatisgarh	
JM-3	JNKVV Centre, Morena	130-132	40	1500-2500	MP	Tolerant to alternaria blight

2(C). Varieties for better oil and seed meal quality

Quality is decided by proportion of fatty acids in oil and amount of glucosinolates in cake. In these varieties erucic acid content is less than 2%, which decreases possibilities of myocardial fibrosis in adults and lipidosis in children. Quality of cake is also good in some of these varieties due to lower amount of glucosinolates.

Variety	Developed by	Maturity duration (days)	Oil content (%)	Average yield (kg/ha)	Recommended area	Major characteristics
Pusa Mustard—21 (LES1-27)	IARI, N. Delhi	137-152	34-40	1428-2277	Delhi, Punjab, Rajasthan, W. UP, Haryana and plains of J & K	Low erucic acid (<2%)
Pusa Mustard-24 (LET-18)	IARI, N. Delhi	134-150	32-39	1995-2044	Delhi, Haryana, Punjab and some parts of Rajasthan	Low erucic acid (<2%)
Pusa Mustard-29 (LET -36)	IARI, N. Delhi	143	37	2169	Delhi, Haryana, Punjab, Jammu and northern Rajasthan	Low erucic acid (<2%)
Pusa Mustard-30 (LES-43)	IARI, N. Delhi	137	36-39	1824	MP, UP, Uttrakhand and eastern Rajasthan	Low erucic acid (<2%)
Pusa Mustard-31	IARI, N. Delhi	144	40	2518	NCR, Delhi	0 erucic acid and total glucosinolates<30m icromol/100g of seed cake
RLC-2	PAU, Ludhiana	142-150	36-38	2039-2342	Haryana, Punjab, J & K, some parts of Rajasthan and Delhi	Suitable for timely sown irrigated conditions, low erucic acid (<2%)
RLC-3	PAU, Ludhiana	145	41.5	1850	Haryana, Punjab, J & K, some parts of Rajasthan and Delhi	0 erucic acid and total glucosinolates <30 micromol /100g of seed cake

2(D). Varieties tolerant to salinity

These varieties give higher yield in the conditions where soils are saline and/or the irrigation water is brackish.

Variety	Developed by	Maturity duration (days)	Oil content (%)	Average yield (kg/ha)	Recommended area	Major characteristics
CS-52	CSSRI, Karnal	135-145	39-43	580-1602	Haryana, Rajasthan and UP.	Suitable for alkaline and saline soils
CS-54 (CS 614-4-1-4)	CSSRI, Karnal	109-147	39-41	1230-2237	Salinity affected areas of Haryana, Rajasthan, UP and Gujarat	Suitable for salinity affected areas

2(E). Hybrids for timely sown conditions

Variety	Developed by	Maturity duration (days)	Oil content (%)	Average yield (kg/ha)	Recommended area	Major characteristics
DMH -1	Dhara vegetable oil company, Delhi	142-152	35-42	1782-2249	MP, UP, Haryana, Punjab, J & K, Rajasthan and delhi	High siliquae density, resistant to white rust
NRCHB-506	DRMR, Bharatpur	127-148	39-43	1550-2542	MP, UP and Rajasthan.	High adaptation
Coral PAC 432	Advanta India Ltd., Sikandrabad	121-132	40-42	1831-2581	UP, Uttrakhand, MP and Rajasthan.	
Coral PAC 437	Advanta India Ltd., Sikandrabad	140-150	39-41	2032-2772	Delhi, Haryana, Punjab, J & K and some parts of Rajasthan	Tolerant to white rust

3. Late sown varieties

These varieties are suitable for those areas, where mustard is sown after harvesting of *kharif* crops and sowing is being done in the month of November. Maturity of these varieties is 110-125 days. Income of marginal farmers can be increased by growing theses varieties.

Selection of variety, method of sowing, water management, recommendations of nutrients are important components for harvesting potential yield in a crop. Short duration varieties are suitable in the areas where crop season is short or farmers have to take more number of crops in a year. Therefore, these varieties are suitable for multiple cropping systems for improving crop intensity. For September sown conditions, we need varieties which possesses high temperature tolerance at early seedling stage so that adequate plant stand can be maintained.

Rapeseed-mustard is sown from mid September to mid November according to the climate of different states or regions. Sowing time is a very crucial factor, therefore, variety should be planted as per the recommendations for sowing time. Varieties tolerant for high temperature are recommended for the areas, where temperature is high at planting time. Moreover, potential yield per unit area can be realised with optimum plant population.

Method of sowing has direct effect on yield of the crop. Sowing should be done in rows at proper depth. Recommended row to row and plant to plant spacing should be followed to maximize the yield. If the soil is salt affected then sowing on ridges is recommended.

Recommended dose of fertilizers should be applied for raising healthy crop. Farmers should apply the nutrients as per the soil health card. Among the oilseed crops rapeseed-mustard require higher dose of sulphur. It increases oil content in the seed and is an essential component of protein, amino acids, glucosinolates and chloroplast. Phosphatic fertilizers should be applied in the form of single super phosphate. If single super phosphate is not available then recommended dose of sulphur should be applied at the time of sowing.

Always use disease free clean seed. Seed should be treated with 2.5g thiram /kg seed to protect from seed borne diseases. Seed should be treated with Metalaxil @1.4 g/ kg seed for protection from white rust disease in the early stage of crop.

The crop should not suffer moisture stress at the critical stages that is at flowering and seed filling.

Variety	Developed by	Maturity duration (days)	Oil content (%)	Average yield (kg/ha)	Recommended area	Major characteristics
CS-56 (CS-234-2)	CSSRI, Karnal	113-147	35-39	1170-1423	MP, UP, Uttra khand and eastern parts of Rajasthan	Bold seed and tolerant to salinity
NRCHB-101	DRMR, Bharatpur	105-135	35-42	1382-1491	MP, UP and Rajasthan	Suitable for irrigated late sown conditions
Pusa Mustard-26 (NPJ-113)	IARI, N. Delhi	115-137	30-41	1481-1895	Delhi, Haryana, J & K, Punjab and Rajasthan	Tolerant to high temperature at maturity suitable for irrigated areas
Aashirwad	CSUA&T, Kanpur	125-130	37-41	1437-1684	MP, UP, Uttrakhand and parts of Rajasthan	
Chhattisgarh sarson	IGKVV,Centre, Jagdalpur	99-128	35-40	1180-1200	Chhatisgarh	
Navgold (YRN 6)	RKVV centre, Navgaon	122-134	34-41	1095-1803	Delhi, Haryana, J & K, Punjab and Rajasthan	
RGN -145	SKVV,Centre, Sriganganagar	121-145	35-39	1448-1640	MP, UP, Uttrakhand and parts of Rajasthan	Suitable for irrigated areas
RRN 505	RKVV,Centre, Navgaon	121-127	40	1208-1506	Rajasthan	
Shatabdi (ACN 9)	Agricultural college, Nagpur	90-105	32-40		Vidarbha, Maharashtra	
RGN-236	SKVV,Centre, Sriganganagar	127	39	1636	Delhi, Haryana, J & K, Punjab and Rajasthan	Suitable for irrigated conditions

To harvest a good crop, the varieties developed by ICAR Research Institutes, SAUs or private research organizations should be planted in the recommended area of adaptation, time and production and plant protection practices. By following the above varietal recommendations the farmers will be able to increase the farm income and improving oil production and productivity.

Improved Agriculture and Allied Practices for Farmers, pp 43-50
New India Publishing Agency, New Delhi, India

4

Scientific Cultivation of Lentil

D. Singh and H.K. Dikshit

Division of Genetics, ICAR-Indian Agricultural Research Institute, New Delhi-110012

Introduction

Lentil is one of the important *Rabi* pulses. It is one of the oldest pulses crops and the most nutritious of the pulses. It has the potential to cover the risk of dryland agriculture. It is also used as a cover crop to check the soil erosion in problem areas. It is mostly eaten as 'dal'. Lentil can also be used as green fodder. Its straw is also used as cattle. Being a leguminous crop, it restores the soil fertility.

Area of cultivation

Lentil is mostly cultivated in the northern and central parts of India. The important lentil growing states are Madhya Pradesh, Uttar Pradesh, Bihar, West Bengal, Orissa, Haryana, Rajasthan, Punjab and Assam. Uttar Pradesh and Madhya Pradesh together contribute 78% towards total national area as well as production.

Climatic requirements: Lentil requires cold climate during growth and warm temperature at the time of maturity. It can tolerate cold and frost winter up to great extent. The optimum temperature for growth is 18-30 degree centigrade. Excessive drought and/or high temperatures during the flowering and pod-fill period also reduce yields.

Soil: Lentil can be grown on a variety of soils such as light loams; alluvial soils and black cotton soils. This crop is also suited to the poorer types of soils, low lying areas such as paddy fields. Lentil does not tolerate flooded or waterlogged soils. Good drainage is required, because even short periods of exposure to waterlogged or flooded field conditions kill plants. A soil pH near 7.0 is best for lentil production. Acidic soils are not fit for growing lentil.

Fertilizer requirement: Similar to other pulses, a small amount of mineral fertilizer (10-20 kg/ha nitrogen and 40-60 kg/ha phosphorus) is either placed in the seed furrow or broadcast and mixed in.

Sowing time: End of October to first fortnight of November is the best time for sowing of lentil.

Seed rate and spacing: A seed rate of 40-45 kg/hectare for large seeded lentil and 30 kg/hectare for small seeded have been recommended as optimum. Sowing should be done in rows 30 centimetre apart. This could be done either by using a ferti-seed drill or by seeding behind desi (country) plough. Lentil seeds should be sown at a shallower depth (3-4 centimetre).

Varieties: The following varieties of lentil have been recommended for cultivation in different states:

Assam: L-9-12, Asha (B-77)

Bihar: Pant L-406, PL-639, PL-209

Gujarat: PL-639, L-4076, L-4147

Haryana: Pant L-639, L-4076, L-4147

Himachal Pradesh: L-9-12

Jammu and Kashmir: T-36, L-9-12

Madhya Pradesh: L-4076, PL-406, B-77

Maharashtra: PL-639, L-4076, IPL-81.

North East Hill region: BR-25, T-36, Pant L-406, PL-639, B-77

Orissa: B-77

Punjab: Pant L-406, Pant Lentil-4, L-4076, L-4147

Rajasthan: L-4076, JL-3, IPL-81

U.P.: Type-26, PL-639, L-4076, L-4147, Narendra Masoor-1, Narendra Masoor-2

West Bengal: C-31, Ranjan (B-256), Asha (B-77)

Rotation

Lentil is grown generally after the harvest of Kharif crops or as the sole crop of the year. The important rotations with lentil are as follows:

Kharif fallow-lentil (rainfed areas)	Bajra-lentil
Paddy-lentil	Jowar-lentil
Maize-Lentil	Groundnut-lentil
Cotton-lentil	

Water Management

The crop is mostly grown in unirrigated areas. It can tolerate drought condition to some extent. Comparatively, higher yields can be obtained by providing one to two irrigation, particularly when winter rains are not properly distributed. First irrigation should be given at 45 days of planting and second, if needed, at pod filling stage. More irrigation may affect the crop performance adversely.

Harvesting and threshing

Lentil should be harvested when the pods ripe but the plants have not become dead ripe. Early varieties are ready to harvest 100-110 days after planting; late varieties are usually ready to harvest 125-135 days after planting. Harvest when plant turns yellow & pods have matured and hardened. Dry lentils for 7-10 days then thresh.

Plant Protection

Wilt: The disease appears in the field in patches at both seedling and adult stages. Seedling wilt is characterized by sudden dropping followed by drying of leaves and seedling death. Adult wilt symptoms appear from flowering to late pod filling stage and are characterized by sudden dropping of top leaflets of the affected plants.

Seed treatment with 0.3 per cent Benlate, use of crop rotations, and resistant varieties such as Pant L 406, Pant L-638 are helpful in controlling the diseases.

Rust: Initial symptoms are yellow spots on leaf surface. The spots develop into oval to elongate reddish-brown powdery. Seed treatment with 0.1 per cent Benlate, spraying the crop with fungicide Dithane M-45 (0.25 per cent) and use of resistant varieties such as L9-12, Pant L-406 is helpful in reducing the disease.

Insect pests

The major insect pests of lentil in the field are aphids (*Aphis craccivora* & *Acyrthosiphon pisum*), leaf weevils (*Sitona* spp.), Lygus bugs, (*Lygus* spp.), and the Cutworm, (*Agrotis ipsilon*).

Yield: Lentil can produce 20-22 q/ha of seeds under normal conditions.

Lentil varieties released in All India Pulses Improvement Programme (1985-2017)

S.No	Name of Variety	Centre responsible for developing	Pedigree	Year of release	Average yield q/ha	Days to maturity	Reaction to major diseases	Area of adaptation	Any other relevant information
1.	K 75 (Malika)	CSAU, Kanpur	Local sel. from Bundelkhand region	1986	14.0	135	-	NEPZ, CZ	Foliage dark green; semi spreading seeds grey mottled large (2.7 g/100 seed wt.)
2.	LH 84-8 (Sapna)	CCS HAU, Hissar	L 9-12 x JLS 2	1991	15.0	135	Res. to rust	NWPZ	Plant semi-spreading; seeds gray mottled bold (2.7 g/100 seed)
3.	Pant lentil 4	GBPUA&T, Pantnagar	VPL 175 x(PL 184 x P 288)	1993	16	140	Res. to rust & wilt	NWPZ	Plant semi-spreading, dark-green ,foliage, small seed
4.	Lens 4076	IARI, New Delhi	PL 234 x PL 639	1993	14	135	Res. to rust	NWPZ & CZ	Darkgreen foliage, semi spreading, large seed
5.	DPL 15 (Priya)	IIPR, Kanpur	PL 406 x L 4076	1995	15.5	135	Res. to rust tol. to wilt	NWPZ	Large seeded
6.	PusaVaibhav (L 4147)	IARI, New Delhi	(L 3875 x P 4) x PKVL	1996	17.8	134	Res. to rust	NWPZ	Small seed
7.	DPL 62 (Sheri)	IIPR, Kanpur	JLS 1 x LG 171	1997	17.20	135	Res. to rust tol. to wilt	NWPZ	Large seed
8.	JL 3	JNKVV, Sehore	Land race Sel. from Sagar	1999	14.33	113	Res. to wilt	CZ	Large seed
9.	IPL 81 (Noori)	IIPR, Kanpur	K 75 x PL 369	2000	12.45	113	Tol. to rust and wilt	CZ	Large seed
10.	KLS 218	CSAU, Kanpur	KLS 133 x L 9362	2005	13.80	120-125	Res. to rust	NEPZ	Small seed
11.	HUL 57	BHU, Varanasi	Mutant of HUL –11	2005	14.0	121	Res. to rust	NEPZ	Small seed
12.	VL 507	VPKAS. Almora	Sel. From ILL-7978	2006	12.4	160-170	Res. to wilt	NHZ	Large seed

13.	VL 126	Almora	LL 498 x LH 84-8	2006	13.0	160-170	Res. to rust	NHZ	Small seed
14.	IPL 406	IIPR, Kanpur	DPL 35 x EC 157634/ 382	2007	17.0	125-130	Res. to rust	NWPZ	Large seed (3.9 g.)
15.	WBL 77	Berhampore (W.B)	ILL 7723 x BLX 88176	2008	14.0	115-120	Res. to rust	NEPZ	Small seed
16.	Pant L 6	GBPUA&T, Pantnagar	Pant L 4 x DPL 55	2009	16-18	125-145	Res. to rust	Uttarakhand	Small seed
17.	Pant L 7	GBPUA&T, Pantnagar	L 4076 x DPL 15	2009	16-18	125-145	Res. to rust	Uttarakhand	Large seed
18.	Pant L 8	GBPUA&T, Pantnagar	DPL 59 x IPL 105	2010	15-16	130-135	Moderately resistant to rust and wilt	CZ	Large seed
19.	IPL 316	IIPR, Kanpur	Sehore 74-3 x DPL 58	2013	14-15	110-115	Tolerant to wilt and rust	CZ	Large seeded
20.	RVL 11-6	RAK College Sehore	JL 3 x DPL 62	2017	11-12	107-113	Tolerant to wilt	CZ	Large seeded
21.	L 4717 (Pusa Ageti Masoor)	IARI, New Delhi	ILL 7617 x 91516	2017	12-13	96-106	Resistant to wilt and Ascochyta blight	CZ	Extra early type

CZ= Central Zone (MP, BundelKhand region of UP)
NHZ= North Hill Zone (Hills of UP, J&K and NE region)
NEPZ = North East Plain Zone (Eastern UP, Bihar, WB and Orissa)
NWPZ= North West Plain Zone (West UP. Punjab, Haryana and parts of Rajasthan)

Lentil varieties released from different states (1990-2017)

S.No.	Name of Variety	Centre responsible for developing	Pedigree	Year of release	Average yield q/ha	Days to maturity	Reaction to major diseases	Area of adaptation	Any other relevant information
1.	JL 1	JNKVV, Jabalpur	Selection from local collection	1991	11-15	110-120			Suitble for rainfed condition
2.	Narendra Masoor 1	NDUAT Faizabad	Precoz x L 9-12	1996	20-22	134-140	Tol. to rust & wilt	MP	Medium seed size
3.	Garima	CCS, HAU Hisar	Pusa 2 x No. 4	1997	12-14	140-145	Tol. to rust	Haryana	Large seed
4.	LL 699	PAU, Ludhiana	PL 639 x PL 77-2	2002	14-15	140-145	Tol. To pod borer moderately resistant to rust & blight	Punjab	
5.	KL 320 (Shekhar 3)	CSAUA&T, Kanpur	PL 405 x P 237	2004	20-22	120-125	Tol. to wilt	UP	
6.	KLB 303 (Shekhar 2)	CSAUA&T, Kanpur	LG 60 x Sagar local	2004	15-18	125-130	Tol. to wilt	UP	
7.	Shalimar Masoor 1	Srinagar Centre, SKUAST-K	Selection from EC 2216	2005	10-11	190-200	Moderately resistant to root marginal rot and resistant to leaf spot and pod blight	Rainfed Karewa land of Kashmir	
8.	Markanday	CSK HPKV, HAREC, Dhaulakuan	Selection from exotic materia	2005	10-12	160-170	Res. To rust, wilt, root rot & l blight	Low and mid hills	
9.	HM-1	CCS HAU, Hisar	K 75 x L 4076	2006	14-16	135-140	Res. to rust	Irrigated areas of Haryana	Small seeded

10.	LL 931	PAU, Ludhiana	LH 90-103 x LL 608	2009	12-13	145-150	Res. To rust &tol. To pod borer	Punjab	
11.	Pant L 6	GBPUA&T, Pantnagar	Pant L 4 x DPL 55	2009	16-18	125-145	Res. to rust	Uttarakhand	Small seed
12.	Pant L 7	GBPUA&T, Pantnagar	L 4076 x DPL 15	2009	16-18	125-145	Res. to rust	Uttarakhand	Large seed
	VL Masoor 129	VPKAS, Almora	VL 101 x VL 1	2009	7-11	145-150	Resistant against wilt and root rot	Uttarakhand	
13.	VLMasoor 514	VPKAS, Almora	VL 501 x VL Masoor 103	2011	8-10	155-160	Moderately resistant to rust and wilt	Uttarakhand	
14.	Shalimar Masoor 2	Srinagar Centre, SKUAST-K	Srinagar Centre, SKUAST-K	2013	12.85	200-205	Moderately resistant to rust and wilt. Resistant to white grub	Suitable upto altitude of 1850 m amsl	Large seed
15.	RVL 31	RAK, Sehore	Selection from Shajahpur local collection	2014	18-19	110-115	Res. To wilt	MP	Large seed High Biomas
16.	KLB 345 (Shekhar 4)	CSAUA&T, Kanpur	Precoz x KLB 231	2014	18-20	110-115	Res. to rust & wilt	UP	Large seed
17.	KLB 2008-4 (Krati)	CSAUA&T, Kanpur	LG 362 x DPL 62	2015	18-20	115-120	Res. to wilt	UP	Large seed
18.	KLS 09-3 (Krish)	CSAUA&T, Kanpur	L 9-12 x WBL 58	2015	18-20	105-110	Res. to wilt & rust	UP	Medium large seed
19.	IPL 526	IIPR, Kanpur	DPL 62 x DPL	2016	10-12	101-110	Tol. To rust and wilt	UP	Medium large seed

20.	Pant lentil 9	GBPUA&T, Pantnagar	Pant L 5 x IPL 105	2016	13-14	113-135	Res. to rust, wilt & Ascochyta blight	Uttrakhand	Small seed
21.	RLG 5 (Keshwanand Masoor 1)	RARI, Durgapura	Sel. From local germplasm	2016	15-16	130	Mod. Resistant to root knot nematode	Rajasthan	Double podded suitable for rainfed condition

Improved Agriculture and Allied Practices for Farmers, pp 51-65
New India Publishing Agency, New Delhi, India

5

System of Rice Intensification (SRI) and Aerobic Rice Culture for Improving Input-use Efficiency in Rice Cultivation

Dinesh Kumar, Nain Singh, Vinod Kumar Tyagi and Shweta Mehrotra

Division of Agronomy, ICAR-Indian Agricultural Research Institute New Delhi-110012

Rice is the most important crop in India occupying about 43.7 million hectares area with a production of 136.5 million tonnes (rough rice/ paddy). As the human population is increasing, so is the demand for rice, yet yields of the crop are leveling out. At present, the rice yield in India is 3.12 t/ha as against the world average of 4.11 t/ha. Not only this, the rice yield in India is exactly half than that of China (6.26 t/ha). The World Development Report 2008 predicts that cereal production would have to increase by 50% by 2030 to meet the escalating worldwide demand. The additional rice for increasing population will have to be produced on less land with less water, less labour, and reduced inputs.The major constraint in rice production in India is the lack of suitable crop management practices and sufficient irrigation facilities. The water crisis is threatening the sustainability of the irrigated rice system and food security in many parts of the world. Despite efforts on a variety of fronts, including genetic engineering of rice strains for improved yield and nutrition, no concrete solution has been found yet. There is a challenge to develop novel technologies and production systems that allow rice production to be maintained or increased in the face of declining water availability. To produce 1 kg of grain, farmers have to put 2 to 3 times more water in rice fields than in those growing other cereals.

The increasing water need for agriculture poses a potential threat to natural water resources and climatic change scenarios in the near future. The great challenge will be to increase food production with less water, particularly in countries with limited water and land resources, by promoting techniques and cropping systems of higher water-use efficiency (WUE).There is also much evidence that water scarcity already prevails in rice-growing areas in India, where rice farmers need technologies to cope with water shortage and ways must be sought to grow rice with less water.The farmers, especially those, who are resource-poor, are losing interest in rice cultivation as its profitability is declining with the rise in input costs. But, system of rice intensification (SRI) and aerobic rice culture appear to be a viable alternative of rice cultivation that may save the expensive inputs, improve soil health/ quality and protect the environment substantially.

System of Rice Intensification (SRI)

The System of Rice Intensification synthesized in Madagascar by Fr. Henride Laulanie′, S.J., has been gaining acceptance around the world because of the remarkable increases in factor productivity that its methods are making possible. Irrigated rice yields are being raised by 50–100%,and often more, without requiring the use of purchased inputs and without changing varieties. The higher yields with SRI practices are achieved with any and all rice varieties while using less water and with little use of mineral fertilizers.Because the resulting plant phenotypes are more resistant to damage by pests and diseases, there is usually little or no need for chemical crop protection. All this has been quite unexpected.While the mechanisms that make these benefits attainable are not yet fully understood, it appears that SRI effects are the result of greater root growth and of increases in soil biological activity that are induced by different cultural practices.

Components of SRI method of rice culture

- Transplanting young seedlings, usually 8 to 12 days old and certainly less than 15 days old.
- Planting single seedlings rather than several seedlings together in a hill.
- Wide space planting (spacing of 25 × 25 cm or wider).
- Planting quickly and carefully to avoid trauma to seedling roots.
- Daily or intermittent irrigation before panicle initiation (PI) after which a thin layer of water (1-2 cm) is maintained. The fields are kept moist and not flodded.
- Hand or mechanical weeding, starting 10 days after transplanting and repeated 2-4 times at 10-12 d intervals until the canopy closes.
- Applying nutrients to soil preferably in organic form such as compost or mulch.

Transplanting of young seedlings prior to the start of the 4th phyllochron of growth, preserves plants' potential for tillering and root growth that is reduced by transplanting later. Direct seeding is an option, however, since what is important is that plant roots should not be traumatized after they start their growth trajectory, on or about the 15th day. Careful transplanting promotes rapid resumption of growth. Wide spacing between plants is maintained with preferably just one plant per hill, and set them out in a square pattern, 25 cm x 25 cm or even wider if soil fertility is good. This gives room for profuse root and tiller growth, achieving 'the border effect' throughout the whole field. Another important aspect in SRI is to keep the soil both moist and aerated, at least during the vegetative growth period, so that roots have access to both oxygen and water. Under continuously hypoxic conditions, rice roots degenerate, with as many as 75% become dysfunctional by panicle initiation.Therefore, the aerobic conditions in SRI promote root growth substantially over submerged situations. In conventional cultivation, seedlings are uprooted, washed, bundled and transported to the main field. Quite often, the main fields are far away, so there is a long time gap between uprooting and planting. SRI practitioners try to avoid any establishment delays, and/or to remove the seedlings with the base medium, thus minimizing any root damage and possible transplanting shock. This might appear to require more time and energy, but in fact it does not, as farmers use far fewer seedlings. Another positive aspect is that it is possible to have the nursery in one corner of the main field so that transfer time can be minimized. Specific practices include frequent weeding, preferably with a rotating hoe that aerates the soil while it prevents growth of weeds by churning them into the soil. Green manures and cover crops are also useful. Correct implementation of the various, interdependent components (SRI practices) requires considerable field experience in which close observation and timeliness are crucial in minimizing labour requirements while maximizing total factor productivity.

Productivity of rice grown by SRI method

The ultimate effect of SRI techniques is reflected in considerably higher grain yields of rice.The potential productivity with SRI is very high, which is more important now as the Green Revolution technologies are showing fatigue. More importantly, the SRI has potential to give higher yields at lower costs. Even when the Andhra Pradesh farmers were unable to implement all aspects of SRI in the first season, just planting younger seedling carefully at wider spacing with somewhat better water management practices resulted in over 2.0t/ha extra yield as compared to the conventional methods using higher inputs.With more experience, still higher yields may be possible.Today there are confirming results from 21 additional countries ranging from China to Peru, with average yields from SRI in the 7-8 tha^{-1} range, and with yields over 15 tha^{-1} reported from at least four countries beyond Madagascar. A summary review of yield results is given in Table 1 (India) and Table 2 (Bangladesh). Evaluation of SRI is made difficult by the fact that it is still evolving, more a set of ideas than a completed technology. Also, data on SRI have not been collected in standard formats because it is being undertaken by a diverse set of NGOs, government agencies, research institutions, universities, farmer groups and individuals with no central support or funding. Of late, there are some researchers at leading institutions in China, India and Indonesia now working on different aspects of SRI. Initial trials have shown high potentials for its acceptance as an improved method of rice production. Some examples are (Husain, 2004):

- In the Sichuan province of China, during the 2004 summer, average yield gains were 3 t/ha over the usual yield of 7.5 t/ha. The highest SRI yield was recorded as 20.4 t/ha in Yunnan province.Water savings of 42% were recorded in Sichuan, while in Zhejiang province, incidence of sheath blight, a major rice disease in the area, was reduced by 70%. The subsequent harvest, almost 13 t/ha, set a record for that province.
- In Andhra Pradesh, India, results of SRI trials by 50 farmers during *rabi* season 2003-04 showed an increase of yield from 7.1 t/ha to 9.7 t/ha with the highest measured yields surpassing 15 t/ha.
- In Cambodia, evaluation of performance of SRI covering five provinces showed average yield gain of 41% with gross profits/ha increasing by 74%. Rather significantly, SRI was found there not to be labour intensive.
- In Sri Lanka, an International Water Management Institute (IWMI) evaluation in two districts found almost a 50% increase in yield, water productivity increase of 90%, reduction in cost of production by 17-27%, and 112% increase in net profit. In Indonesia also, the Agency for Agricultural Research and Development (AARD), based on evaluation of SRI practices for three years, made SRI part of its new national strategy for integrated crop resource management to restore growth in the rice sector.

A field experiment was conducted during 2004-2007 at Indian Agricultural Research Institute to evaluate the comparative performance of system of rice intensification (SRI) and conventional rice culture (CRC) at varying nutrient management practices. Newly released scented variety of rice, *i.e.*'Pusa Sugandh - 5'(Pusa-2511) was transplanted in first week of July every year at a spacing of 25 cm x 25 cm in SRI and 20 cm x 10 cm in CRC. One and two seedlings/hill were transplanted in SRI and CRC, respectively. Under SRI, the plots were kept moist and aerated throughout the crop period, however, under CRC a continuous submergence was maintained. On an average, the SRI crop needed about 1/3 to 2/5 less water than the CRC. The results of the study showed a non significant difference in grain yield of rice under both (SRI and CRC) the systems of rice cultivation in first and third year (Table 3). However, in second year, CRC produced significantly more rice yield than SRI.The nutrient management practices affected the grain yield of rice significantly.

Table 1: Results of SRI trials in Andhra Pradesh state, 2003

Region of Andhra Pradesh	No. of trials	Range of trial yields (t/ha)	Farmer trials with yields > 10 t/ha	Yield advantage (t/ha)
Rayalseema region	10	7.7-15.5	6	4.73
Telangana region	40	4.2-16.2	10	2.50
Coastal region	84	3.2-14.3	17	1.15
Total Andhra Pradesh State	134	3.2-16.2	33	1.87

Source: Uphoff (2003)

Table 2: Yield from SRI and conventional practices, Bangladesh, boro season, 2003

Particular	SRI method	Conventional method
Average weight (grams/1000 grains)	24	21
Unfilled grain (number/100 grains)	14	23
Average yield (tha^{-1})	7.7	5.4
Average weight of straw (tha^{-1})	6.4	4.6

Source: BRAC (2003)

Table 3: Comparative performance of SRI and conventional rice culture in rice-lentil cropping system

Treatment	Rice (2004)	Lentil (2004-2005)	Rice (2005)	Lentil (2005-2006)	Rice (2006)	Lentil (2006-2007)
Method of cultivation						
System of rice intensification	5.31	1.27	3.12	1.35	3.83	1.48
Conventional rice culture	5.28	1.28	5.40	1.34	3.41	1.46
CD (P=0.05)	NS	NS	0.20	NS	NS	NS
Nutrient management						
Control	3.98	1.29	3.98	1.18	2.60	1.31
NPK	5.32	1.25	5.32	1.44	3.50	1.42
FYM	5.25	1.29	5.25	1.27	4.32	1.47
½ FYM + ½ NPK	6.07	1.28	6.07	1.37	3.90	1.67
Compost	5.03	1.28	5.03	1.42	3.50	1.50
½ Compost + ½ NPK	6.10	1.27	6.10	1.37	3.90	1.47
CD (P=0.05)	3.2	NS	0.30	0.81	0.86	0.18

Phenotypical Changes

SRI changes in the management practices for plants, soil, water and nutrients produce different and more productive phenotypes (Table 4).

Table 4: Phenotypical differences in rice plants and grains associated with SRI practices

Character	Measured Differences
Tillering	30–50 tillers plant^{-1} are common with SRI methods, and twice this number or more can be attained with good use of the methods and with improved soil.
Root Growth	Enhanced and vigorous root growth
Grain Filling	Positive correlation between number of panicles and size of panicles, contrary to the typical rice phenotype grown with continuous flooding, for which the correlation is negative; usually reduced % of unfilled grains.
Grain Weight	A 5–15% increase is common, contributing to higher yield; this can occur without an increase in grain size, which represents greater grain density.
Reduced Time for Maturation	Often 1–2 weeks less time for same variety to ripen when SRI methods are used.
Grain Quality	10–15% higher outturn of milled rice from SRI paddy, due to fewer unfilled grains (less chaff) and fewer broken grains during milling (less shattering), reflecting higher grain density. A Sichuan Agricultural University evaluation in China found a 16% increase in outturn of milled rice, and also less chalkiness.

Reported Differences	
Resistance to Pests and Diseases	Widely reported by farmers in different countries, saying that pests and diseases do not cause enough damage to warrant use of biocides.
Resistance to Drought, Wind and Rain Damage, and Cold Temperatures:	Larger root systems appear to account for the resistance to drought and cold as well as to cyclone and typhoon damage damage as observed in India, China and Sri Lanka; lodging is rare with SRI plants despite larger panicles, even under extreme weather conditions.

Enhanced root growth

SRI encourages vigorous root growth of rice. Tests of root-pulling resistance (RPR) have shown that conventionally grown clumps of 4 to 6 rice plants required, on average, 22.00, 35.00 and 20.67 kg plant^{-1} at the stages of panicle initiation, anthesis, and maturity, respectively. In contrast, for single SRI plants at these respective stages, RPR was 55.19, 77.67 and 53.00 kg plant^{-1} (Barison, 2002). This represents a 10-fold difference in the amount of force per plant. The increased root volume, number and weight helps the roots to absorb more nutrients, withstand adverse soil conditions and increase the availability of nutrients, etc. With their larger root systems, SRI rice plants tiller more profusely and are less subject to senescence and to pest or disease damage. With SRI, there is a positive correlation between the number of panicles per plant and the size of panicles (number of grains). The roots of rice plants grown in flooded fields remain near the surface, about 75% in the top 6 cm (Kirk and Solivas, 1997). This enables the plants to capture dissolved O_2 in the irrigation water, with root growth into lower soil levels limited by the hypoxic conditions. When the rhizosphere soil is kept continuously saturated, most of the root system will degenerate by the time of flowering. Controlled comparison studies have shown 78% of rice roots degenerated when grown in continuously flooded soil compared with almost no degeneration of roots in well-drained, aerated soil (Kar et al., 1974). Larger roots in SRI grown rice provide more water and nutrients to the canopy; a bigger canopy produces more photosynthates that benefits the roots.

Water savings in SRI method

Water requirements with SRI are usually reduced by about half since paddies are not kept flooded during the entire crop cycle. Water requirement is much reduced during the vegetative growth phase, and only a minimum quantity of water is kept in the field during the reproductive phase. This saving in water will become increasingly important, especially during drought year, in the agricultural sector.

Seed savings

Because many fewer plants are grown, the seeding rate is only 5-10 kg/ ha. This is a benefit particularly for hybrid rice where seed cost can be a barrier for adoption.

Increased factor productivity and profitability

The returns to farmers' land (yield per hectare), labour (income per hour), water (output per m^3) and capital (profitability), all increase at the same time as SRI practices enable plants to get the benefits of greater supplies of carbon, nitrogen and oxygen.

Biological support for rice production

SRI has had liability of appearing 'too good to be true'. But its validity is becoming more and more evident. Fewer external inputs can produce more output, as using less water (avoiding soil hypoxia) and chemical fertilizer (affecting microorganisms) enhances soil biological activity. Fewer chemical sprays leave more beneficial organisms in the canopy and soil to control pests. With reduced external inputs, there is more input to crop production processes of beneficial gases as well as biological mineralization of nutrients in the soil. The beneficial and even necessary contribution of root exudates to supporting larger and more diverse microbial populations is well established. However, its implications for agricultural production have been little considered.

There are a number of biological processes well documented in the literature that could be boosting rice yields with SRI management of plants, soil, water and nutrients when soils are alternately flooded and drained, this gives both aerobic and anaerobic bacteria as well as mycorrhizal fungi an opportunity to contribute to plant growth. Under these conditions, there is increased biological N fixation. Plants with larger root systems and canopies are able, indirectly, to produce more N in their root zone by exuding photosynthates made from atmospheric CO_2 into the rhizosphere. These photosynthates support larger bacterial populations on the roots that in turn are 'grazed' by protozoa which then (because their own C:N ratio is lower) excrete excess N onto the plant roots which utilize this biological waste product. Many species of bacteria and fungi produce phytohormones in the rhizosphere — auxins, cytokinins, ethylene, etc.- that regulate and promote root growth.

Limitations

Certainly, there are some limitations on the use and adoption of SRI. The most objective one is the need of having good water control to get best results. Water savings can only be made and soil can only be kept aerated enough to support the growth and functioning of aerobes in the soil if farmers can apply limited amounts of water rather than keep their paddies continuously flooded. Many farmers do not have such water control as they operate in field-to-field (cascade) systems of distribution. Once the benefits attainable from SRI methods are better known and well documented, investments to improve water control, even sacrificing a little field area, should become attractive to improve capabilities for water management. Initially, SRI methods require more labour, as they need to be learned and mastered. In fact, this requirement is reduced in subsequent seasons, and SRI labour demand eventually becomes less than with conventional cultivation. The major requirement for SRI is motivation and skill. Farmers need to become more conscientious and knowledgeable managers of their plants, soil, water and nutrients. This is a cast at the outset, since time and effort are needed to disseminate SRI techniques correctly to farmers who may not have much confidence in them.This will become easier and quicker as a positive reputation of SRI spreads and human resource development where farmers increasingly engage in making their own observations, evaluations, and decisions becomes over time a benefit offsetting any cost. Farmers who already know how to grow irrigated rice and who are motivated to learn SRI are able to do this with very little instruction. Particularly if the information is passing farmer-to-farmer, where credibility can be maximized, SRI adoption requires relatively little investment. It is important that methods be presented with an eye to minimizing labour effort, cutting corners and economizing so that the yield per unit of investment is increased.

Aerobic rice culture

Traditional lowland rice with continuous flooding has relatively high water inputs. Because of increasing water scarcity, there is a need to develop alternative systems that require less water. “Aerobic rice” is a new concept of growing rice: it is high-yielding rice grown in non-puddled, aerobic soils under irrigation and high external inputs. To make aerobic rice successful, new varieties and management practices must be developed. Crops such as wheat and maize, in contrast, are grown on non-puddled, aerated soils. Irrigation is applied when the soil becomes too dry, and the quantity of applied water is sufficient to bring the soil to field capacity. Savings in irrigation water and increases in water productivity might be possible if rice could be grown under aerobic soil conditions such as wheat and maize. However, new varieties must be developed if this concept of "aerobic rice" is to be successful. Upland rice exists but is targeted at unfavorable environments without access to irrigation, with low external inputs, and hence with a low yield potential. Aerobic rice should be responsive to high inputs (water, nutrients) to reach high yields under non-flooded conditions. The realization of water savings combined with high yields, however, depends on good water management. The basic understanding of crop-water relationships in aerobic rice is missing and irrigation management recommendations have to be developed.

In recent years in India, there have been concerns related to shortages of labour, which causes transplanting costs to rise and delay the planting of the rice crop. Also, there have been restrictions on the availability of water early in the season, a further cause of delay. To overcome these constraints, aerobic rice appears to be a good alternative. Rice, like wheat or maize, can be directly seeded through dry seeding. Dry seeding of rice can be done by drilling the seed into a fine seedbed at a depth of 2–3 centimetres. Weed management is a critical factor in aerobic rice. Timely application of herbicides and two or three hand weedings may provide effective control. The success of the aerobic rice depends entirely on the availability of rain or irrigation water, besides proper stand of crop. If sowing is done in a properly prepared land, proper stand of crop can be achieved. A field with fine tilth facilitates the seed to come in contact with the soil moisture after drilling and enables the seed to germinate quickly and uniformly. Thus, an ideal preparation of the land will help to achieve a uniform stand, facilitate weeding and fertilizer practices. Therefore, with number of ploughings of the field and timely sowing, the aerobic rice crop generally gives better yield. Seeding under dry situations is done in three different ways - viz. (i) drilling i.e. sowing in the furrow behind a plough, (ii) dibbling and (iii) broadcasting. But drilling method is the ideal one. The light soils which generally come into conditions quickly, any method can be adopted. Seeding with drilling method has got a greater advantage over other methods, because of the uniformity of the stand and the control of the population of the plants per unit area. Heavy soils which do not come in conditions quickly, other methods except broadcasting are not feasible. It has been found that drilling or dibbling always gives considerably better yields than broadcasting system. It eliminates the use of seedlings and related operations such as seedling nursery preparation, care of seedlings, pulling, bundling, transporting, and transplanting. Research findings suggest that following practices can be adopted for success of aerobic rice.

Seed Depth and Soil Moisture

i) Optimum depth of seeding is 3-4 cm and the seed should be covered with soil for proper germination and to avoid bird damage. A pre-sowing irrigation is necessary if sufficient moisture is not available in the field. After seeding, planking is necessary. Seeding in presence of residues of previous crops help to conserve soil moisture longer to improve the crop stands and yields.

ii) Seeding of rice must be done at a soil moisture content sufficient for proper germination.Surface mulch help retain soil moisture longer to improve emergence of seedlings and reduce weed menace.

iii) Precision land leveling helps in good germination and reduces mortality of young tender rice seedlings due to uneven distribution of water during an irrigation event.

Cultivar choices, sowing time, seed rate and spacing

For aerobic rice, it is very important to develop varieties that may produce high yields under aerobic situation.The drought-tolerant upland varieties are often non-responsive to high inputs such as nutrients and produce very less yield. Similarly, the existing lowland varieties developed for puddle-transplanted-continuous situation, show high yield loss under aerobic conditions.Hence, development/screening/identification of new rice varities/ genotypes, which may yield high under aerobic situation, is necessary to get high yield. Furthermore, the choice of varieties would also depend on crop rotations being followed by the individual farmers. The existing varieties could be suitably used till the breeders in India are able to develop the varieties for aerobic situation. Early to medium short duration cultivars, hybrids and scented rices that have early vigour and smothering effect on weeds perform better. Cultivars such as Sarbati, PB-1; HKR 47, IR-64; Sugandha-4 or 5, Basmati-370, CSR-31, Prabhat, Sarjoo-52, Pant-4, PR-111, PR-115, NDR-359, Saroj, PHB-71, KRH-2 (coarse grain) and PRH-10 (fine grain). etc. have been found very promising. Sugandha, Prabhat and Hybrid rice compete very well with early weeds. aerobic rice crop matures earlier than transplanted rice by 1-2 weeks.Therefore, aerobic rice can be seeded from June 1-15 in northwest Indo-gangetic plains (IGP) to vacate fields by early November. In eastern IGP, the aerobic rice crop can be established from middle of May to first half of June depending upon the rice ecologies and crop rotations. Timely rice crop establishment improves rainwater use efficiency by 40-50% and enhances the total productivity of cropping system up to 30 percent in Gangetic plains. The recommended seed rates are 60-80 kg/ha for manual sowing with plough and 40-60 kg/ha for sowing with tractor-drawn seed drill. Sowing is done in rows with a row-to-row spacing of 25 cm.

Fertilizer Use

External use of fertiliser inputs should be as per local recommendations. Generally, 150 kg N, 60 kg P_2O_5 and 40 kg/ha K_2O is recommended for aerobic rice. It is generally variable between specific crops and cropping systems and soil conditions. P and K should be applied in full as a basal dose with 1/3 nitrogen at seeding time. Rest of the nitrogen should be applied in two-equal splits at 30-35 and 60-65 days after seeding. A LCC (leaf colour chart) value of 3 and 4 is used for scented and other rice cultivars, respectively for N application. LCC should be used to avoid excess N application. In low lying and flood prone areas, it is better to deep band place 80% of recommended N as a basal dose at sowing time. Also, iron deficiency is a serious problem in aerobic rice cultivation. It could be overcome by spraying 2.0% iron sulphate may be done starting at tillering in two weeks interval (Singh and Chinnusamy, 2006).

Weed Management

Productivity of the aerobic rice largely depends on timely weed management practices. Yield losses from uncontrolled weed growth in aerobic rice are much higher compared to transplanted rice. Management of weeds very early in the rice season hold the key for a successful aerobic rice crop. If it rains before seeding, or it takes more time to seed after pre-sowing irrigation, most weeds would germinate early. Pre-germinated weeds can be knocked down with glyphosate (Roundup) or by 1-2 very shallow ploughings (stale seed bed method).

Stale Bed technique of weed control

In the stale bed technique, irrigation is applied to facilitate the germination of weeds, which can be killed by nonselective herbicides (Grammoxone or Glyphosate at 0.5%) two days before seeding. Alternately, weeds can be eliminated by 1-2 shallow ploughings. Once weeds are knocked down, rice crop can be established and second flush of weeds can be removed manually.

Herbicides and Weed Management

Herbicides are becoming increasingly popular amongst farmers. Many farmers are however not fully aware about their proper and correct use against diverse weed flora. Surface mulch act as a good herbicide and if possible, must be retained in main fields. If weed management is to be primarily based on use of chemical molecules (herbicides), information in ensuing sections may prove useful: (i) Pendimethalin (Stomp) at 1.0 kg a.i. ha applied immediately after seeding effectively controls grassy weeds for 25 days in relatively dry soils. (ii)

Post-sowing, herbicide molecule, Pretilachlor-1 with Safner (Sofit / Erase-N) at 500 g a.i. ha is also very effective in control of grassy weeds if applied within 3 days but requires wet soil moisture regimes for few days to be effective. (iii) For the control of broad leaf weeds 2,4-D at 500 g a.i. ha^{-1} or Almix (Chlorimuron + Metsulfuron) at 4 g a.i. ha^{-1} can be applied after 21 days. (iv) Post emergence application of tank mixture of Whip Super (Fenoxaprop ethyl) and Sunrise(-1 Ethoxysulfuron) at 50+18 g a.i. ha^{-1} controls both grassy and broad leaf weeds effectively. (v) Repeat application of herbicides (vi) after 25-30 days of seeding will effectively control the second flush of grassy weeds.

Irrigation Management

After germination of rice, irrigation can be delayed for around 7-15 days depending on soil texture and water table conditions. Delayed irrigation facilitates deeper rooting and makes seedlings resistant to drought. Since elimination of puddling practice significantly reduces soil cracking, and aerobic rice crop does not require ponded conditions, irrigation frequency of 3-7 days can be safely practiced for a aerobic rice crop, after disappearance of water from the field. Under limited water supply and drought situations, irrigation can be delayed up to 10-15 days. However, watering is crucial once tillering has begun.

Green Manuring and Surface Retention of Crop Residues

New multi-crop zerotill ferti-seed drills with punch or disk type openers allow seeding in presence of loose residues. In combine harvested areas, crop residues must be spread evenly before planting for a uniform germination and crop establishment. Surface retained residues serve as physical barrier to emergence of weeds, moderate the soil temperature in summers and winters, conserve soil moisture, add organic matter and nutrients to the soil on decomposition. Traditionally, farmers grow green manure crops before rice culture and incorporate it by puddling before transplanting rice seedlings. This means an additional need for irrigation water for *Sesbania* crop *perse* and fuel, costs for incorporating it. Since there is little water in the reservoirs during peak summers, farmers have not been able to take full advantage of green manuring in rice growing season. In ''Brown Manuring' practice, both rice and *Sesbania* crops are seeded together and allowed to grow for 30 days. Subsequently co-culture *Sesbania* crop is knocked down with 2,4-D at 500 g/ha. Co-culture technology described above reduces weed population by nearly half without any adverse effect on rice yield. *Sesbania* surface mulch decomposes very fast to supply N and other recycled nutrients.

Summary

Rice yields all over the world have leveled out under the present system of flooded cultivation. We need to be looking for alternatives to existing practices with an open mind. SRI is still evolving and it is hoped that the scientific community will collaborate in further refining the SRI practices and working out the scientific reasons for the reported higher productivity. Planting young seedlings carefully and at wider spacing gives the plant more time and space for tillering and root growth. Careful water management keeping the field wet and not flooded gives better yield as it supports healthy root growth. This practice should be encouraged everywhere as the whole world is facing water shortages. Weeding rice fields with a rotary weeder helps by churning the soil and incorporating the weed biomass as it aerates the root zone. Therefore, seeing the overall benefits, the SRI needs to be evaluated further and refined to suit the local environments.

Traditionally, rice is transplanted after puddling and wheat is sown after pulverizing the soil in Indo-gangetic plains. This reflects an edaphic conflict in traditional soil management for rice and its consequent deleterious effects on the soil environment for the succeeding wheat and other upland crops. The basic question is, can rice culture be like wheat so that it can benefit cropping system productivity and sustainability? Aerobic rice, which removes puddling and drudgery of transplanting the young rice seedlings provides an option to resolve the edaphic conflict and enhance the sustainability of rice-wheat cropping system. Puddling requires lots of scarce water at a time when there is little water in the reservoirs, destroys soil structure and adversely affects soil productivity. In addition, direct seeding (aerobic) of rice can overcome the problem of seasonality in labour requirement for rice nursery raising and transplanting operations. Non-development of ground water in *Kharif*, late onset of monsoons and drudgery of operations often delays rice transplanting which leads to late vacation of fields, forcing farmers to plant wheat after the optimum sowing time. In eastern India, farmers waste almost 40% of rainwater in preparatory tillage before transplanting rice, ending in terminal droughts and consequent low yields and total system productivity. Aerobic rice facilitates timely establishment of rice and succeeding winter crops. Unlike puddled fields, aerobic rice fields do not crack and thus help save irrigation water. In raising a successful aerobic rice crop, weed management is an issue that requires serious attention by farmers. With timely farming operations, this problem can be tackled easily.

References

Barison, J. 2002. Nutrient-use efficiency and nutrient uptake in conventional and intensive (SRI) rice cultivation systems in Madagascar. Master's thesis, Crop and Soil Sciences Department, Cornell University, Ithaca.

Doberman, A.2004. A critical assessment of the system of rice intensification (SRI). *Agricultural Systems*, **79** : 262-281.

Husain, A.M.M. 2004. Evaluation of the System of Rice Intensification (SRI) Trials in Bangladesh: Results from PETRRA Sub-projects. Paper presented at the World Rice Research Conference (WRRC) Held in Tsukuba, Japan November 3-7.

Kar, S. et al. 1974.Nature and growth pattern of rice root system under submerged and unsaturated conditions, *Il Riso* (Italy), **23**: 173–179.

Kirk, G.J.D. and Solivas, J.L.1997. On the extent to which root properties and transport through the soil limit nitrogen uptake by lowland rice, *Eur. J. Soil Sci.*, **48**: 613–621.

Satyanarayana, A., Thiyagarajan, T.M. and Uphoff, N. 2007. Opportunities for water saving with higher yield from the system of rice intensification. *Irrigation Science* **25**:99–115.

Sheehy, J. E., Peng, S., Dobermann, A., Mitchell, P.L., Ferrar, A., Yang, J., Zou, Y., Zhong, X. and Huang, J. 2004. Fantastic yields in the system of rice intensification: Fact or Fallacy? *Field Crops Research*, **88**: 1-8.

Singh, A.K. and Chinnusamy, V. 2006.Aerobic rice prospects for enhancing water productivity. *Indian Farming*, 56(7): 58-61.

Stoop, W.A., Uphoff, N., Kassam, A., 2002. A review of agricultural research issues raised by the system of rice intensification (SRI) from Madagascar: opportunities for improving farming systems for resource-poor farmers. *Agric. Syst.* **71**: 249–274.

Surridge, C. 2004. Rice cultivation: Feast or famine? *Nature*, **428**: 360-361.

Uphoff, N. 2003. Higher yields with fewer external inputs? The system of rice intensification and potential contributions to agricultural sustainability. *International Journal of Agricultural Sustainability*, **1** (1): 38-50.

Improved Agriculture and Allied Practices for Farmers, pp 67-83
New India Publishing Agency, New Delhi, India

6

Package of Practices for Sugarcane Cultivation

Dinesh Kumar and Shweta Mehrotra

Division of Agronomy, ICAR-Indian Agricultural Research Institute, New Delhi-110012

Sugarcane is the most important sugar crop contributing more than 62% of the world's sugar production. With increasing human population in the world, sugar requirement has also gone up. Molasses is an important by-product of sugar industry. It is used for the production of alcohol by the distillaries. Alcohol serves as a base material in pharmaceuticals and several other industries. The manufacture of gur (jaggery) and khandsari as alternative sweetening agents in India from sugarcane also plays considerable role in the Indian economy, particularly in rural area. Nearly 40 % of the sugarcane produced is utilised for manufacture of gur and khandsari. Sugarcane is bound to play a greater role in the Indian economy in the years to come by offering a stable income to the farmers by way of cane price, by providing employment to rural masses, by realising excise duty, Cane Cess/ purchase tax and foreign exchange earning.

Acreage and production

Sugarcane is grown in tropical and sub tropical regions. Brazil has largest area and production of sugarcane.India is second to Brazil in sugarcane production. It is followed by Cuba and China. Other important cane growing countries are Pakistan, Mexico, South Africa, Australia, Indonesia etc. So far, as average sugarcane productivity is concerned Peru ranks first in average sugarcane yield. To meet the sugar requirement of increasing population in future, it is essential that sugarcane production is stepped up. Since there is no scope at present to bring additional land under sugarcane, productivity of sugarcane, the targets have to be achieved only through increased productivity.

Area and production of sugarcane in India

At present sugarcane is cultivated throughout the country except in hilly tracts in Jammu & Kashmir, Himachal Pradesh and North Eastern parts. The sugarcane growing areas are broadly classified into two regions viz., sub-tropical and tropical. Subtropical belt comprises of U.P., Bihar, Punjab, Haryana, Rajasthan, M.P., Orissa, Assam and West Bengal. It accounts for about 70% of the area but gives 50% of the total production of sugarcane in India. Tropical belt comprises of Maharashtra, Gujarat, Andhra Pradesh, Tamil Nadu, Karnataka and Kerala. Broadly there are five agroecological zones in India where sugarcane is grown.

North - Western Zone

North - Central Zone

North - Eastern Zone

Peninsular Zone

East Coast Zone

Of all the states, U.P. has the highest area under sugarcane followed by Maharashtra, Other important sugarcane growing state are Bihar, A. P., Tamil Nadu, Karnataka, Gujarat and Punjab.

Species of Sugarcane

Sugarcane belongs to the grass family and is classified in the tribe Andropogonae under Gramineae. The name of genus *Saccharum* is derived from Sanskrit word *Sharkara* meaning sugar. At present, the genus comprises three cultivated and two wild species. They are:

1. *Saccharum officinarum* Linn.
2. *Saccharum barberi* Jesw.
3. *Saccharum sinense* (Roxb) Jesw.

Wild species

1. *Saccharum spontaneum* Linn
2. *Saccharum robustum* Brand and Jesw

Some important characteristics of these are described below:

S. officinarum :- The species is characterised by thick culms in various colours with broad leaves. The inflorescence is big and showy. The clones are termed as chromosome number is $2n = 80$. The clones are noted for their high sucrose content and low non-sugars. The varieties are susceptible to diseases and pests.

S. barberi Jesw :- The species is characterised by thin stalks with narrow or medium leaves. Chromosome number varies from 2n = 82 to 124.The inflorescence is small. Well adopted to the sub-tropical region. Clones are tolerant to diseases and pests.

S. sinense (Roxb) Jesw :- The species is supposed to have originated from south east China and spread to India. Its canes are thicker and softer than those of *S. barberi* with broader leaves. The chromosome number is 2n = 118. The clones are susceptible to moisture stress and to mosaic and red rot diseases.

S. spontaneum L :- Clones of this wild species are small and bushy. Tillering is profuse and the clones exhibit marked resistances to moisture stress low temperature and number of diseases and pests. The chromosome number is 2n = 40 to 128.

S. robustum Brand and Jesw :- The wild species is endemic to New Guinea and neighbouring islands. The stalks are medium in thickness with broad leaves. The clones have a strolling habit with high fibre content. The chromosome number is 2n = 60 and 80.

Sugarcane Varieties

Before the introduction of the improved hybrid sugarcane varieties during the early part of the century, the clones of *S. officinarum* were grown in the tropical region of the world and the indigenous Indian varieties classified under *S. barberi* and *S. sinense* in the North Indian Belt. By the start of twentieth century the hybrid varieties replaced original varieties, and there was a spurt in yield per unit area. Important sugarcane varieties released for subtropical and tropical zones are given in Table 1.

Soil and Preparatory Tillage

Sugarcane can be successfully grown on various types of soils varying from sand to heavy clays provided climate, water, drainage, nutrient, conditions, soil texture and depth for water retention and root development are favourable and each type of soils is given its form of management. Moderately heavy and medium deep (1-2m) loams are better suited than heavier and shallow soils. The soil must be of good depth and drainage, with no salt and compactness. Ensuring a good seedbed is essential and is more so in sugarcane than most other crops, because the soil is prepared once every 2 to 3 years. The successful raising of plant and subsequent ratoon crop depends to a considerable extent on the seed bed provided to the plant crop. A poor preparatory tillage results in poor plant and ratoon crop. The soil should therefore be well prepared in improving soil tilth which contributes to good germination, stand and final yield of the crop.

Root system of sugarcane plant consists of a temporary non-extensive assemblage of sett roots, which are replaced by extensive system of shoot roots. The latter is more important for absorption of water and nutrients. The rooting pattern of different varieties by and large indicates maximum concentration at 50 cm depth. Hence it is essential that preparatory tillage is done at deeper layers for better spread of roots. Experiments have shown that deep tillage increases the cane yield and profit. After the harvest of previous crop, the field is deep ploughed with a soil turning plough. All large clods are broken and levelling done to facilitate irrigation and drainage. Subsoiling is done in Hawaii to a depth of 60 cm.

Planting Seasons

The duration of sugarcane crop in India ranges from 10 to 18 months. A twelve month crop is most common. Time of planting is governed mainly by weather conditions. Spring and autumn plantings are two important planting seasons.

Spring planting

In northern India, spring planting is done in February - March while in peninsular India it is done in January-February. Spring planted crop is known as *Suru* in Maharashtra and *Eksali* in Gujarat and Andhra Pradesh.

Autumn planting

In northern India, autumn planting is done in September- October while in Bihar and peninsular India, it is done in October-November. Autumn planting is also known as preseasonal planting in Maharashtra and Gujarat. The preseasonal crop matures in 13 to 15 months and supplies sugarcane during early crushing period.

Adsali planting

In Maharashtra and Karnataka, *adsali* planting is done in July- August and the crop matures in 16 to 18 months. Because of extended growing season there is increase in yield as well as sugar recovery. Biggest advantage of *adsali* is that it passes through only one summer season. Now-a-days, area under *adsali* planting is declining because of less availability of irrigation water.

Late Planting

Because of wheat crop taken on large area in *rabi* season in North and Central India, planting of sugarcane is delayed until harvesting of wheat in

March-April. Research results have proved conclusively that delay in planting causes marked yield reduction. Verma and Yadav (1986) reported that delay in planting from February to March reduced cane yield by 16 %. The loss increased to 20 % when delays are prolonged until April. The results of All India Co-ordinated Research Project on sugarcane revealed 24 tonnes per ha reduction over spring planting. Due to late planting, crop duration is reduced. This causes reduction in irrigation. For late planting varieties Co S 767, Co 1148, CoLK 7710, CoLK 8102, Co S 802, B. O. 91, CoS 7918 and CoS 88216 were found suitable.

Plant Material

Sugarcane is propagated commercially by vegetative methods which involve the planting of sections of the stem of immature cane. The planting material is known as seed pieces or setts. The true seed of sugarcane is unsuitable for producing commercial crop. The buds on the setts germinate to give plants.The cane used for planting should preferably be well grown immature plants of about 10-12 months old. It is excellent practice to establish a special cane nursery for quality seed production. Plant material should be healthy, rich in nutrient status and free from pest and disease incidence.

Planting Methods

In Australia and Hawaii, sugarcane is planted mechanically. In other countries planting is only partly mechanised. Manual/hand planting is followed is India. Planting is done in fairly deep furrows in which cane cuttings of 40-50 cm. are laid out and lightly covered. Indian Institute of Sugarcane Research, Lucknow has developed several planting techniques to sustain greater population of shoots per unit area. Three important systems of planting are.

Flat system which is common in U.P., Punjab, Haryana and Rajasthan.

Trench system which is followed in parts of U.P., Bihar, West Bengal, A. P., Tamil Nadu, Orissa and Kerala.

Pit method followed on slopy lands.

Recently, sugarcane cutter – planter with two discs and a ground wheel has been developed at IISR, Lucknow. It is about 40 % economical in form of money and energy for planting as compared to conventional method.

Seed Rate and Spacing

Sugarcane is planted mostly by opening a furrow with a ridger. Row to row spacing varies from 60 to 90 cm. In case of *adsali* planting, 120 cm row spacing is followed. Seed rate of 30000 setts of three buds, each is considered as optimum seed rate. In case where 2 budded setts are used, the number of setts will be above 80,000/- per ha.

Seed Treatment

To prevent the incidence of diseases, setts should be treated/dipped in a solution of carbendazim fungicide. If pest incidence is there, then diamethoate is to be added to the solution of fungicide for simultaneous dipping.

Optimum Shoot Population

At maximum photosynthetic efficiency of leaves and under ideal conditions of growth, sugarcane could yield 364 tonnes per ha of green stalks. Yields of this order have actually been harvested under ideal conditions of Hawaii and some other parts of world where optimum growing conditions with regard to sunshine, temperature, humidity, moisture and nutrition are ensured.Among the three yield attributes of sugarcane, number of canes and cane length could be altered by modifying the micro environment and providing optimum conditions for plant growth. It is difficult to change cane girth as it is genetically controlled.To get maximum possible yields of sugarcane about $1.8–2.0x10^5$ millable canes per ha each of two kilogramme weight are required. However, at present $1.0x10^5$-$2.0x10^3$ millable canes per ha each having 0.6 to one kilogramme weight are obtained. Therefore, there is a considerable scope to improve productivity by increasing number of millable canes per unit area.

Sugarcane Fertilization

Fertilization of sugarcane has been studied in India by number of research workers. Recommendations given on N, P_2O_5 and K_2O rates in different states one presented in the Table 1.

Table 1: Rates of fertilizer application recommended in different states of India for sugarcane.

State	Fertilizer Nutrients (kg/ha)		
	N	P	K
Maharashtra			
Suru	250	115	115
Pre-seasonal	350	170	170
Adsali	400	170	170
Andhra Pradesh Coastal districts	122-224	-	-
Telangana region	112-400	112	-
Tamil Nadu Coastal belt	110	-	-

Lift irrigation belt	90	-	45
Jaggery areas	70	-	45
Karnataka Mandya	247	99	124.00
Belgaum	247	74	186.00
Tungabhadra project area	247	74	74.00
Heavy rainfall area	185	124	124.00
Orissa	200	100	60
Gujarat	250	125	125
Kerala Pandalam and Thiruvella area	165	165	82.00
Chittur area	225	75	75
Newly cleared forest area	15	75	90
West Bengal	160	60	60
Uttar Pradesh	150	45	-
Assam	135	70	60
Bihar North Bihar	120	85	30
South Bihar	140	70	30
Rajashtan Kota and Udaipur	150	50	On soil-test basis
Sriganganagar region	100	50	On soil-test basis
Madhya Pradesh	120	34	24

Biofertilizers in sugarcane

Inoculation with *Azospirillum* and 75% of inorganic N increased the cane yield considerably. The increase was spectacular in sandy loam soil and moderate in red loam and deep clay soils. Results of 106 on-farm trials in T.N. indicate following: In well drained soils, where moisture is not a limiting factor, *Azospirillum* was superior. In semidry loamy and sandy soils, *Azotobacter* is a better source.

Water Management

As a general world average, 12000 to 13000 m^3 of water per ha per year is estimated to pass through sugarcane plants. Considering the efficiency of water use of 45 to 70 %, a quantity of 20000 to 30000 m^3 (200 to 300 cm ha^{-1}) of water is required in order to obtain satisfactory cane yields.

Scheduling of irrigation

Sugarcane requires adequate moisture supply in the root zone. Moisture stress results in setback in the growth which reduces yield and juice quality. Excessive moisture during grand growth period also affects the sugarcane root development and causes physiological drought. Hence, the crop should be given optimum moisture supply during peak growth phase.

Experiments conducted at Coimbatore indicated that there can be saving of water when irrigation was given at 50 % depletion of available soil moisture. Another irrigation experiment at Cuddlore indicated that irrigation at 50 per cent depletion at tillering and growth phases and 75 percent at maturity phase is advantageous in giving higher cane yields. In general, sugarcane crop should be irrigated at 10 days interval during summer and 15 to 20 days interval during winter. Water should be withheld about a month before harvests for quality improvement.

Irrigation methods

Different methods which are practised for irrigating sugarcane are:

surface irrigation – flooding / furrow method

Overhead irrigation – Sprinkler

Drip irrigation

Of the above methods, furrow method is most common method in India.

Management practices for moisture stress situations

In majority of the areas in India, the crop suffers due to shortage of irrigation water supply. The situation is much aggravated during summer when water is very much scarce. Loss of water through transpiration and evaporation from soil surface results in stunted growth and poor yields. For ameliorating the effect of moisture stress, following techniques have been developed.

i) Inducing hardiness: Calcium is known to induce hardiness and develop better root system. Soaking of seed material in saturated (80 kg lime + 400 litre of water) lime solution for two hrs. has been found advantageous.

ii) Use of antitranspirants: Cane yield can be improved by spray application of Kaoline during drought period.

iii) Potash application: Adequate potash is found necessary to protect the crop from water stress. Application of potassium at 60 kg K_2O/ha along with trash mulching in the furrows and removal of non-functional leaves have been found useful to the crop

iv) Foliar spray of urea and potash and trash mulching: The beneficial effect of urea with potash spray as well as mulching during formative phase was seen. Lakhdive et al. (1983) studied the effect trash mulching on irrigation requirement and it has been reported that trash mulching decreased number of irrigation from 30 to 20.

v) Drought tolerant variety : Research undertaken has led to the identification of Co 740 as a suitable variety for cultivation under drought condition from the point of view of cane yield and sugar recovery. In addition to Co 740, Co 285, Co 997, Co 6304 and Co 1148 were identified to be drought tolerant.

Interculture

Interculture refers to post planting operations done to crop to maintain the soil around the young plant in the right physical condition and free from weeds. Cultivation including hoeing, weeding, earthing are important intercultural operations. In a number of sugarcane growing states of India, prior to emergence of shoots, blind hoeing is done, which facilitates emergence of young sprouts. Subsequently two to three inter-row cultivations are done. In Maharashtra, a local modern implement dufan is used for interculture. At 4 ½ month stage or when cane has formed 2-3 internodes earthing up is done with the help of a ridger. After earthing up, furrows are converted into ridges, and ridges into furrows. Earthing up operation gives support to the shoots and crop does not lodge later on.

At 6 to 8 month age, crop attains a good height and becomes prone to lodging when high velocity winds are blown. Lodging of the crop causes loss in yield and sugar. It is, therefore, necessary to protect the crop from lodging. In subtropical belt, the crop may be bound by taking leaves from two adjacent clumps in a row twisting them into a rope and then tying the two clumps together. This operation is repeated after a month or so. This time, already tied clumps from two adjacent rows are brought together and tied by twisting their leaves.In tropical belt tying and propping of the crop is done when the crop attains a height of 130 to 150 cm. Another operation is carried out when the crop grows to a height of 200 cm or more.

Weed Management

Sugarcane being initially a slow growing crop, faces an acute competition from weeds.Due to favourable conditions like frequent irrigation, heavy fertilization, coupled with wider spacing weeds appear in rank growth and smother cane crop.Weed competition caused considerable yield losses. The loss is mainly due to restriction on the tiller production.To attain economical cane yields, weeds must be removed during critical period of competition, which falls in between 40 to 70 % days of planting.

The experiments have clearly shown that the suppression of sprouting and subsequent growth of weeds using pre-emergence herbicides was more efficient than allowing the weeds to grow and their removal by hand weeding. Several herbicides have been evaluated as pre- and post-emergence applications individually and in combination. Under field conditions, atrazine at 1.75 kg a.i./ ha has been found most effective as pre-emergence application in controlling most of the weeds except *Cyperus*. Application of Metribuzine as pre-emergence or early post-emergence proved quite promising in eliminating weed competition on sugarcane. Also, substitued urea herbicides namely diuron and linuron are in use for control of weeds in sugarcane. Diuron at 2.5 to 3.5 kg/ha when applied as pre-emergence proved to be more stable in its weed control potential.

Integrated Weed Management

Being a widely spaced crop, there is an ample scope of using cultural/ mechanical and chemical methods in combination so as to reduce dependance on either of these methods. In planted and ratoon crop, trash mulching followed by post emergence application of either contact or systemic herbicides has been found effective. Combination of herbicide use and interculture is also ecomonical and effective and it brings labour economy.

Intercropping systems

Growing two or more crops simultaneously on the same field is known as intercropping/companion cropping. Crop intensification is in both temporal and spatial dimensions. Until recently, sugarcane was cultivated as a single crop. But rapidly increasing population, insufficient food, limited scope for extension of cultivation to new areas, diversified needs of small farmers for food and cash etc. have forced the adoption of the system of intercropping. In Taiwan, intercropping in sugarcane is a standard practice, where it is done with rice in wet lands and with sweet potatoes, groundnut, soybean, cotton and maize in uplands.Much of the space between two rows of sugarcane remains unutilized for an initial period of 100 to 120 days, because of slow growth. Companion cropping offers opportunity for profitable utilization of available space.

Suitable intercrops

The crops suitable for companion cropping should have following characteristics. The duration of intercrop should be less than four months to avoid competition with sugarcane. Intercropping should not be competitive with the main crop in respect of sunlight, air, water and nutrients. The growth habit of intercrop should be erect type and should have dissimilar root system. The intercrop should not host insect and disease which might affect sugarcane adversely. Maximum cane yield was obtained when potato/onion were grown as companion crops with autumn planted sugarcane followed by potato/wheat, toria/onion. The best crops identified with spring planted sugarcane were mungbean and uradbean. Mungbean, sunflower or soybean as an intercrop in sugarcane field have been found to be profitable.

Pest Management

Sugarcane crop is infested by a number of pests from emergence till harvest. It is broadly estimated that about 10 % of the crop is lost annually due to pest incidence.Recently, major emphasis is given to integrated pest management rather than chemical control alone. In integrated pest management, the principal objectives are to maximise natural control forces like natural enemies, resistant varieties, together with cultural practices and to resort to use of chemical as the last effort. Important pests of sugarcane are stalk borer, shoot borer, top borer, pyrilla, scale insect, mealy bugs, termites etc.

Borers

The larvae/caterpillar of borers bore through shoot/stalk causing considerable damage. Following major species of borers have been listed in Table 2.

Table 2. Major species of borers

S.No.	Borer	Scientific names	Destribution
1.	Shoot borer	*Chilo infuscatellus*	All states
2.	Pink borer	*Sesamia inferens*	All states
3.	Green borer	*Raphimetopus oblutella*	North India
4.	Root borer	*Emmalocera depressela*	A. P., Karnataka, North India
5.	Top borer	*Scirpophaga excerptalis*	All states
6.	Stalk borer	*Chilo auricilius*	North India
7.	Gurdaspur borer	*Acigona steniellus*	Haryana, Punjab, U.P.
8.	Plassey borer	*Chilo tumidicostalis*	North East part

Highly susceptible varieties like Co J 64 promote infestation of top borer while Co C 671, favours internode borer. Resistant varieties have tight leaf sheath and hard rind. Tolerant varieties known for tolerance to different borers are as under:

1. Tolerance to Co 975, CoJ 64, Co 1148, internode borer Co 62198 and Co C 772
2. Tolerance to Co J 67, Co 1007, Co 1158 top borer Co 745, Co 1111, Co 1330, Co 7224
3. Tolerant to shoot borer Co 617, Co 758, Co 798.

Cultural practices

Advancing the time of planting in December – February reduces the severity of shoot borer infestation. Alternative hosts like *Saccharum spontaneum* for stalk borer may be destroyed from field bunds together with trees/bushes. Sugarcane may be rotated with other crops.

Mechanical control: Collection and destruction of borer infested shoots / dead hearts.

Biological Control: Species of *Trichogramma* have been tried extensively but failed to control shoot borers because of unfavourable weather during summer which is adverse for parasite. The tachinid parasite *Sturmiopsis inferens* was found effective for control of shoot borers. Another effective parasite is *Isotime javensis* introduced from North India to Tamil Nadu and Karnataka for control of top borers. It establishes rapidly in new areas even when top borer population is low.

Pyrilla

Pyrilla perpusilla is a sucking pest of sugarcane. The pest is active from April to October. The pest can be controlled by spraying 0.05 per cent malathion 50 EC. Biological control of pest with ectoparasite *Epiricania* (*Epipyrops*) *melanoleuca* is very effective. Release of 5000 cocoons or 5 lakh eggs per ha gives good suppression of the pest.

Scale insects

Scale insects such as *Melanaspis glomerata* cause damage by sucking cell sap of the stem. Infested canes become shrivelled, growth is stunted and juice quality is impaired. The incidence of these pests is heavy in peninsular India. Dipping seed material before planting in 0.1 per cent malathion emulsion is effective preventive measure. Pest can be controlled by spraying malathion 0.1 per cent spray after detrashing.

Two indigenous predators, *Chilocorus nigritus* and *Pharascymnus horni* were released for control of scale insects in Andhra Pradesh. The predators have considerable feeding potential. The parasite, *Botryoideclava bharatiya* has been established in Andhra Pradesh and Tamil Nadu.

Disease Management

Of the 147 diseases listed as affecting sugarcane crop over the world, 84 have been reported in India. There are seven diseases of economic importance in India namely:

Fungal disease	:	Red rot, smut, wilt and pineapple
Virus disease	:	mosaic
Bacterial disease	:	Ratoon stunting
Micoplasma	:	Grassy shoot

Red rot

Red rot has been endemic in areas like north Bihar and eastern U.P. and has been greatest obstacle in sugarcane cultivation. Excepting Maharashtra and Karnataka the disease occurs in all the states.Varieties which are susceptible show the red coloration of inner tissues throughout the length of the stalk. The fungus also produces lessions on the midrib and reddish areas in the leaf sheath. The varieties resistant to the disease and currently in cultivation in India are: Co 1148, Co 1336, Co 6498, Co 623 99, Co 6304, Co C 671, Co S 659, Co S 698, Bo 70.

Smut

In India, smut is prevalent in all the states and is more infectious in the southern states. The losses due to smut up to 29 % has been reported in plant cane. Ratooning of infected cane is known to increase the infection. The characteristic symptom of the disease is the production of a long whip like structure from the terminal bud of the stalk. Resistant varieties to smut are Co 449, Co 527, Co 853, Co 1148, Co 1336, Co 62175, and Co 6806. Besides the use of resistant varieties, rouging of disease affected clump is an effective method of control. The standard hot water treatment of 50 °C for 2 hrs. partly controls smut.

Ratoon stunting disease (RSD)

The disease occurs throughout the sugarcane growing areas of the country. It affects plant as well as ratoon canes but the effect is more pronounced on ratoons. Affected crops look stunted in growth. Maintenance of heat treated seed nurseries is the best control measure for the disease.

Grassy shoot disease (GSD)

The diseare has been recorded throughout India. Affected shoots show excessive and stunted tillers with narrow leaves. The canes formed are stunted with leaves being yellow or white. The whole stool appears to be like a big bunch of grass.

Ripening

Ripening of sugarcane refers to rapid synthesis and storage of sucrose in the stalks. Accumulation of sugar in the stalk begins soon after completion of elongation phase when glucose produced during photosynthesis is not utilized for vegetative growth but stored in stalks largely as sucrose. When concentration of sucrose exceeds 16 percent in the juice and purity increases over 85 per cent, the cane crop is said to be matured. As the crop advances in maturity, its water content decreases, sucrose content increases and reducing sugars decrease. Both organic and in-organic non sugars also decrease. At peak maturity, sucrose content is at maximum and reducing sugars and non-sugars are at minimum.

Ripening is influenced by a number of factors, of which, climate, varieties and nutrition are most important. Cool dry weather is the most effective single factor in inducing ripening. During maturity phase, bright sunshine during day and a day temperature of 28–30°C and low temperature (12–14°C) during night coupled with low relative humidity favour ripening process and accumulation of sucrose. The RTD (relative temperature disparity) around 55 and above temperatures during maturity phase is highly favourable for sugarcane ripening, sugar accumulation and quality.

Ratooning of Sugarcane

Sugarcane is unique among the agricultural crops in the sense that a number of succeeding cane crops are raised from single planting. The crop raised from planting cane setts is called as plant crop. After plant crop of sugarcane is harvested, the underground portions of stem stubble are left in the field. The buds of the stubbles sprout and give rise to a succeeding crop referred to as ratoon crop. A practice of taking a ratoon crop is known as ratooning. Ratooning is an important integral component of sugarcane production system.

Ratooning has following three advantages: i) Expenditure on seedbed preparation planting material and planting operation is reduced ii) Ratoons may give equal or more yield than plant crop, and iii) Ratoons mature earlier than plant cane.

Number of ratoon crops taken varies. In Cuba, 10-11 ratoon crops are taken. In South Africa and U.S.A., 4 to 6 ratoons are raised whereas in Hawaii, Brazil and Australia 2 to 3 ratoons are taken. In India, ratoon crop generally yields less than the plant crop and its yield goes down with successive ratooning. Very little attention is paid to its culture and manuring. In India, generally one ratoon is recommended. On an average 30 to 40% of the total sugarcane area is under ratoon crop. Due to poor yields of ratoon, the average productivity of sugarcane is low.

Ratoon management for higher productivity

- A good plant crop stand invariably gives a good ratoon crop. Plant crop should be raised by following improved package of practices, namely, selecting ratooning variety, judicious use of manure and fertilizers, appropriate plant protection etc.
- Plant crop should be harvested in February-March to ensure favourable temperature for the regrowth of ratoon sprout.
- Sugarcane varieties differ in then ratooning ability. Generally early maturing varieties are poor ratooners. Thin or medium thin varieties are better ratooners than thick varieties.
- Harvesting close to ground level is desirable for a good ratoon crop. If harvesting is done unevenly, stubble shaving should be done.
- After harvesting of plant crop, the trash left in the field should be burnt to clean the field and destruction of eggs and pupae of insect pests.
- Shoulder breaking or off barring is essential for decaying stubble roots and formation of new shoot roots.
- Gap filling is highly essential. For this purpose young sprouts from the stubbles are desirable or single bud setts of matching age may be used.
- Ratoons are less efficient utilizers of N than plant crop. Therefore comparatively higher dose of nitrogen is required. Response to phosphate and potash is more consistent. Ratoon crop should be fertilized judiciously.
- Crop should be irrigated at 60% depletion of available soil moisture. Thus ratoon crop tolerates relatively more moisture stress.
- Ratoon crop needs more plant protection care than a plant crop. Grassy shoot disease, ratoon stunting disease and smut are the major diseases associated with ratooning. Preventive measures should be taken. Ratoon chhorosis occurs in calcareous soil which may be controlled by spraying 0.5 % $FeSO_4$ solution.

Sugarcane varietal scene over the decades

Name of variety	State (s) for which recommended	Key characteristics			
		Maturity	Cane Yield (t/ha)	Sucrose (%)	Resistantto
Co 85004	Gujarat, Maharashtra, Karnataka, Kerala, Interior of Tamil Nadu & A.P., Madhya Pradesh &Chattisgarh	Early	90.5	19.5	Smut
Co 86032	-do-	Midlate	102.0	20.1	Smut
Co 87025	-do-	Midlate	98.2	18.3	Smut
Co 87044	-do-	Midlate	101.0	18.3	Smut
Co 8371	-do-	Midlate	117.7	18.6	Smut
CoM 88121	-do-	Midlate	88.7	18.6	-
Co 91010	-do-	Midlate	116.0	19.1	Smut
Co 94008	-do-	Early	119.8	18.3	Red rot
Co 99004	-do-	Midlate	116.7	18.8	Red rot
Co 2001-13	-do-	Midlate	108.6	19.03	Red rot, Smut, Wilt
Co 2001-15	-do-	Midlate	113.0	19.37	Red rot, Smut
Co 0218	-do-	Midlate	103.77	20.79	Red rot
Co 86249	Coastal Tamil Nadu & Andhra Pradesh and Orissa	Midlate	104.2	18.7	Red rot, Smut
CoC 01061	-do-	Early	110.8	17.4	Red rot
CoS 91230	Punjab, Haryana, Rajasthan, Central & Western U.P. &Uttarakhand	Midlate	68.2	18.8	Red rot
CoPant 90223	-do-	Midlate	73.3	18.5	Red rot, Smut
CoH 92201	-do-	Early	70.0	18.2	Red rot

CoS 95255	-do-	Early	70.5	17.5	Red rot
CoS 94270	-do-	Midlate	81.5	17.1	Red rot
CoH 119	-do-	Midlate	82.8	17.5	Red rot
Co 98014	-do-	Early	76.3	17.6	Red rot
CoS 96268	-do-	Early	69.8	17.9	Red rot
CoPant 97222	-do-	Midlate	88.2	18.2	Red rot
CoJ 20193	-do-	Midlate	75.9	17.9	Red rot
CoS 96275	-do-	Midlate	80.8	17.3	Red rot
Co 0118	-do-	Early	78.2	18.45	Red rot, Smut, Wilt
Co 0238	-do-	Early	81.08	17.99	Red rot, Wilt
Co 0124	-do-	Midlate	75.71	18.22	Red rot
Co 0239	-do-	Early	79.23	18.58	Red rot
Co 87263	Earstern Uttar Pradesh, Bihar, West Bengal and Jharkhand	Early	66.3	17.4	Red rot, Smut
Co 87268	-do-	Early	78.9	17.5	Red rot, Smut
Co 89029	-do-	Early	70.6	16.3	Red rot
BO 128	-do-	Midlate	69.2	17.6	Red rot, Smut
CoSe 95422	-do-	Early	67.8	17.7	Red rot
CoSe 92423	-do-	Midlate	70.1	17.5	Red rot
CoSe 96234	-do-	Early	64.1	17.9	Red rot
CoSe 96436	-do-	Midlate	67.1	17.7	Red rot
CoLk 94184	-do-	Early	76.0	18.0	Red rot
Co 0232	-do-	Early	67.82	16.51	Red rot, Smut, Wilt
Co 0233	-do-	Midlate	67.77	17.54	Red rot, Smut, Wilt
Co 0232	Assam	Early	67.82	16.51	Red rot, Smut, Wilt
Co 0233	-do-	Midlate	67.77	17.54	Red rot, Smut, Wilt

Section 2: Horticultural Production Technology

Improved Agriculture and Allied Practices for Farmers, pp 85-92
New India Publishing Agency, New Delhi, India

7

Protected Cultivation of Horticultural Crops

Mam Chand Singh and Praveen Kumar Singh

Centre for Protected Cultivation Technology, ICAR-Indian Agricultural Research Institute, New Delhi-110012

India has made a considerable progress during the last three decades in horticulture production in India producing 300.6 million tonnes of horticultural produce from 24.85 million hectares in 2016-17. The average productivity of different horticultural crops is quite low as compared to the best yields at national and world level, indicating the scope to enhance the yield in different horticultural crops by many fold along with its quality specially in vegetables. Unlike vegetables, flower cultivation is growing at a very fast speed and registered a marked growth along with employment generation through protected cultivation. According to NHB (National Horticulture Board), during the year 2014-15 the area under floriculture production in India was 248.51 thousand hectares with a production of 1,658 thousand tonnes loose flowers and 484 thousand tonnes cut flowers. However, India's total export of floriculture was Rs. 548.74 crores/ 82.05 USD Millions in 2016-17. This is because of an appealing gap between potential yield and the actual yield of various vegetable crops. In many vegetable crops, the yields realized are even less than 50 % of their potential yield. Due to fluctuating market prices and poor storage infrastructure available for horticultural crops, farmers are not able to get their production cost. Though, there are some crops, fetches high prices during off-season. Several biotic and abiotic stresses are major factors, responsible for low productivity and poor quality in several horticultural crops under their open field cultivation during rainy and post rainy season more specifically, in vegetables. The increasing demand of off-season and high quality vegetables in various markets of the big cities has drawn the attention of the vegetable growers for diversification from traditional system.

Protected cultivation of horticultural crops offers the best choice for diversification from traditional agriculture production system for a number of reasons. Production of crops under protected conditions has great potential in augmenting production and quality of vegetables, flowers and in some fruit crops in main and off season and maximizing water and nutrient use efficiency under varied agro climatic conditions of the country. This technology has great potential especially in peri urban agriculture, since it can be profitably used for growing high value vegetable crops like, tomato, cheery tomato, coloured peppers, parthenocarpic cucumbers; flowers like cut flowers, chrysanthemum, rose, gerbera, carnation and lilium; fruits likc strawberry, grapes etc. and for off season cultivation of vegetables and their healthy and virus free seedlings production.

Protected cultivation in partially modified environment structure is useful in combating both biotic and abiotic stresses that limit the productivity and quality of horticultural crops. This requires careful planning and attention, including timing of production and harvest to coincide with high market prices, choice of varieties adopted to the off season environment, and able to produce economical yields of high quality produce etc.

Large number of farmers, who are living in peri urban areas of the country can successfully diversify their traditional agriculture by adopting or using various levels of protected cultivation technologies for production of horticultural crops looking to their resources, availability of emerging markets of usual and unusual off season horticultural produce, year round demand of high value vegetables like standerd tomatoes, coloured peppers and parthenocarpic cucumbers etc. High quality nursery raising in vegetables is another area, where improvement over the traditional system of nursery raising is required. Under Indian conditions, all kind of protected technologies may not be economical and useful to the farmers, because of their very high initial, running and maintenance cost, but some protected technologies are simple and highly profitable under Indian conditions and more specifically for peri-urban areas, which can be adopted by Indian farmers for production of different horticultural crops in the following manners.

1. Plug tray Seedling Production Under Protected Conditions

To ensure high productivity and high quality of the produce, raising of high quality seedlings through use of good quality seeds at right time and at a appropriate place is one of the cheapest but most important way. Most of the Indian farmers are raising their vegetables, flowers and fruits like papaya seedlings under open field conditions, which is always inferior in quality, as the seedlings are infected with virus when raised in open during rainy and post rainy season. On one side soil borne fungus and nematodes create severe problem for raising the seedlings in soil media in open fields during hot summers and rainy season but on the other hand the very high cost of hybrid seeds in vegetables has also warranted the farmers to improve or change their traditional nursery raising method to increase the productivity and quality of vegetables. Plug tray nursery raising in vegetable crops has already working as a industry in several European countries, Israel, USA, Morocco, Turkey, Japan and China. It is rapidly increasing in India also under this system seedlings are raised in plastic multicelled plug-trays in artificial soil-less media in especially designed greenhouses or other protected structures. A large number of virus free healthy seedlings of different vegetables can be raised in a small area of green house in plastic plug-trays by using soil-less media for growing vegetables either for main season or for their off season cultivation. With the use of this technology it is now almost possible to raisehealthy vigorous seedlings of different cucurbits and corn, otherwise it was not possible in the traditional system of nursery raising. The farmers or unemployed agriculture graduate youths of our country can very successfully start nursery raising as an agri-entrepreneur in major vegetable growing pockets of the country. By this way the vegetable growers will get the virus free or off-season healthy nursery as per their requirement and it will also generate additional employment in agriculture sector. Therefore, this is the first and most important step for enhancing productivity and quality in horticulture through protected vegetable cultivation.

2. Off-Season Cultivation of Vegetable Crops Under Plastic Low Tunnels

In most parts of our country the farmers are growing various vegetables during their main season of cultivation, but the prices of those vegetables are very low and sometime the vegetable growers are even not getting back the cost of cultivation of the vegetables. But the same vegetables are sold on very high price during their off-season in several cities of our country. The demand of off-season vegetables is increasing day by day in several big and medium sized cities of the country, which provide wide scope of vegetable production through off-season cultivation of some vegetables mainly the cucurbits in peri-urban areas of the country. Plastic low tunnel technology is a simple and profitable technology for off-season cultivation of cucurbits during the winter season in northern plains of our country. Crops like summer squash can be grown as a complete off-season crop, whereas other cucurbits like muskmelon, round melon, bottle gourd, cucumber, bitter gourd, watermelon can be advanced by 30-40 days over their normal growing season.

Plastic low tunnels are flexible transparent coverings that are installed over single or multiple rows of vegetables to enhance the plant growth by warming the air around the plants in the open field during winter season when the temperature is below 80°C. Plastic low tunnels are often used to promote the growth of plants during the period of winter season. Low tunnels are supported above the plants by using hoops of GI wire and a clear or transparent plastic of 20-30 micron is covered/stretched over the hoops and the sides are secured by placing in soil. The plastic is vented or slitted during the growing season as the temperature increase with in the tunnels. The farmers can grow different varieties of summer squash (round fruited, long fruited) which is a emerging crop along with cultivation of netted muskmelon varieties in place of traditional varieties. Bitter gourd and round melon are two other crops with increasing demand and which usually fetches very high price during off-season and can be grown successfully by using the plastic low tunnel technology. This technology is highly suitable and profitable for the farmers living in northern plains of India.

3. Off- Season Vegetable Cultivation Under Walk in Tunnels

Walk in tunnels are the temporary structures erected by using G. I. pipes & transparent plastic. Walk in tunnels are used for complete off season cultivation of vegetables like bottle gourd, summer squash, cucumber etc. during winter season (Dec. – mid February) the basic objective & utility of walk in tunnels is to fetch high price of the complete off season produce to earn more profit per unit area. The ideal size of a walk in tunnel can be of 4.0 m width and 30m length (120 m^2) and total cost of fabrication may be Rs.15000-18000/-.

Objectives & Utility of the walk in tunnels

1. Off season cultivation of vegetables for earning more income.
2. Walk in tunnels are only erected over the crop during the peek winter months of December - mid February and there after the structure is removed from the crop.
3. Since, the plastic is used only for two months (December to mid February) therefore, life of the plastic can be 8-10 years.
4. These temporary plastic structures are suitable for off season vegetable cultivation in northern plains and low hills.

4. Nursery raising for advancing vegetables ahead to their normal season

If the vegetable growers are not so resourceful to use plastic low tunnels, they can advance these cucurbits even under their open field cultivation. For advancing these crops, first seedlings of these crops are raised under protected conditions in plastic pro-trays in artificial media. Because under this system of nursery raising the root development of the seedlings is vigorous so that there are no chance of damage to the root or shoot parts of the seedlings. After raising the seedlings these cucurbits can be directly transplanted in open after first week of February, when there are no further chances of frost in northern plains of India. By this way these crops can be advanced by 20-30 days over their normal and traditional system of cultivation of these crops by seeds.

5. Insect Proof net Houses for Virus free Vegetable Cultivation

Usually the farmers are growing their vegetable crops like tomato, chilli, sweet pepper, okra etc. under open fields. But during rainy and post rainy season it is very difficult to grow these crops successfully due to leaf curl and yellow vein mosaic and other viruses, respectively. These viruses are mainly spread by insect vectors like whitefly or aphids. The population of white fly after on start of monsoons is very high and it remains in the open environment upto end of October depending upon the temperature. The farmers are using several insecticides for several sprays to control these vectors, even they could not control these vectors and their tomatoes, cucumber, chilli or okra crops are highly infected with viruses. The second most common and most severe problem in tomato, brinjal and okra is the fruit borer against which also the growers are using huge amount of insecticide even they are unable to control this insect. The only way to control the virus and fruit borer is to put a mechanical barrier between the crops and open environment and this is possible with the use of insect proof net of 40 or 50 mesh in form of net houses or insect proof net covered walk in tunnels. By this way the growers can directly reduce the use of insecticides and they can grow virus free crops of tomato, chilli, sweet pepper and okra during rainy or post rainy season. But for growing these crops under insect proof net houses, it is pre-requisite to raise virus free healthy seedlings of these crops either in the greenhouse or by covering the nursery beds with insect proof net. The farmers can erect these insect proof net houses by using half inch size GI pipes after bending them in half circle shape. Other insect proof net houses can also be made by covering all sides and top with insect proof net of 40 or 50 mesh, but the net should be UV stabilized. Under these net house crops like sweet pepper, tomato chilli or okra can be grown successfully without infestation of viruses or other insects like fruit borer etc. and the growers can save the huge amount spent on pesticides.

6. Vegetable Cultivation Under Shade net Houses

Shade nets are perforated plastic materials used to cut down the solar radiation and prevent scorching or wilting of leaves caused by marked temperature increases within the leaf tissue from strong sunlight. These nets are available in different shading intensities ranging from 25% to 75%. Leafy vegetables and ornamental greens are recommended to be grown under shade nets whose growth rates are significantly enhanced compared to unshaped plants when sunlight is strong. The basic objective of shade net is to reduce radiation and temperature up to some extent during critical summer months (May-Sept). Black colour shade nets are most efficient in reduction of temperature compared to other colours like green, white or silver etc. as the black colour in the maximum absorbent of heat. Mostly leafy vegetables like beet leaf and green coriander are preferred to be grown under shade nets, but it is also suitable for growing early cauliflower and radish cultivation during June to September months.

7. Low Cost Polyhouses for Vegetable Cultivation

Low cost polyhouses are the other structures, which can be used successfully and efficiently for off-season vegetable production or for virus free healthy nursery raising purposes. These low cost polyhouses can be erected by using GI pipes of 1.0 and 1/2 inch. The 1.0 inch pipe is useful erecting the side walls upto 6-7 feet in height and on the top the ½ inch size bended GI pipes are fixed in the 1.0 inch size pipes. The width of these structures is 5 to 5.5 feet. The top of the structure is covered with plastic of 180-200 micron, whereas the sidewalls are first covered with insect proof net of 40 or 50 mesh up to 5-6 feet height from the ground after placing it in soil and a rollable plastic is fixed over the net. During summer one can roll the all side plastic for cross ventilation in the structure and it can be opened during the winter season to increase the inside temperature of the structure.These structures can be used for off-season or virus free cultivation of vegetable or for year round virus free healthy nursery raising of different vegetables. These structures can be made with in Rs. 300-350 per m^2 cost and the farmers themselves can made these structures according to their requirement at village level by using rural artisans.

8. Naturally Ventilated Greenhouses/polyhouses for High Value Vegetables and Cut -Flower Production.

Naturally ventilated greenhouses are the protected structures where no heating or cooling devices are provided for climate control. These are simple and medium cost greenhouses which can be erected with a cost of Rs.900-950 per sq. meter and these greenhouses can be used successfully and efficiently for growing year round parthenocarpic slicing cucumber, off season muskmelon, tomato and sweet pepper crops for 8-9 months duration. These structures are having a manually operated cross ventilation system as and when required.

Looking to the year round, increasing demand of high quality parthenocarpic slicing cucumber in upmarkets of the metro and other big cities of the country, this is one of the most suitable and profitable crop for cultivation under naturally ventilated green houses in peri-urban areas of the country. Three successful crops of cucumber can be grown in a naturally ventilated greenhouse in a period of one year. Muskmelon is the second crop, which can be successfully cultivated for its complete off-season availability, which can fetch very high price of the off-season produce in the up markets of the metro and other big cities of the northern parts of the country.

Similarly high value vegetables like standard tomatoes, cherry tomatoes and coloured peppers are three crops which can be grown for long duration (8-10 months period) under naturally ventilated greenhouse conditions. In metro cities like Delhi NCR, there is year round demand of these high quality vegetables in the up markets viz. five star hotels, shops of embassies or high commissions of various countries situated in Delhi. They are ready to pay very high price for the high quality produce; therefore, it may be a profitable venture, if this technology is adopted around metro cities of the country. Greenhouse vegetable production is a highly intensive enterprise requiring substantial labour and 24 hour-a-day commitment, which restrict the adoption of this technology. But now the time has come when the vegetable growers in various parts of the country can use the naturally ventilated green house technology for cultivation of high value vegetables for high profits.

Among the flowers, cut flower production under protected structures has increased 480 % during the last 10 years. However, the area under floriculture has gone up to 2,50,000 ha registering about 400% increase which includes approximate 1,26,235 ha under cut flowers inclusive of 5,600 ha under protected structures. India has to achieve the ambitious export target of Rs.1,000 crores per annum over the next 5 years with an ever increasing domestic demand for quality cut flowers and foliage potted plants. The scientific Institutions and industry need to prioritized with economies of scale, range of flowers and their products, suitable varieties, quality control and certification, cold chain management and transit facilities. Protected cultivation, although is in limited area (approximate 5% of the total crop area), its contribution to the flower trade is significant. At present there are about 240 export oriented units in operation covering an area of approximately 3600 ha. These units are growing mostly roses. However, the science of producing cut flowers especially in such crops that are responsive to the photoperiods and light acclimation to produce a large number of cut stems per unit area are being subjected to and have been reported to produce 20 to 35 % more yield with desired quality. The application of LED lighting technology in Europe and US has proven such changes in the cut flower industry and production of compact plants at large scale. Rose, chrysanthemum and gerbera are highly remunerative while cultivated in climate controlled greenhouse and fetch handsome income at cost benefit ratio of 1:3.6, 1:6.20 and 1:3.72, respectively.

Keeping in view the increasing demand of off season and high value vegetables, fruits and flowers in several big cities of the country, there is a urgent need for diversification from the traditional agriculture by production of high value horticultural crops under different protected conditions for increasing their productivity and quality for getting high returns. Nursery raising under protected cultivation can be adopted as a agri entrepreneur in major vegetable growing areas of the country by unemployed youths who are graduate in agriculture or post graduate in horticulture. Low cost protected technology like plastic low tunnels or walk in tunnels can be used for off season vegetable cultivation for getting high returns from off season produce. Similarly insect proof net houses can be used on a large scale for safe vegetable cultivation by way of minimizing the use of pesticides in vegetable cultivation and virus free seed crop production in large number of vegetables.

Improved Agriculture and Allied Practices for Farmers, pp 93-102
New India Publishing Agency, New Delhi, India

8

Hybrid Vegetable Seed Production

B.S. Tomar, Girish Kaddi and Kalyanrao

Division of Vegetable Science, ICAR-Indian Agricultural Research Institute New Delhi-110012

India has emerged as the second largest producer of vegetables 176.17 mt after China 573.93 mt in the world and is gradually striding towards achieving food and nutritional security at national level. The average productivity of vegetable crops is 17.6 t/ha (NHB, 2015) which is comparatively lower than the productivity of China (23.4t/ha), Spain (39.3t/ha) and USA (32.5t/ha). The lower productivity could be attributed to limited use of quality seeds, inefficient production management, impact of biotic stress and climatic factors during crop production. Among the limiting factors limited availability of quality seed is one of the important factors. Since, the performance of other inputs is also depend on the quality of seed. At present the seed replacement rate among vegetable crops is around 25%, whereas highest in cabbage (>90%) followed by tomato, capsicum, bitter gourd, muskmelon, bottle gourd etc.The use of hybrid seed is in increasing due to higher productivity, tolerance to biotic and a biotic stresses and better market value than the open pollinated varieties but the higher price of the hybrid and their availability is still a constraint in adopting the use of hybrid seeds among small and marginal farmers.

Role of hybrids in vegetable productivity

The average productivity of vegetable during 1991-92 was 10.46t/ha which has increased to 17.6 during 2013-14 (Table 1). The tremendous improvement in vegetable productivity was realized after the introduction of new seed policy, 1988 which came into operation 1989 and has allowed the import of hybrids under OGL from different countries. As a result the hybrids in number of crops were introduced by NSC (cabbage; Green Boy and Green Express) and by private sector companies (tomato, sweet pepper, summer squash and parthenocarpic cucumber).

Table 1: Vegetable productivity at a glance (1991-92 to 2013-14)

Year	Area (000ha)	Production (000MT)	Productivity (t/ha)
1991-92	5593	58532	10.46
2001-02	6156	88622	14.39
2011-12	8989	156220	14.4
2012-13	9205	162180	17.4
2013-14	9396	162890	17.6

The impact of the use of hybrids in improving the crop productivity especially cabbage, tomato, brinjal and onion has been realized in many vegetable crops (table 2). This steady increase in productivity has been achieved mainly by the use of quality seeds with built in inbred and hybrid vigor, coupled with the application of modern technology and sound government policies.

Table 2: Comparative productivity of selected vegetable crops

Crop	2001-02	2011-12	2012-13	2013-14
Tomato	14.7	20.6	20.7	21.2
Brinjal	16.6	18.3	18.6	19.1
Cabbage	15.6	21.6	22.9	22.6
Cauliflower	14.8	18.8	19.6	19.8
Onion	14.2	16.1	16.0	16.0
Okra	8.5	12.1	12.0	11.9

Development of hybrids in vegetable crops

Vegetable seed industry in India with a history of more than 80 years was quite slow to pick up the potential of hybrid seeds. Wide range of scepticism pertaining to the high cost of hybrid seeds prevailed in the country up to 1973. The NARS also laid major emphasis to exploit inbred vigor rather than hybrid vigor in vegetables. It was in 1973 when Bangalore based Indo-American Hybrid Seed Company from the private sector launched its first successful hybrid in tomato and capsicum in India, amply demonstrating to the Indian farmers, good returns possible for the money invested in hybrid seeds.

The emphasis has been laid out in the development of hybrids and testing under the All India Coordinated Improvement Project on vegetables during mid 1980's and has resulted the identification and releases 151 F1 hybrids in various vegetable crops. There are number of F1 hybrids developed by public sector organization which are popular and multiplied by national and state level seed corporations (table 3) and many private sector companies through MTA.

Table 3: Popular public sector hybrids in vegitable crops under seed production in India

S.No.	Crop	Hybrids	Source
1.	Tomato	Pusa Hybrid-1, Pusa Hybrid-2, Pusa Hybrid-4, Pusa Hybrid-8, Pusa Divya (Kt-4)	IARI, Delhi
		Arka Rakshak, Arka Ananya, Arka Samrat, Arka Shreshta, Arka Vishal, Arka Vardan, Arka Abhijit	IIHR, Bengaluru
		Kashi Abhiman	IIVR, Varanasi
		Pant Hybrid-1, Pant Hybrid-2, Pant Hybrid-10, Pant Hybrid-11	GBPUAT, Pantnagar
		Rajashree, Phule Hybrid-1	MPKV, Rahuri
2.	Brinjal	DBHL-20, Pusa Hybrid-5 (Long), Pusa Hybrid-6 (Round), Pusa Hybrid-9, Pusa Anupama(Kt-4)	IARI, Delhi
		Arka Navneet	IIHR, Bengaluru
		Kashi Sandesh	IIVR, Varanasi
3.	Chilli	CH-1, CH-3	PAU, Ludhiana
		Arka Meghana, Arka Harit, Arka Sweta	IIHR, Bengaluru
		Kashi Early, Kashi Surkh	IIVR, Varanasi
4.	Sweet-pepper	Pusa Deepti, KTCPH-3	IARI
5.	Cucumber	Pusa Sanyog	IARI
6.	Bitter-gourd	Pusa Hybrid-1, Pusa Hybrid-2	IARI, Delhi
	Bottle-Gourd	Pusa Hybrid-3	IARI, Delhi
		Kashi Bahar	IIVR, Varanasi
		Pant Sankar Lauki 1	GBPUAT, Pantnagar
		Narendra Sankar-1	NDUAT, Faizabad
7.	Musk-melon	Pusa Rasraj	IARI, Delhi
		Punjab Hybrid-1	PAU, Ludhiana
8.	Pumpkin	Pusa Hybrid-1	IARI, Delhi
9.	Summer squash	Pusa Alankar	IARI, Delhi
10.	Water-melon	Arka Jyoti	IIHR, Bengaluru
11.	Cauli-flower	Pusa Kartik Sankar, Pusa Hybrid-2, Pusa Snowball Hybrid-1	IARI, Delhi/ R.S Katrain
12.	Cabbage	Pusa Cabbage Hybrid-1	IARI, Delhi

13.	Carrot	Pusa Vasudha, Pusa Nayanjyoti	IARI, Delhi/R.S Katrain
14.	Onion	Arka Lalima, Arka Kirtiman, Arka Bhima	IIHR, Bengaluru
15.	Okra	Kashi Bhairav	IIVR, Varanasi
16.	Ashgourd	Pusa Shreyali and Pusa Urmi	IARI, Delhi

Areas suitable for Hybrid seed production

The temperature, photoperiod, rainfall pattern, amount and distribution greatly affect flowering, seed set, seed development and maturation, therefore due consideration should be given at the time of selection of site for hybrid seed production to the above factors. Besides the climatic factors the availability of skilled labour, irrigation facility, mode of transport and security should be considered. Thus the various locations in northern Karnataka (Ranebannur, Haveri, Dharwad, Koppal, Belgum), southern Karnataka (Bangalore rural, Ramnagar, Chikkaballapur), Andra Pradesh (Vijaywada, Guntur, Karimnagar, Anantpur), Maharastra (Jalna, Aurangabad), Gujrat (Rajkot, Jamnagar) have been commercially exploited for hybrid seed production of tomato, brinjal, chilli, cucumber, pumpkin, capsicum, okra, gourds, melons and onion. The hybrid seed production of tropical carrot, radish and cauliflower is produced in south-west irrigated and northern part of Haryana. There are various locations (Sproon valley, Kullu valley, Kalpa valley, Kashmir valley) which are suitable for hybrid seed production of temperate vegetable hybrids. However the temperate locations are not being exploited due to shift in commercial vegetable production or security reasons.

Method/systems of hybrid seed production

The hybrid seed of vegetables crops are being produced by different methods and systems commercially in different vegetable crops (table 4). Hybrid seed in tomato, brinjal, capsicum and chilli are produced through hand emasculation and pollination, bottle gourd, bitter gourd and pumpkin through protection of female flower and hand pollination, cucumber through natural pollination in case of gynoceious seed parent, onion, cabbage and cauliflower hybrid seed are being produced by utilizing the CMS and SI system respectively where as the hybrid seed of summer squash is produced through the use of ethephon for inducing the pistilate flower and natural pollination. The functional male sterility (Psaij) has been exploited for hybrid seed production of tomato cv Pusa Divya under polyhouse condition by Manjunath (2009).

Table 4 Method/systems of hybrid seed production

Hybrid seed production methods	Crop
Hand emasculation and manual pollination	Tomato, Brinjal, Sweet pepper, Okra, Hot pepper
Pinching of staminate flowers and hand pollination	Bitter gourd, Bottle gourd, Pumpkin
Removal of staminate flower + emasculation + hand pollination	Watermelon and Muskmelon
Functional male sterility and hand pollination	Tomato, Brinjal
GMS + bee pollination	Chilli
CMS + natural pollination	Capsicum, Onion, Cabbage, Carrot, Radish
Self Incompatibility and natural pollination	Cauliflower, Broccoli
Gynocecism and natural pollination	Cucumber
PGR and natural pollination	Squash's
Detasseling + wind pollination	Sweet corn, Baby corn

Development and maintenance of parental lines

High yielding plants are selected from varietal population for hybrid seed production. Population of breeder seed may be used for this purpose. After selection of plants may be raised in single plane rows. Further, selected lines are produced by selfing either naturally or artificially to attain homozygocity, fixation of desired characters and uniformity. The degree of selfing or inbreeding depends on the response of parental lines to selfing, and degree of uniformity required in hybrids. Phenotypic uniformity of an F1 hybrid depends upon genetic homozygocity in the inbred parental lines.In cross pollinated vegetables like carrot, cabbage, cauliflower and onion which exhibits severe reduction of vigor upon selfing, inbreeding is restricted to three to four generations only. In cucurbitaceous vegetables inbreeding can be practiced to attain homozygous breeding lines without significant loss in vigor. Self pollinated vegetables like brinjal tomato, pea and beans do not loss vigor upon selfing. Cole crops permit inbreeding to fifth generation with low seed production, but onion and carrot scarcely be taken to third cycle of selfing and in leek inbreeding depression may be as high as 35% at a single cycle of selfing. This loss of vigor results in low yields of hybrid seed and difficulties in maintaining the inbreed. In this kind of crops some variation in inbred (early generation inbred) and as well as in hybrids may be allowed. Second approach may be the development of hybrids by three way or double cross. In case of diseases resistant hybrid, heterogeneity offer greater insurance against new pathogen or physiological races, therefore little variability is allowed in the parental lines.

Maintenance of male sterile and gynoceious parental line

The genetic male sterility has been reported and commercially exploited for hybrid seed production of muskmelon cv Punjab Hybrid 1 (MS -1 x Hara Madhu). The male sterile line is maintained by sib pollination from the fertile counter part and seed is collected from the sterile plant only. The CMS line is maintained by crossing with isogenic line which is having recessive genes in nucleus and fertile cytoplasm. This system of maintenance is applied in cabbage, onion and broccoli. In case of cole crop the SI system is also utilized and maintained through bud pollination or enhanced level of Co2 (500ppm for 18-24 hours depending upon the level of SI) for temporary induction of self compatibility. Similarly the gynoceious line of cucumber could be maintained with three application of GA3 (1000ppm) at fortnight interval from true leaf stage or by single spray of silver nitrate solution (600mg/lt) before first flower opening or silver thio-sulphate (25-50ppm) at 2-3 leaf stage to produce staminate flower (Gautam *et. al.* 2006).

Isolation distance for multiplication of parental lines and hybrid seed production

In order to produce genetically pure seed of parental lines and to achieve high genetic purity during field production/ multiplication plot should strictly adhered to the prescribed isolation distance (table 5) within the isolation distance no other varieties of same crop and wild relative should not be present.

Table 5: Isolation distance for multiplication of parental lines and hybrid seed production

Crop	Maintenance for parental lines (m)	For hybrid seed production (m)
Brinjal	200	100
Tomato	200	100
Okra	400	200
Onion	1000	500
Melons and gourds	1500	1000
Cabbage, Cauliflower	1600	1000
Carrot	1000	800
Pepper (hot/sweet)	400	200

Planting ratio and methods of planting

The planting ratio is determined by the mode of pollination and system of hybrid seed production. The planting ratio between female and male is higher when pollination is arranged by hand or by wind. Generally, the planting ratio in case of cauliflower is determined weather the one of the parent or both the parental line containing SI system. If seed parent is SI and pollen parent is self compatible than planting ratio kept 2:1 and in the event that both seed and pollen parent is SI than planting ratio is advice to kept 1:1 for harvesting higher seed yield. In case of hand pollination block planting (tomato, brinjal, chilli, capsicum, bitter gourd, bottle gourd and pumpkin) is advised where as regular planting of female and male is practiced when pollination is carried out by natural/ augmented population of honey bee (cabbage, cauliflower, carrot, cucumber, summer squash and onion) or by wind (sweet babycorn corn and spinach).The planting ratio and method of planting generally followed in hybrid seed production is presented (table 6)

Table 6: Planting ratio and methods of planting for hybrid seed production

Crop	Planting Ratio (M:F)	Method of planting
Brinjal, Tomato	1:5	Block planting
Okra	1:4 or 1:5	Block planting
Onion	1:4 or 1:8	Continuous/Regular planting
Cauliflower, Cabbage	1:1 or 1:2	Continuous/Regular planting
Spinach	2:6 or 2:14	Continuous/Regular planting
Carrot	1:2 or 1:4	Continuous/Regular planting
Radish	1:1	Continuous/Regular planting
Watermelon	1:1 or 1:2	Block planting
Cucumber	1:2 or 1:3	Continuous/Regular planting
Summer squash	1:5	Continuous/Regular planting
Bottle gourd, Bitter gourd	1:4	Block planting or Continuous/Regular planting
Pumpkin	1:3	Continuous/Regular planting
Muskmelon	1:2	Block planting or Continuous/Regular planting
Sweet pepper	1:4	Block planting
Hot pepper	1:2	Block planting or Continuous/Regular planting

Pollination management

Pollination management refers to optimum time of emasculation (bisexual flower), time of pollination, collection of pollens from anther, method of pollination and repeated pollination. The emasculation is generally done day before the anthesis preferably in afternoon (3-5PM) followed by protection from chance pollination by covering with white butter paper bag/ non moisture absorbent cotton swab. On the same day the male flower is also protected to avoid genetic contamination from unwanted source and robbing of pollen. The time of pollination is affected by season, location and very crop to crop. The optimum time of pollination in different vegetable crops is presented in table 7.

Table 7: Optimum time and method of pollination

Crop	Time of pollination	Method of pollination
Brinjal	9-11 AM	Hand pollination
Tomato	7-111 AM	Hand pollination
Okra	7-9 AM	Hand pollination
Onion	8-11AM	Insect pollination
Muskmelon and watermelon	7-9AM	Hand pollination/ Insect pollination
Bitter gourd	7-9AM	Hand pollination
Bottle gourd	1-3PM	Hand pollination
Pumpkin	5-7AM	Hand pollination
Summer Squash	7-9AM	Insect pollination
Cucumber	7-9AM	Hand pollination/ Insect pollination
Cabbage, Cauliflower	9-11AM	Insect pollination
Carrot	9-11AM	Insect pollination
Pepper (hot/sweet)	9-10AM	Hand pollination

Hand pollination is most common method of pollination employed for hybrid seed production of many vegetable crops.hand pollination method is recommended for higher number of fruit set, higher number of seed per fruit and seed yield per plant by Pham Duc Tuan (2006)and Vishwanath (2007) in bottle gourd and pumpkin respectively. Repeated pollination which involves removal and putting of covering bags increase the chance of damage to the gynaceium and some of the flower withered next day hence have limited scope of application but contrary to the above Manjunath (2009) has suggested the repeated pollination in hybrid seed production of tomato and recorded higher fruit set (78%), increased fruit weight (36.8g), higher seed yield per fruit (102mg) in poly house under Delhi condition.

Hybrid seed production under protected structures

High value low volume vegetable hybrid seed can be produced successfully under insect proof nets and naturally ventilated polyhouse condition. In hybrid seed production genetic purity and seed health of parental line is a matter of concern because of non availability of adequate isolation and attack of various diseases and viruses. Singh *et al.* (2009) has reported the production of high quality seed of the parental lines of pumpkin Pusa hybrid-1 under insect proof net house condition and compared with open field condition and observed significant difference for fruit development attributes (table 8), seed yield attributes (table 9) and seed quality attributes (table 10) in female and male parental lines under insect proof net house conditions.

Table 8: Effect of growing conditions on fruit development characters in parental lines of pumpkin Pusa Hybrid-1

Character/ Parental line	Growing conditions		Character/ Parental line	Growing conditions	
Female	Open Field	Insect proof Net house	Male	Open Field	Insect proof Net house
Fruit weight (g)	2.80	5.31*	Fruit weight (g)	1.46	3.27*
Fruit length (cm)	20.67	24.49*	Fruit length (cm)	10.18	11.64*
Fruit width (cm)	18.37	23.70*	Fruit width (cm)	17.89	22.75*
Fruit thickness (cm)	3.12	3.30	Fruit thickness (cm)	2.4	2.93*
Cavity size (L x W) cm^2	179.83	306.44*	Cavity size (L x W) cm^2	90.26	126.91*

* Significant at 5 % level

Table 9: Effect of growing conditions on seed yield attributes in parental lines of pumpkin Pusa Hybrid-1

Character/ Parental line	Growing conditions		Character/ Parental line	Growing conditions	
Female	Open Field	Insect proof Net house	Male	Open Field	Insect proof Net house
Total seed	376.40	389.83*	Total seed	304.35	387.20*
Un-Filled seed	16.80	6.83*	Un-Filled seed	10.57	18.06*
Filled seed	359.60	382.75*	Filled seed	293.78	368.46*
Seed yield (g)	61.88	78.98*	Seed yield (g)	42.20	65.03*
100 seed wt. (g)	14.19	15.28*	100 seed wt. (g)	13.36	17.79*

* Significant at 5 % level

Table 10: Effect of growing conditions on seed quality attributes in parental lines of pumpkin Pusa Hybrid-1

Character/ Parental line	Growing conditions		Character/ Parental line	Growing conditions	
Female	Open Field	Insect proof Net house	Male	Open Field	Insect proof Net house
Germination (%)	99.30	100	Germination (%)	99.78	100
Root length (cm)	18.84	20.71*	Root length (cm)	17.53	21.99*
Shoot length (cm)	17.13	21.17*	Shoot length (cm)	16.79	20.04*
Fresh weight (mg)	9.42	15.65*	Fresh weight (mg)	10.08	18.12*

Dry weight (cm)	0.78	0.84*	Dry weight (cm)	0.76	1.00*
Seed vigour Index (I)	3571.82	4188*	Seed vigour Index (I)	3424.44	4203*

* Significant at 5 % level

Sujata (2008) has standardized the hybrid seed production technology of tomato cv. Pusa hybrid-4 in polyhouse under Delhi conditions and also compared the seed yield and quality attributes with open field conditions.

Future Thrust

Since the demand of vegetables is bound to increase with the increase in the population and to overcome the problem of malnutrition among the Indian population especially in children's and women the vegetables can play an important role being majority of Indian population is vegetarian in diet. There is an urgent need to develop high yielding F1 hybrids coupled with biotic resistance and standardization of hybrid seed production technology of various hybrids in vegetable crops. Secondly the new locations for hybrid seed production of vegetables should be searched since the existing locations is over exploitation and built up of soil borne pathogens are posing serious problem in successful hybrid seed production.Thirdly the change in environment behaviour is also a matter of concern thus to ensure higher yield and better quality of hybrid seed there is a need of explore the possibility of low cost insect proof net house and naturally ventilated polyhouse.

References

National Horticulture Board (2015). Indian Horticulture Database. Ministry of Agriculture, Government of India.

Singh,Balraj,Tomar, B S and Shailja Thakur (2009).Quality seed production of parental lines of pumpkin (*Cucurbita moschata* L.).Paper presented in 4th International Cucurbitaceous Symposium held at HAU, Changha (Hunan), China during September20-24, 2009.

Vishwanath (2007).Studies on floral attributes and methods of pollination on seed yield and quality in pumpkin (*Cucurbita moschata* L.).cv. Pusa Hybrid-1.M. Sc. Thesis submitted to PG School ,IARI, New Delhi-110012.

Pham Duc Tuan (2006). Study on hybrid seed production in bottle gourd (*Lagenaria siceraria*) under open and poly house condition. M. Sc. Thesis submitted to PG School ,IARI, New Delhi-110012.

Manjunath, C (2009). Studies on functional mal sterility in Tomato (*Lycopersicon esculentum* Mill.) and its application in hybrid seed production. PhD Thesis submitted to PG School, IARI, New Delhi-110012.

Sujatha Patta (2008). Studies on standardization of hybrid seed production technologygy in tomato (*Lycopersicon esculentum* Mill.) PhD Thesis submitted to PG School ,IARI, New Delhi-110012.

Gautam, P. L., Hari Har Ram and Singh, H. P. (2006). Cucurbits breeding and production technology, Proceedings of national seminar on cucurbits, 22-23 Sept, 2005, Pantnagar.

Improved Agriculture and Allied Practices for Farmers, pp 103-110
New India Publishing Agency, New Delhi, India

9

Product Diversification for Improved Profitability-Production of Special Flowers

T. Janakiram

Assistant Director General, Horticulture, ICAR, New Delhi-110 012

The floriculture has become an important sector all over world. In India, floriculture has been associated with culture and heritage since very ancient time. In recent decades there has been increasing in demand of floriculture products with increasing income. It is souring industry in Asian countries including India. Floriculture is an emerging area with great potential both in the domestic as well as export market.In India, commercial floriculture is ongoing development but have a long tradition of various types of flowers.Floriculture was being cultivated in 3.5 lakh hectares in the country and 250 varieties of flowers exist in the country.Besides potted plants, roses, carnation, orchids and chrysanthemum are important cut flowers, which are popular in our country. In the increasingly competitive international cut flower and pot plant market, new and novelty ornamental plants plays an important role in stabilization and expanding market share. Always novelty has been considered as an important attribute in ornamental industry. The consumer choice is also changing for product quality level depending upon purpose of purchase, as well as for higher levels of services and a wider and deeper assessment suggesting ***product diversification***. The increasing pace of agricultural transformation in India has been characterized by widening share of commercial crops in the overall basket. Among others, horticulture has shown great promise because of the availability of high-value domestic and international market against the low production base so far. Understandably, it has received special attention of policy-makers and being termed as the sunrise industry within agriculture. India, with its varied agro-climate is ideally suited for growing several types of flowers in different seasons. Although traditional ornamental crops will continue to play key role in the floriculture trade, a distinct trend towards increasing the share of new crops is clearly evident in recent years often referred to as **'specialty flowers'**, these are cut flowers, which are not widely known and also grown in small pockets but have high prospects in the international market. There are over 60 types of new ornamentals, which have been reported to hold potential for commercial utility. Some progressive growers in India have already initiated the process of introducing new ornamental crops in their fields.These new products normally fetch high prices than the traditional crops for a certain period, but quite often the prices drop when market is saturated. By that time new products should be ready to enter the market. In this context introduction of new ornamental crops is therefore continuous effort.

Flowers have always been an integral part of Indian culture. But, floriculture as a business proposition has not received the attention it deserves. This industry is still in a nascent stage and the potential remains untapped. Floriculture includes not just cultivation of a wide variety of flowers but also extensive research and creation of healthier, stronger seed varieties and plugs. Globally, more than 140 countries are involved in the cultivation of floriculture crops. India's floriculture business is worth Rs 9 billion. With rich and varied climatic conditions and abundance of cheap labour, it is most suitable for floriculture.Organised floriculture in India is 15 years old. Of the 110,000 hectares under floriculture in the country, only 500 hectares come under organised floriculture. In the early 1990s, floriculture was recognised as a sunrise industry and was accorded 100 per cent export-oriented status. This enabled new floriculture units that were producing exclusively for exports to get several fiscal benefits, including income tax holiday and exemption from certain import duties. Yet, only a fraction of the total floriculture area of around 2.5 lakh hectares is under high-tech flower production in the protected environment of green houses. Much of the floriculture activity remains confined to growing fresh flowers in open fields for the domestic and export markets.Several formidable constraints prevent flower growers from realising adequate returns - high post-harvest losses, poor down-the-line infrastructure and inefficient marketing are some of the significant factors. A study by Hossain and Rahman (1994) reveals that the capital investment in flower business has been increasing, the total number of shop exhibits an increasing trend, merchandizing patterns have been diversified, the scale of flower made products is rising and all sales are for cash. The positive attitude of respondents toward flowers, multipurpose uses of flowers, increasing uses of flower made products and an unsatisfied demand of some flowers are all positive indicators of the business boom. Hassan (1996) showed producers, marketing intermediaries and traders secure significant profit from flower production and marketing.

Diversification

Many flower producers in India are considering alternative crops to increase their incomes Diversification into alternate flower crops, which have domestic consumption, export demand, are the urgent need of the hour for the industry and ***specialty flowers*** seems to fit into this crop diversification. Field production of specialty fresh cut flowers could be a profitable alternative crop. Our country has a climate advantageous to production of many flowers,

Perennial flowers that are high in demand, such as heliconia, red ginger, bird of paradise, ornamental gingers flourish in our climate and soils. National, regional, and local markets exist for high quality specialty cut flowers. Some of these specialty cut flowers are grown as back yard ornamental plants across the country but in the recent times the market conditions, technology and new varieties made them commercially popular in flower industry.

The following factors determine for diversification towards new ornamental crops

- Local markets are saturated with all traditional flowers; therefore expansion is possible for many more new crops.
- The limiting factor in agricultural production is water and floriculture has high return per cubic water.
- India has a suitable climate for various flower crops production.

The ideal crop for diversification

An ideal cut flower crop would have the following characteristics. No single species or variety of plant material will meet all of these criteria. They are presented here to provide a means of evaluating the relative desirability of producing a crop of a specific plant.

- Low cost of production—materials and labor
- High value and unlimited demand
- High production per square foot of bed space
- Extended production and marketing season
- Long productive life
- Ability to sell fresh and to sell surplus as a dried or preserved floral product
- Long stems and Post harvest vase life of at least 7 days

- Resistance to diseases and pests of all types
- Resistance to heat and drought stress
- Easy harvest and handling
- Aesthetically pleasing and/or fragrant flowers, foliage, or stems

Since diversification is viewed as an important factor to survive in the floriculture trade, some of the potential specialty flowers were discussed.

Heliconia

1Heliconia belonging to Heliconiaceae are the most attractive of all exotic tropical flowering plants. They are known as lobster's claws, parrots flower, parrot plantain and false plantains. The heliconias are gaining importance due to their diversity in both colour and form and has good potential as commercial cut flower. Its brilliant colour, exotic form, long straight peduncles and excellent post harvest characteristics make it an outstanding flower for the florist trade. *H. psittacorum* and *H. latispatha* x. *H. psittacorum* are excellent for floral arrangements. It is well adapted to all major agro climatic zones of the country can be cultivated at a height of 3000 to 4000 meter and grows well in the North Eastern part of India.

Major heliconia- producing nations include Barbodos,Hawaii,Brazil and Venezuela.They are cultivated in the Netherlands and Germany. India has an annual production of about one lakh stems which account 1% of the total floral production of the country. Fifty per cent of the production of the heliconias comes from West Godavari district of Andhra Pradesh.This crop is gaining importance as a commercial flower crop in Karnataka and Kerala in recent times

Heliconias thrive in the open area and soils rich in organic material. The bract color may be slightly more intense under light shade than full sun for some *H. psittacorum* selections. The plants thrive well under Jorhat condition in Assam (sub – humid tropical climate).When planted in January, the shoot emerges within 2 months and emergence of inflorescence takes place within 6 months. Full blooms are obtained 6-7 months after planting, the peak flowering period being from September to December. In the succeeding years, it flowers in April and continues up to December. However, flowers are produced almost throughout the rest of the year. During severe winter, partial shading of leaves occurs and flowering is arrested from January to March.

Soils: Heliconias thrive. However, a slightly acidic, well- drained medium with high fertility is preferred. Chlorosis develops readily on plants grown in soils with a high pH.

These are propagated byrhizomes and planted at a spacing of 40 x 40cm for rhizomes (weighing 40 g). But most other species may require higher spacing.

Suggested spacings (rows): H. *psittacorum* (0.75 -1 M), H. *hirsuta*, H. *metallica*, H. *angusta,* H. *aurantiaca* small H. *stricta* forms, H. *vaginalis* (1.2 -1.5M), H.*rostrata*, H.*angusta* flava, H.x 'Golden Torch' taller H. *stricta* form H. *latispatha* (, 1.5 -2 M) H. *caribaea*, H. *bihai,* H. chartacea, H. *wagneriana*, H.*collinsiana*, H.*bourgaeana*, H.champneiana, H. platystachys, H. indica. (2-2.5 M)

Heliconias require substantial amounts of water and should be grown in a well – drained site as poor drainage causes major root problems. Water stress in heliconias is indicated by longitudinal rolling of the foliage.

Heliconias need rich compost during active growing period. Well rotten farmyard manure at the rate of $4kg/m^2$ should be applied at the time of soil preparation and 20g each of N, P_2O_5 and K_2O/m^2 at the time of planting rhizomes. A top dressing with N@ $20g/ m^2$ is done two months after planting. Micro- element, spray is recommended three to four times each year as a foliage spray for better leaf color.

Bird of paradise

Bird of paradise (*Strelitzia reginae*) is an evergreen perennial herbaceous plant It is highly prized as cut flower due to brilliant colour and unusual appearance of the spectacular flowers. It is a native of South Africa and belongs to family Musaceae.

Bird of paradise grows 90 to 180 cm high and flowers, which emerge one at a time from the spathe, consist of three brilliant orange sepals, standing up like the crest on a bird's head, and three shimmering peacock blue petals, two of which are joined together to form an arrow like nectar.

The orange and blue flowers on long stiff stems will bloom intermittently all year long

Common mode of propagation is by seeds; division of clumps and separation of offsets or suckers from the mother plant. Seedling plants take longer time to come in flowering than vegetatively propagated plants. Presently tissue culture is widely practiced to propagate this crop.

Crop grows well under tropical and subtropical conditions and needs full sun to light shade with intermediate to warm temperatures. In the subtropical conditions the crop can be cultivated under greenhouse. For better quality flowers the minimum and maximum temperature requirement is 10-13^{0}C and 20-22^{0}C respectively. Well-drained sandy loam soil rich in organic manure is best suited for growing the crop.

Generally February, August and September months are optimum time for planting. The plants can be cultivated in open fields and pots. Pits of 90 cm^2 are dug and filled with soil and FYM. Care should be taken to avoid damage to roots while planting.

A spacing of 90cm between the plants and density of 4000-4500 plants/ ac is recommended. For early flowering and better root development, a liquid nutrient solution containing 6g Single Super Phosphate and 3g/l Potassium nitrate should be sprayed for three months with an interval of 10 days. At the time of flowering a spray of liquid nutrient solution containing 5g of DAP/l or 3g potassium sulphate and 4g of urea /2 l can be recommended. It is a moisture loving plant. The first watering is done immediately after planting followed by irrigation once in a week depending on season and growing conditions.

Flower stalks are usually pulled rather than cutting in the field at the commercial stage of harvest, which is indicated by, split of bract at the top, emergence of first flower and exposure of orange sepals. At the packing shed, stems are trimmed to a standard length ranging from 60 to 120 cm and the flowers are washed to remove field dust, insects, and exudates. Flowers are packed singly or bunched in five, with or without sleeving Approximately 180 stems cushioned with moist shredded newspaper are packed per box. Bird of paradise flowers may be stored in water at 8^0 C for a week or less, Storage under dry condition with 90 to 95 per cent relative humidity is best for longer shelf life

Zingiberaceous cut flowers

Zingiberaceae family is characterized by many strikingly beautiful plants, which are valued for their display and cut flower value by virtue of their post harvest longevity. The red gingers.Torch gingers and ornamental Curcumas are the most prominent among them.These plants are adapted to partially shaded conditions and are ideal choice as intercrops in coconut gardens.

Red ginger (*Alpinia purpurata*)

The red ginger plant is a tropical ornamental plant with bright red flower bunches. It is also known as fire ginger. The flower bunch is a compact spike or open raceme, about 30 cm long consisting of numerous large, boat shaped bright red bracts and much reduced flowers which are white. Because of their long lasting qualities and bold effect, they are excellent cut flowers. Jungle

King and Eileen Mc Donald are cultivars with commercial cut flower value. There is a pink type, presumed to be a variant of this red type.

Alpinias require a rich soil and cool shady situations.They are propagated by divisions of rhizomes or from small plantlets produced in bract axils of old inflorescences.Micropropagartion is also possible

Flowers are produced throughout the year, the highest production being in June and July. Flowers are harvested with long stalk of 50-80 cm length. They last for 2-3 weeks after removing from plant. One of the treatments recommended for extending vase life of red ginger flowers is preconditioning at 40°C for 15 minutes, keeping in a bucket of water at room temperature for one hour and then a hot water treatment at 50°C for 12-15 minutes.

With the fast changing scenario in floriculture trade, the availability of different climatic zones, good climate and soil, cheap labour and land and skilled manpower. It is certain that our growers can do extremely well in competitive sector by taking steps for diversification in the fast changing scenario

- Keep information of new products.
- Experiment with test plots, before expanding production.
- Understand that a high-quality product is essential for success.
- Set prices to ensure profits. Price cutting without lowering costs can have drastic effect on profitability.

Conclusion

Floriculture is an age old farming activity in India having immense potential for generating gainful self-employment among small and marginal farmers. In the recent years it has emerged as a profitable agri-business in India and worldwide as improved standards of living and growing consciousness among the citizens across the globe to live in environment friendly atmosphere has led to an increase in the demand of floriculture products in the developed as well as in the developing countries worldwide. The production and trade of floriculture has increased consistently over the last 10 years. Floriculture is a viable and profitable alternative for the new generation of farmers. Though the quality of Indian cut flowers has been gaining popularity globally, domestic flower growers may not be able to meet the growing domestic demand owing to lack of re-plantation activities. Replantation of cut flower plants is not taking place in India as importing the planting materials is too expensive. High freight rates add to the crisis.

References

K. Radha Manohar & Igathinathane, (2007) Greenhouse Technology and Management B.S.Publication, Hyderabad 500 095 A.P.

S.K. Bhattacharjee, (2006) Advances in Ornamental Horticulture, Volume 6) Pointer Publishers Jaipur 302 003(Rajasthan) India.

Access from electronic source www nhb.gov.in , (2008) Indian Horticulture data base, Ministry of Agriculture Government of India, Gurgaon

Access from electronic source www.apeda.com (2008) Ministry of Agriculture Government of India, New Delhi

Hassan, M. (1996). A Study on Production and Marketing of Tuberose in Some Selected Areas of Bangladesh. An M.S. Thesis, Department of Cooperation and Marketing, Bangladesh Agricultural University, Mymensingh. Bangladesh.

Hossain, M.B.and M.M.Rahman. 1994. The Potential of Flower Marketing in Dhaka City. Bureau of Business Research, University of Dhaka, Bangladesh.

Improved Agriculture and Allied Practices for Farmers, pp 111-122
New India Publishing Agency, New Delhi, India

10

Improved Cultivars and Production Technology of Important Fruit Crops

V.B. Patel and A.K. Singh

Division of Fruits and Horticultural Technology, ICAR-Indian Agricultural Research Institute, New Delhi-110012

Fruit production is gaining momentum in the country. With the production about 90.18 million tonnes of fruits from 6.30 million ha in 2015-16, India is the second largest producer of the fruits in the world after China. During the past three decades, international trade in fruits has seen a rapid expansion.The newer markets are now opening for Indian produce e.g. Japan and USA have opened the route for Indian mangoes. At domestic level, the increase in demand of fruits for consumption both in fresh and the processed form has been evident. The income and livelihood security from this sector are increasing. The government has also supported the horticulture sector particularly area expansion and productivity enhancement of fruits through technological interventions like drip irrigation, fertigation, emphasis on integrated nutrient management, production of high quality planting material etc. Selection of an ideal variety and availability of plant material from a reliable source is the basis for achieving high yield and quality uniform produce. Lack of clonal uniformity not only leads to the problems in crop management but also the differences in maturity may cause delay in commercial picking. A number of improved fruit crops' varieties have been developed during past few years, which ranged from the high yielding to stress resistant to high quality produce. In fruits and plantation crops more than 150 high yielding varieties suitable for various purposes have been developed. A list of some popular cultivars and new cultivars in major fruit crops has been given here (Table 1).

Table 1: Improved cultivars in fruit crops

Fruit Crop	Cutlivars/Hybrids
Mango	**Northern India** : Bombay Green, Dashehari, Langra and Chausa. **Central India** : Langra, Dashehari, Chausa, Safeda and Sunderja. **Western India** : Alphonso, Kesar, Mankurad, Vanraj and Pairi. **Southern India** : Baneshan (Banganpalli), Neelam, Pairi, Mulgoa, Rumani, Bangalora, Raspuri, Badami and Suvarnarekha. **Eastern India** : Langra, Kishenbhog, Zardalu, Fazli, Himsagar, Bombai, Maida and Gulabkhas. **New Cutivars** : Mallika, Amrapali, Pusa Surya, Pusa Arunima, (IARI, New Delhi); Dashehari 51, Ambika, CISH M-2 (CISH, Lucknow), Arka Puneet, Arka Arunna, Arka Anmol, Arka Neelkiran (IIHR, Bangalore) Ratna and Sindhu (FRS, Vengurala), Pusa Lalima, Pusa Peetamber, Pusa Shreshth and Pusa Pratibha
Banana	Dwarf Cavendish, Giant Cavendish, Virupakshi, Harichhal, Nendran, Poovan, CO-1, Gandevi., FHIA -01, FHIA -03, CO-2, H-1 and H-2
Grape	**Western and Southern India :** Arkavati, Arka Kanchan, Arka Hans, Arka Shyam, Arka Neelmani, Tas-e-Ganesh, Sharad Seedless, Sonaka and Dilkush. **Northern India :** Pusa Urvashi, Pusa Navrang, Pusa Seedless, Perlette, Beauty Seedless, Centennial Seedless etc.
Citrus	**Mandarin:** Nagpur, Coorg, Khasi, Darjeeling and Kinnow. **Sweet orange :** Mosambi, Sathgudi, Valencia Late, Jaffa, Malta Blood, Blood Red and Hamlin. **Grape fruit :** Duncan, Foster and Marsh Seedless. **Lime :** Kagzi, Pramalini, Vikram,Jai Devi, PKM-1 and Sai Sarbati. **Lemon :** Kagzi Kalan, Eureka, Lisbon and Assam Lemon.
Guava	Allahabad Safeda, L-49, Lalit, Shweta, Arka Mridula, Arka Amulya, Safed Jam and Kohir Safeda
Papaya	H-39, Cp-81, CO1-7, Surya, Pusa Delicious, Pusa Majesty, Pusa giant, Pusa Dwarf and Pusa Nanha
Litchi	Swarna Rupa, Sabour Madhu, Sabour Priya
Pomergranate	Hybrid Ruby, Ganesh, Arakta and Bhagwa
Ber	Umran, Kaithli, Banarasi Karaka, Gola, Seo and Goma Kirti
Aonla	Chakaiya,Krishna, Kanchn, NA-6, NA-7 and NA-10
Beal	NB-5, NB-9, CISH B-1, CISH B-2, Pant Aparna, Pant Shivani, Pant Sujata
Apple	Sunehari, Lal Ambri, Chaubatia Princes, Amrich, Amred and Royal Ambri
Custard apple	Arka Sahan
Sapota	Kalipatti, Cricket Ball, PKM-2 and DHS
Kiwifruit	Abbott, Allison, Bruno, Hayward and Monty
Peach	Shan-e-Punjab and Flordasun (low-chilling)

Production Technology of Important Fruit Crops

Soil Management

Soil plays a vital role in fruit production, because none of the fruit can be grown in each and every type of soil. Specific fruits require specific soil for their optimum growth, development and productivity per unit area. Tropical and sub-tropical fruits may be grown on a wide variety of soils. Sandy loam soils with adequate drainage and water holding capacity are the best for most of the tropical and sub-tropical fruits. A large part of the cultivated area has turned problematic due to different reasons but fruit growing can be a remunerative option if practiced scientifically. A list of fruit crops is being given for different problematic soils

i) **High tolerance** (ESP 40 to 50, ECe 12 to 15 dSm^{-1}) e.g. date palm, *ber* and wood apple.

ii) **Medium tolerance** (ESP 30 to 40, ECe 9 to 12 dSm^{-1}) e.g. *aonla,* tamarind, *jamun, mahua, lasoTa, karonda, phalsa, khirni* and custard apple.

iii) **Weak tolerance** (ESP 20 to 30, ECe 6 to 9 dSm^{-1}) e.g. guava, citrus, mango and *bael.*

iv) **Susceptible** (ESP<20, Ece <6 dSm^{-1}) e.g. banana, papaya and pineapple.

Besides, salinity and sodicity, moisture stress is one of the major constraints over wide range of the soil situations. Plants having xerophytic characteristics, viz. deeper root system, deciduous nature, reduced foliage, sunken or covered stomata, waxy coating or hairy leaves etc., minimize evapo-transpiration and make plant amenable for their cultivation under moisture stress situations. Fruits, like *ber, aonla,* tamarind, wood apple, cashew nut, custard, apple, *karonda, mahua* and some indigenous fruits like *kair (Capparis decidua), khejri (Prosopis cineraria),* drumstick *(Moringa oleifera), lasora (Cordia dichotoma), khirni (Manikara hexandra),* have xerophytic characteristic and may be cultivated under moisture stress situations.

Climatic Requirement

Mango is found growing all over the country, up to an altitude of 4,000 feet, but it does not fruit well above 2,000 feet. Banana is grown principally in the South and East. Citrus fruits grow in most parts of the country, except at high elevations. Guava is common in U.P. and Bihar and papaya in all frost-free areas. Grapes can be grown in most places, except at very high altitudes and in wet and humid areas. Litchi and loquat can be grown in the sub-tropical submontane areas. At the foot of the Nilgiri Hills in South India, several exotic fruits have been grown successfully. These include mangosteen, durian, avocado and *rambutan. Ber* can be grown anywhere in the plains and the date-palm can be cultivated in the dry desert areas. Other fruits successful in India include *phalsa (Grewia subinaequalis),* pomegranate, fig, pineapple, custard-apple, *jamun (Syzygium-cuminii),* jack-fruit, *sapota (Acharas zapota),* cape gooseberry, mulberry, passion fruit, *aonla (Emblica officinalis)* and tree tomato *(Cyphomandra betacea).*

The variety of climate and different types of the fruits, which can be grown in India, are indeed impressive. However, the climate in some places is very rigorous and only very hardy varieties can thrive. For example, in the plains of North India the climate is not really sub-tropical, but semi- tropical. The fruit trees being perennials remain in the field all through the year and they must successfully withstand the desiccating hot weather in May and June, frequent frosts during winter and water-logged conditions along with the incidence of all kinds of diseases and insect pests during the monsoon season. Failure to withstand any of these conditions will result in the death of plants. Annual crops have the advantage that they can complete their life-cycle in the favourable period of a few months confined to just one season.

Irrigation Management

Irrigation is an important factor, which determines the success of the fruit growing. Moisture requirements of fruit plants are calculated on the basis of their ability to synthesize the quantity of dry matter at the expense of the quantity of water transpired. The water scarcity is evident in most of the fruit growing regions of India particularly during the period of peak water requirements. Therefore there is need of judicious use of water and minimizing the losses in conveyance and application. Water harvesting and use of drip/ micro irrigation has been standardized for some fruit crops. It has been observed that by using drip irrigation alone there is increase in yield by 52% in banana, 23% in grapes, 50% in sweet lime and 45% in pomegranate. A range of techniques standardized for enhancing water use efficiency in fruit crops is given in Table 2.

Table 2: Techniques standardize for improving water use efficiency in fruit crops

Crop	Techniques to improve water use efficiency
Arid fruits	Drip irrigation-25% saving in irrigation water Double ring method of irrigating anola, water saving 50-60 percent
Banana	Water needs 900-1000 mm per year, Drip irrigation, 30 per cent water saving & 40% increase in yield
Cashew	Crescent bund treatment of terrace increased the soil moisture content by 14.8% and reduced the run-off by 4.6 l/s/ha.80 litres of water per tree once in four days from December to March through drip was found efficient compared to hose irrigation with 200 litres.

Citrus	Drip irrigation, requirement of water reduced by 40%
Grapes	Tas-A-Ganesh grafted on rootstock for better WUE-25% water savingSubsurface irrigation methods saved 25% water Mulching and spray of antistress combination reduced 25% irrigation requirement
Pomegranate	Drip irrigation, saving of water is 50%
Temperate fruits	Irrigation at 80% ET increased fruit set in apple Use of black polythene/ dry grass as mulch saved water by 40-70% & increased yield of apple.

(*Source:* Recent Initiative in Horticulture, 2010)

Fruit crops require proper drainage.Un-seasonal heavy rains may lead to disease incidences such as collar or stem rot in papaya, Panama wilt and leaf spot diseases of banana and pests like weevils and aphids are predominant in wet and humid conditions.Similarly, the tea mosquito bug causes severe damage in cashew nut under such adverse conditions.Plants under drip irrigation show early establishment and uniform bearing. Slight drought stress before flowering in fruit crops enhances the number of flowers. Heavy irrigation after fruit set must be avoided to check excessive fruit drop.

Mulching

Covering of plant basin with organic waste materials, black polyethylene strips or emulsions, is termed as mulching. Mulching reduces the evaporation by interrupting radiation falling on the soil surface and thus delays the drying of soil and reduces soil thermal regime during daytime. It also reduces the weed population and improves the microbial activity of soil by improving the environment around the root zone. Continuous use of the organic mulches is helpful in improving the soil organic matter content, which in turn improves the water holding capacity of the soil. Experiments indicated that in *aonla, ber* and guava, mulching of the tree basin with F.Y.M., paddy straw and sugarcane trash have shown positive response in maintaining the optimum moisture regime, weed control, improving the physical and chemical properties of the sodic soils and thus inducing better tree vigour. However, use of the inorganic mulches is expensive and it does not incorporate organic matter to the soil. Leguminous crops are preferred over cereals for mulching as they add nitrogen. Natural sod of grasses growing in the inter-space between rows, reduce the incidence of spongy tissue in 'Alphonso' mango. Organic and plastic mulches were found effective in pineapple and papaya as they check the excessive growth of weeds. Weeds take away the Ca, which hamper the storage quality of the fruits.

Nutrient Management

Nutrients are one of the major inputs accounting for nearly 35 per cent of the cost of cultivation.The aim of nutrient management is to maintain the soil fertility and plant nutrient supply to an optimum level for sustaining the desired crop productivity through optimization of the benefits from all possible sources of plant nutrients in an integrated manner. Initially, the soil chosen for plantation has to be brought to an optimal level of fertility, and nutrient supply of the soil has to be understood depending on chemical analysis. To achieve this, depending on primary character of the soil, heavy green manuring, incorporation of organic matter and other amendments have to be made by pit making and pit filling with appropriate mixtures. Sufficient quantity of manures and fertilizers should be incorporated before planting. A general recommendation is that in each pit besides plant protection chemicals against white ants, 50kg organic manure (FYM), 2.5kg single super-phosphate and one kg muriate of potash should be incorporated. Depending on soil reaction and salt concentration, other materials such as lime, gypsum and pyrite can also be mixed. This is based on soil test recommendation. The range of different nutrient availability in soil is given below:

Table 3: Optimum soil fertility range for production of fruit crops

Parameter	Low< (less than)	Medium	High> (more than)
Soil pH (1:2.5)		6 to 7	
ECE (dsm^{-1})		<2.0	
Organic carbon (%)(Walkley and Black)	0.5	0.50-0.75	0.75
Available N (kg/ha)(Alkaline $KMNO_4$)	280	280-560	560
Available P_2O_5 (kg/ha) (Olsen's)	28	28-56	56
Available K_2O (kg/ha)(Ammonium oxalate, pH 7.0)	140	140-280	280
Sulphur (ppm) (Heat soluble)	10	10-20	20
Zinc (ppm) (DTPA Extractable)	0.5	0.5-1.0	1.0
Iron (ppm) (DTPA Extractable)	5.0	5-10	10
Copper (ppm) (DTPA Extractable)	0.2	0.2-0.4	0.4
Manganese (ppm) (DTPA Extractable)	0.5	5-10	10
Molybdenum (ppm) (Ammonium oxalate Extractable)	0.05	0.05-1.00	1.0
Boron (ppm) (Hot water soluble)	0.1	0.1-0.5	0.5

To optimize the soil fertility level and to keep the plant nutrient status; above the critical level, a number of recommendations for the fruit crops are available. Based on the averages of these recommendations and the results available under the All India Coordinated Research Project on fruit crops, the recommendations are being made for different fruit crops are presented in Table 4. Besides these recommendations, application of zinc and iron has become common. Spray of boron has helped in *aonla* and mango. Therefore, regular applications of the micro-nutrients are necessary to get an optimum production per unit area.

Table 4: Manure and fertilizer recommendations for some tropical and sub-tropical fruits at their full bearing age

Fruit	FYM (kg/plant	Nutrient recommendation (g/plant/annum)			Remarks
		N	P_2O_5	K_2O	
Mango	75	775	500	700	-
Banana	10	180	180	270	Green manuring before planting
Citrus (acid lime)	50	900	250	500	-
Guava	60	360	180	180	-
Grape	25	270	450	900	In Karnataka
Thompson Seedless	25	900	450	750	In Maharashtra, application of lime in acid soil is common
Litchi	50	600	200	600	-
Papaya	10	250	250	500	-
Pineapple	25 t/ha	12-16	2-4	10-12	-
Pomegranate	30	600-700	200-250	200-250	-
Ber	50	500	200	300	-
Sapota	50	1,000	500	500	-
Aonla	50	1,500	750	1,000	-
Date palm	30	1,000	500	500	-

Leaf is the principal site of plant metabolism, and changes in nutrient supply are reflected in the composition of the leaf. The concentration of nutrients in the leaf at specific growth stage is related to the performance of crop. The leaf nutrient status required for optimum production in important fruit crops are presented in Table 5. Efforts should be made to maintain these levels in leaves before flowering.

Table 5: Optimal leaf nutrient status in important fruits

Crops	Nutrients									
	N%	P%	K%	Ca%	Mg%	S%	Bppm	Feppm	Mnppm	Znppm	Cuppm
Mango	1.23	0.06	0.54	1.71	0.91	0.12	-	171.0	66.0	25.0	12.0
Banana	3.29	0.44	3.11	2.12	0.24	-	-	-	-	-	-
Citrus	2.50	0.14	0.90	4.20	0.43	0.25	-	90.0	112.5	62.5	10.5
Guava	1.70	0.50	1.30	2.1	0.46	-	10.0	-	-	70.5	75.0
Grape	1.75	0.56	1.66	0.94	-	0.20	-	-	125.0	92.0	-
Litchi	1.47	1.03	—	-	-	-	-	-	-	-	-
Sapota	1.66	0.08	0.80	0.83	0.48	0.07	-	100	39.3	15.7	6.7
Papaya	1.66	0.50	5.21	1.81	0.67	0.38	-	-	-	-	-
Pineapple	1.40	0.16	3.70	-	-	-	-	-	-	-	-
Pomegranate	2.50	0.19	1.47	-	-	-	96.0	199.0	196.0	55.0	-
Ber	2.60	0.25	1.82	1.21	0.45	-	-	-			

To meet the requirement of nutrients recommended for different fruit crops, it could either be through inorganic source or organic source, but each one has its limitations. General consensus is that it is desirable to meet fifty per cent of nitrogen through organic sources and fifty per cent through fertilizers for a good performance, otherwise continuous use of inorganic fertilizers is detrimental to the soil health. Addition of the organic matter would not only provide the needed nutrient including micro-nutrient at a slow pace for continuous growth, but it would also improve the physical condition of soil. Furthermore, it would improve the aeration of soil, providing a good scope for better root growth and increased production.Fortunately, in fruit crops from the very beginning, application of organic manures has been in practice and all our recommendations have organic component.It should be pursued.The evidence that the organic manures and biofertilizers reduce the need of chemicals supplying specific micro-nutrient and to minimize the risk of inducing deficiencies of many other micro-nutrients also commands itself a sound practice for a sustainable horticulture based on the low external chemical input.Besides micro-nutrients, biofertilizers viz. VAM, *Azotobacter, Azospirillum,* P-solublizing bacteria, etc. are now gaining popularity in fruits production. Root or soil inoculation with biofertilizer inoculants should be done at the planting stage. In case of the pomegranate, nitrogen use efficiency has been reported to be increased through the use of 250g *Azotobacter* culture with 100g nitrogen causing net saving of 200 g nitrogen per plant.

Canopy management and High Density Planting

Canopy in a fruit tree refers to its physical composition comprising of stem, branches, shoots and leaves. While, the canopy density is determined by the number and size of leaves, its architecture is determined by the number, length and the orientation of stem, branches and shoots. Canopy management of fruit trees deals with the development and the maintenance of their structure in relation to the size and shape for the maximum productivity and quality. The basic concept in the canopy management of perennial trees is to make the best use of the land and the climatic factors for an increased productivity in a three dimensional approach. Some of the basic principles in the canopy management are as follows:

- Maximum utilization of the light.
- Avoidance of the build up of micro-climate congenial for diseases and pests.
- Convenience in carrying out the cultural operations.
- Maximizing the productivity and quality.
- Economy in obtaining the required canopy architecture.

Economic production can only be achieved when plants are of desired canopy size, i.e.height and spread. Appropriate training and pruning systems help in building a strong framework and to maintain a balance between vegetative and reproductive growth of any plant. High density or ultra high density plantation is the current buzzword with the growers for its known merits. In mango dwarf rootstock 'Vellai Collumban', were found to be ideal for Alphonso mango. Dwarf hybrids/selections viz.Amrapali in mango; Dwarf Cavendish and Basrai in banana; Pusa Dwarf and Pusa Nanha in papaya and Kew in pineapple may be ideal for high density plantations.Planting under high density has resulted in yield realization of 86.60 to 174.39 t ha^{-1} in banana spaced at 1.2 m x 1.2m and 2.0 m x 2.0 m having 6,944 to 7, 500 plants ha^{-1}. In mango, 2.5-3 m x 2.5-3 m spacing is recommended for high density planting. In guava, 10' x 10' (1,076 plants ha^{-1}) in a square system of planting has been found highly successful. Papaya panted at 1.2 m x 1.2 m spacing. Similarly, Troyer Citrange, a dwarf rootstock has been found suitable for high-density plantations of Kinnow mandarin. In guava, *Psidium friedrichsthalianum* and Aneuploid No. 82 have shown to impart dwarfing effect in Allahabad Safeda. Technologies for meadow orcharding in guava, high density planting, canopy management and rejuvenation of old and senile orchards of mango, anola and guava are developed.

Use of bio-regulators

Ever since, the discovery that the plant hormones have the ability to influence several physiological processes, these compounds are applied to bring about specific physiological /biochemical effect governing vegetative or reproductive growth stages. Auxin namely, indole-butyric acid (IBA) are commercially and extensively being used in propagation of different fruit crops through cutting, air-layering or stooling. Soaking of rootstock species seeds in gibberellic acid (50 to 200 ppm) improves germination so as to get higher number of seedlings for grafting/budding. For floral induction and deblossoming growth regulators are extensively used in different fruits. In mango, 200 ppm NAA should be applied during October followed by de blossoming in January to minimize floral malformation. NAA at 20 to 40 ppm may be applied at 25 days interval after fruit set to control fruit drop in mango. For managing alternate bearing in mango, deblossoming by applying 50 or 100 ppm ethephon during 'on year' followed by application of 100 ppm TIBA at fortnightly interval after September and 500 or 1,000 ppm cycocel (CCC) before flowering season during 'off year' can induce flowering. Similarly, soil application of paclobutrazol (PP333) @ 5 g i.e. per plant has also helped in inducing flowering during 'off season' in mango to sustain the yield. Calcium carbide (1.0 g/per plant) or 25 ppm NAA induce uniform flowering in pineapple. In citrus, off season deblossoming is brought about by the application of 500 ppm ethrel or 200 ppm 2,4-0. In lime, it is being done during June-July.

For promoting growth and development of fruit, several bioregulators are found very effective. In banana, 100 ppm GA_3 twice at 25 and 50 days after shooting improved fruit size considerably. In grape, pre-bloom application of 10 to 20 ppm GA_3 helped in increasing the bunch and berry size. Reduction in fruit drop can be brought about by the application of 2-4-D (20-40 ppm) in citrus and 100 ppm SA+100 ppm NM in grape *etc*. Furthermore, ethrel sprayed at 50 ppm enhances fruit set in cashew nut.Whereas, in sapota, 10 or 20 ppm 2,4-D can improve fruit set while SAOH at 100 ppm or planofix (NAA) at 300 ppm resulted in higher fruit retention. Application of growth retardants, like cycocel (CCC), Maleic hydrazide and paclobutrazol (PP333) have shown to reduce vine vigour and increase fruitfulness in some grape cultivars.

Integrated pest and disease management

Fruits being rich in nutrients are easily prone to the attack of diseases and pests. Further, edaphic and weather factors constantly pose a threat for biotic stress. Fruit growers apply multiple sprays of agro-chemicals to get instant control for any pest or disease, although, these operations also kill the beneficial organisms, i.e. biological/natural control agents.Biopesticide/ fungicide with low or negligible toxicity should form the part of integrated pest and disease management practices.Annual life cycle and population dynamics of pest/ diseases should be managed by using safe agro-chemicals (low residual toxicity), botanics and bio-pesticides in a schedule under a joint eradication programme comprising growers of the entire region.In mango, pests like mango hopper can be controlled by spraying carbaryl (0.2%) at panicle emergence, whereas, mango mealy bug may be controlled by tieing alkathene sheet (300 gauge) 30 cm wide on the base of main trunk. Annual deep ploughing of soil near the trunk destroys larvae and pupae of fruit fly and mealy bug.The incidence of banana weevil and aphids may be reduced by avoiding high humidity regions.The diseases of mango viz.,anthracnose*(Colletotrichum gloeosporoides)*,stem-end rot *(Botrydip/odia theobromae)* and soft brown rot *(Nathrasia mangiferae)* may be controlled by spraying systemic fungicides, i.e. benomyl, carbendazim and copper oxychloride in a combination. Wilt and virus infected plants of banana and papaya should be uprooted and burnt to check the spread of viral diseases.

Maturity standards/harvesting stage

Developing fruit undergoes an array of physical, biochemical and physiological changes until it reaches maturity. Fruit maturity at harvest is directly associated with the market to which the produce has to be dispatched. Normally, growers determine the maturity stage for any fruit by adjudging the change in shape, size, colour, aroma and disappearance of astringency (Chadha, *et al.*, 2005). Following criteria have been identified to judge the maturity and harvest stage to enhance the post-harvest life of some tropical and sub-tropical fruits (Table 6).

Table 6: Criteria for judging harvesting maturity in some tropical and sub-tropical fruits

Fruit	Physical characters	Chemical characters
Mango	Light green peel colour, clear and enlarged lenticels, reflexed shoulder and desired specific gr. (1.02).	Starch (%), flesh colour
Banana	Light skin colour, loss of angularity, drying of outer leaves and days from emergence of inflorescence (100-115 days)	Pulp/peel ratio(1.35 to 1.40), starch content
Papaya	Appearance of yellow patches/ streaks and days after flowering (135-145 days). Darkening of seed colour, gelliness of sarcotesta	Yellowing of flesh
Pineapple	Greenish-yellow skin, smooth surface, drying of bracts/projections	TSS (12-14^0 Brix), acid (0.5-0.6%)
Sapota	Development of dull orange skin colour, disappearance of brown scale and easy fall of adhering stigma upon touching	Slow release of milky latex

Conclusion

Fruits add greater variety to diet in terms of colour, texture, flavor and also provide much needed fibres, minerals and vitamins. A large number of improved fruit varieties and range of agro techniques have been developed suiting to different climatic conditions. Some notable examples are high density planting in mango, pineapple, banana, citrus, apple etc, canopy management, integrated nutrient management etc.The techniques for enhancing water productivity through micro-irrigation have amply demonstrated in major fruit crops which not only saves water but capable of enhancing yield to the tune of 50- 200 per cent. Development of efficient nutrient management schedule including fertigation, standardization of leaf nutrient standards paved the way for enhancing nutrient use efficiency and improving quality of fruits. The use of latest varieties and technology certainly help in achieving higher productivity and quality of produce and finally enhancing the income of producers.

References

Indian Council of Agricultural Research. (2005) Pawar Inaugurates new National Research Centre on Pomegranate. 1 Sept. 2006. http://www.icar.org.in/pr/25092005.htm >.

Janick, J. (2007) Fruits of the bibles. *HortScience* 42:1072–1076.

Janick, J. and Paul, R.E., editors. (2008) The Encyclopedia of Fruits and Nuts . Cambridge, MA. CABI

Singh, D.B., Samadia, D.K. and A.R.P. Kingsly. (2006) Conservation, characterization and evaluation of pomegranate germplasm under arid ecosystem of India. p. 15. In: 1st International Symposium on Pomegranate and Minor Mediterranean Fruits, Abstracts contributed papers, 16-19 Oct, Adana, Turkey.

Improved Agriculture and Allied Practices for Farmers, pp 123-134
New India Publishing Agency, New Delhi, India

11

Modern Fruit Production Techniques for Higher Yield and Better Income

M.K. Verma, S.K. Singh, V.B. Patel and Chavlesh Kumar

Division of Fruits and Horticultural Technology, ICAR-Indian Agricultural Research Institute, New Delhi-110012

Indian horticulture has made a rapid stride, converting the country from food scarce to self sufficiency. The horticulture, which includes fruits, vegetables, spices, flowers and medicinal and aromatic plants, has proved beyond doubt its potentiality for gainful diversification. One of the significant developments is that horticulture has moved from rural confine to commercial production, and this changing scenario has encouraged private sector investment in production system management. Since the growing of fruit crops is highly rewarding to the farmers in terms of returns per unit area, the horticulture sector is expected to contribute significantly for the food and nutritional security, employment opportunity and poverty alleviation.

In recent past, several innovations and new initiatives have been made both in research and development for improving production, productivity, quality and marketing and to identify critical gaps in various emerging areas in fruits & horticulture technology. Some of the brief details are discussed on under:

High yielding superior fruit varieties for commercial cultivation

Mango: The mango varieties have been developed as a national priority at IARI, which are highly popular amongst the fruit growers of India. The important features of these hybrids are summarized below:

Amrapali (Dushehari x Neelum): This hybrid was released in 1979. Trees are dwarf and suitable for high density plantation (2.5 m x 2.5 m). It is highly regular and precocious and bears small to medium (150-180 g) fuirts and having light greenish yellow in colour at maturity. Pulp plentiful (74.8%) and deep orange red in colour and fibreless.TSS (22.8%) with high carotenoid content (16,830 μg/ 100 g of pulp). Late maturing and fruits matures by 4th week of July under north Indian conditions.

Mallika (Neelum x Dashehari): First mango hybrid released in 1971 by IARI for commercial cultivation after planned hybridization. Trees are semi-vigorous having moderate spreading and fairly regular bearer. Fruits are medium to large (250-400 g) in size and have attractive apricot yellow colour. Pulp is fiberless and firm with high recovery (75%). Taste and flavour of fruits are excellent with high TSS (24 °Brix) and has good keeping quality. Fruits mature during 3rd-4th week of July. It is suitable for table and processing purposes.

Pusa Arunima (Amrapali x Sensation): This mango hybrid was released in 2002. It is regular bearer, semi-vigorous and suitable for closer planting (6 m x 6 m). Fruits mature by first week of July under north Indian conditions. Fruits are medium to large (230 to 250 g) having attractive red blush on the shoulders. It has medium TSS (19.5° Brix) and long shelf-life (10-12 days) at room temperature after ripening

Pusa Surya (Syn. Eldon): Released in the year 2002. Trees are semi-dwarf and suited for close planting (6 m x 6 m). Fruits are medium to large (260 to 270 g) in size having attractive golden apricot- yellow colour at maturity. Pulp plentiful (70%) with medium TSS (19 °Brix) with long shelf-life (10 to 12 days) at room temperature after ripening. Fruit ripens by mid-July under north Indian conditions.

Pusa Pratibha (Amrapali x Sensation): This unique hybrid was released in 2012. It is regular bearer and suitable for closer planting (6 m x 6 m). The red peel colour on golden yellow background makes it very appealing. The fruits are medium (181 g) in size and have high pulp content (71.1%). It has medium TSS (19.6%) and has good flavour with better shelf-life (7 to 8 days) at room temperature. Fruits start maturing from in first week of July under north Indian conditions.

Pusa Shreshth (Amrapali and Sensation): It is a unique hybrid released in 2012. The plants are semi-vigorous and are suitable for closer planting (6 m x 6 m). It has regularity in bearing, attractive elongated shape, red peel and orange pulp. The fruit size is medium (228 g) having attractive red peel colour and higher pulp content (71.9%). It has medium TSS (20.3%), good flavour and better shelf-life (7 to 8 days) at room temperature. Fruits start maturing from in first week of July under north Indian conditions.

Pusa Lalima (Dushehari x Sensation): This hybrid was released in the year 2012. Plants are semi vigorous and can be planted at at 6 m x 6 m distance. The fruit size is medium (209 g) and higher pulp content (70.1%). Fruits are having attractive red peel colour on greenish background. It has medium TSS (19.7%), and good shelf-life (5 to 6 days) at room temperature. It is an early maturing hybrid. Fruits start maturing by first fortnight of June under north Indian conditions.

Pusa Peetamber (Amrapali x Lal Sundari): This hybrid was released in the year 2012. It is regular bearer and suitable for closer planting (6 m x 6 m). Fruits are having attractive oblong shape and have attractive yellow peel colour. Fruits are medium (213 g) in size with juicy pulp (73.6%). It has medium TSS (18.8%) and good shelf-life (5 to 6 days) at room temperature. It is moderately resistant to mango malformation and major insect pests of mango.

2. Sweet Orange

Pusa Sharad: A selection having medium vigorous trees, thorn present, dense canopy, suitable for planting at a distance of 5 m x 5 m. Average fruit weight 227.56 g, round in shape, juice recovery 50.12%, medium thick peel, higher TSS (9.20°brix) and low acidity (0.77%). It yields higher than Jaffa and Valencia. Suitable for cultivation under plains of North and central India as well sub Himalayan regions of Uttarakhand.

Pusa Round: It is a selection and released for commercial cultivation in both northern plains and Sub-Himalayan regions of Uttarakhand. It has uniform bigger sized fruits and higher juice content (48.26%). It yields higher than Jaffa and Valencia. TSS (10.14%); ascorbic acid content (38.06 mg/100 ml juice), moderate acidity (0.92%).

3. Lemon

Kagzi Kalan: It is an excellent lemon, ideally suited for north Indian conditions. The plant is semi-vigorous, low spreading prolific bearer and starts bearing after three years of planting. The first crop is ready for harvest from the month of May till August and the second crop is harvested in December-January. The fruits are medium to large in size and are extremely juicy. The fruits are oblong or round, peel yellow, smooth, thin, seedless; pulp greenish-yellow, in about 10 segments, juicy, moderate acidic. Fruits often borne in large terminal clusters unprotected by the foliage.

4. Acid Lime

Pusa Abhinav: It is a clonal selection identified from diverse population of acid lime. It has dense canopy, heavy yielder, fruits round in shape, medium (34.6 g), fruit length (41.0 mm), diameter (37.2 mm), peel thickness (1.1 mm); seeds (16.0/fruit), fruit yellow at ripening; juice content (64.5%); TSS (8.2%); acidity (7.7%). It produces fruits round the year with peak harvesting during August-September and March-April. It is moderately susceptible to citrus canker. It is suitable for cultivation under Indo-gangetic plain and trans-gangetic plains of India.

Pusa Udit: It is a clonal selection. It has dense canopy, heavy yielder, fruits round in shape, medium in size (37.9 g), fruit length (43.9 mm), diameter (39.3 mm), peel thickness (1.1 mm), seeds (8.3/fruit), fruits bright yellow at ripening, juice content (43.3%), TSS (8.5%), acidity (6.9%). It produces round the year crop with peak harvesting from August-September and February-March. It is moderately susceptible to citrus canker. It is suitable for cultivation under Gangetic plain and trans-gangetic plains of India.

5. Grape

Pusa Seedless: It is a selection made from Thompson Seedless. It resembles its parent Thompson Seedless with regard to most of the characters but its berries are more elongated. It was recommended for commercial cultivation in 1976. It is highly responsive to GA_3 application. The berries have high TSS (22-24^{0}Brix) content. It is suitable for both table purpose and raisin making.

Beauty Seedless: It was introduced by this Horticulture Division of IARI from California, USA and was evaluated and recommended for commercial cultivation in north India in 1972. The berries ripen by the end of May. The vine is medium in vigour; bunches medium to large, long shouldered and compact with blueish-black coloured, spherical, medium sized berries. It is a prolific bearer but the keeping quality is very low.

Pusa Navrang: This hybrid is a cross between Madeleine Angevine and Rubired and was released 1996. It is an early ripening (1st week of June), basal bearer and teinturier variety containing red pigment both in peel and pulp. The bunch is loose, medium in size with round and medium sized berries with TSS ranging from 18-20%. It is ideally suited for coloured juice and wine making and is resistant to anthracnose disease.

Pusa Urvashi: This hybrid (Hur x Beauty Seedless) was released in 1996. It is early ripening (1st week of June) and a basal bearer. Its bunch is loose and medium in size with seedless greenish-yellow berries. The berry TSS varies from 20 to 22%. It is suitable for table purpose and raisin making.

Pusa Aditi: It is a hybrid between 'Banqui Abyad' x 'Perlette'. It is early maturing (1st week of June), seedless having large round berry (2.7 g), yellowish-green in colour, firm pulp with good TSS (19.3°Brix). Vine is moderately vigorous, and is a spur pruned variety. It is tolerant to anthracnose and powdery mildew. Grape bunches and berries are GA_3 responsive. The fruits are good for table purpose and making juice. The average yield is about 12-15 t/ha.

Pusa Trishar: It is a three way cross hybrid between ('Hur' x 'Bharat Early') x 'Beauty Seedless'. It is early maturing, suitable for sub-tropical conditions, semi-vigorous and spur-pruned, having moderate tolerance to anthracnose, powdery mildew and termite. It matures by 10th June. Berries are round (2.15 g), yellowish-green in colour, with good TSS (18.4°Brix) and firm pulp. Grape bunches and berries are GA_3 responsive. The fruits are good for table use and juice making purpose. The average yield is about 14-16 t/ha.

6. Papaya

Pusa Dwarf: This variety is dwarf with oval round fruits of medium size having weight of 1 to 1.5 kg. It starts fruiting at a height of 35 cm above ground level.

Pusa Nanha: This variety was evolved through mutation. It is exceptionally dwarfing (1m) and fruiting starts at 30 cm bright from ground level. It is suitable for planting in pots and under high density planting system.

Improved Fruit Production Technologies

High density orcharding-Amrapali Mango: Mango is generally being cultivated at wider spacing (10-15 m) wherein only 80-100 trees could be planted in a hectare of area. Under this system lot of available area/land could not be utilized efficiently which causes a loss of revenue from the available land are results in low productivity. High density orchard is accommodate more than 375- 450 and more trees per hectare. High density plantations are the consequences of continuous decline in the availability of cultivated land, rising energy and land cost together with the mounting demand for horticultural produce. Besides an increased number of trees/ha, a high density orchard must come into bearing within 2-3 years after planting, through using precocious rootstocks. As tree density increased, profitability increased up to approximately 2500 tress per hectare. The result of medium density plantation is most encouraging in mango. Dwarf and compact trees in high density orchards, not only provide high yield and net economic returns per unit area in the initial years but also facilitate more efficient use of fertilizers, water, plant protection measures, weed control, and easy efficient harvest management practices.

Table 1: Comparative details of high density in mango over traditional system in terms of estimated investment, net returns etc.

Particular	Traditional system	High density planting
Spacing (m)	10 x 10	2.5 m x 2.5 m
No. of plants/ha	100	1600
Cost of establishing orchards	Rs. 40,000	Rs. 75,000
Annual maintenance cost	Rs. 40,000	Rs. 45,000
Age of stable yield	8 to 10 years	7 to 8 years
Production (kg/ha)	10,000 to 12,000	18,000 to 22,000
Sale of produce (whole sale) @ Rs. 10.0 per kg	Rs. 1,00,000 to 1,20,000	Rs. 1,80,000 to 2,20,000
Net Return	Rs. 40,000 to 50,000	Rs. 1,20,000 to Rs 1,30,000

***High density orcharding-Kinnow*:** Kinnow mandarin is very potential and highly remunerative fruit crop. Vigorous rootstock like Jatti Khatti and Rough lemon are very common for Kinnow cultivation under traditional system. However, Troyer Citrange rootstock offers dwarfing statute to kinnow. Therefore, high density orcharding is also possible using Troyer Citrange rootstocks at closure planting (1.8 x 1.8 m) accommodates 3086 trees/ha.

Table 2: The comparative advantage of high density in kinnow over traditional system is given

Particulars	Traditional system	High density
Spacing (m)	6 x 6	1.8 x 1.8
No. of plants/ha	278	3086
Cost of establishing orchards	Rs. 35,000	Rs. 75,000
Annual expenditure (Rs.)	Rs. 25,000	Rs. 50,000
Age of stable yield	5 years	4 years
Production (kg/ha/year)	10,000 to 12,000	22,000 to 25,000
Sale of produce (whole sale) @ Rs. 9 per kg	Rs. 90,000 to Rs 1,08,000	Rs. 1,98,000 to Rs. 2,25,000
Net Return	Rs. 65,000 to Rs. 80,000	Rs. 1,48,000 Rs. 1,75,000

***High density orcharding-Guava*:** In recent past, guava becomes a remunerative fruit crop due to nutritional awareness and its use for processing. High density orcharding for guava has been developed through use of dwarfing rootstocks, methods planting, in combination with judicious training and pruning. Indian Agricultural Research Institute has developed a dwarfing rootstock 'Pusa Srijan'. Grafting of 'Allahabad Safeda' on 'Pusa Srijan' rootstocks gave a dwarfing statute facilitates closure planting. This reduces about 50% of the tree vigour of 'Allahabad Safeda'. In the conventional system of planting trees generally plated at wider spacing (6 x 6 m) accommodates 278 trees/ha. Whereas, on Pusa Srijan these can be planted at much closure spacing (3 x 3 m) accommodated more than three time trees (1111) per hectare. When trees attain the full canopy, needs regular annual pruning to facilitate proper sunlight penetration as well as air ventilation. This is good for getting higher yield per unit area.

Table 3: The comparative advantage of high density in guava over traditional system is given in below table.

Particulars	Traditional system	High density
Rootstock	Allahabad Safeda	Pusa Srijan
Spacing (m)	6 x 6	3 x 3
No. of plants/ha	278	1,111
Cost of establishing orchards and maintenance	Rs. 40,000	Rs. 55,000
Annual expenditure (Rs.)	Rs. 25,000	Rs. 30,000
Age of stable yield	4 years	4 years
Production (kg/ha/year)	10,000 to 14,000	20,000 to 25,000
Sale of produce (whole sale) @ Rs. 8.0 per kg	Rs. 1,20,000 to Rs 1,40,000	Rs. 2,00,000 to Rs. 2,50,000
Net Return	Rs. 60,000 to Rs. 70,000	Rs. 1,15,000 Rs. 1,65,000

***High density orcharding-Papaya*:** Development of dwarfing varieties in papaya given an opportunity to grow them at very closure spacing. In past, non-availability of such cultivars forced to papaya growers to grow them at wider spacing, which causes a loss of available land for unfruitful purpose. These were generally being planted at wider spacing (2.4 m x 2.4 m) as compared to closure planting (1.25 m x 1.25 m) using dwarf varieties like 'Pusa Nanha'. This variety was developed by Indian Agricultural Research Institute and catering the needs of growers by supplying true-to-type seeds. Generally, plants starts bearing just 30 cm above from the ground surface and total length of tree reached not more than 1.5 m from ground. Therefore, these are generally recommended to plant at 1.2 m x 1.2 m accommodated 6400 plants per hectare. Another variety 'Pusa Dwarf' is also developed by this Institute. This variety is also dwarf in statute and but little vigorous over 'Pusa Nanha'. This is planted at 1.5 m x 1.5 m spacing accommodates 4444 plants per hectare.

Table 4: The comparative advantage of high density over in papaya traditional system is given in below table.

Particular	Traditional system	High density
Spacing (m)	2.4 x 2.4	1.25 x 1.25
No. of plants/ha	1,736	6,400
Cost of establishing orchards (Rs.)	Rs. 70,000	Rs. 1,05,000
Annual Expenditure (Rs.)	Rs. 40,000	Rs. 75,000
Age of stable yield	1 years	1 years
Production (kg/ha/year)	45,000 to 50,000	80,000 to 90,000
Sale of produce (whole sale) @ Rs. 7.0 per kg	Rs. 3,15,000 to Rs 3,50,000	Rs. 5,60,000 to Rs. 6,30,000
Net Return	Rs. 2,15,000 to Rs. 2,50,000	Rs. 3,40,000 to Rs. 4,10,000

Management of mango malformation: Mango malformation is a serious threat to mango cultivation in many sub-tropical areas in the world. It is now well established that *Fusarium mangiferae* causes the symptoms of mango malformation. This melody is a serious threat to mango industry. Productivity in malformed mango trees can be improved by a single spray of NAA (200 ppm) in the first week of October followed by deblossoming once at the bud burst stage during January. It has become the most widely accepted control measure for floral malformation.

***Control of biennial bearing*:** To solve the problem of alternate bearing in mango, the use of paclobutrazol (cultar PP333) was recommended for the first time in India. Application of paclobutrazol (PP 333) @ 22.5 ml/10 L of water at pea stage of mango fruit was found to induce new shoot growth in 65% of the fruited shoots which accounted for fruiting in the next season. This treatment was found to be the most effective in Langra, which is considered to be a highly alternate bearer.

***Control of fruit drop in mango*:** Causes attributed to the problem of fruit drop in mango, like lack of self- pollination, low stigmatic receptivity, and defective perfect flowers, poor pollen formation, presence of self-incompatibility, lack of moisture and incidence of pests and diseases etc. are the some of intensive work carried out in the Division. Application of 2,4-D (10 ppm) holds good for controlling pre-harvest drop in most of the mango cultivars.

***Control of fruit drop in citrus*:** Pre-harvest fruit drop is a serious problem in different types of citrus. Among the various growth regulators, 2,4-D was found to be the most effective in checking fruit drop. Thus, the recommendations for application of 2,4-D (8, 15, 20 ppm) for Valencia, Mosambi and Jaffa cultivars, respectively during October got accepted commercially.

Control of citrus granulation: It is a serious physiological disorder of citrus in north India. It is highly pronounced in Mosambi sweet orange followed by mandarins. Studies indicated that the respiration rate in the affected fruits increased by 7-15% while disease and pectin esterase activities decreased, which led to reduction in sugar and increase in polysaccharides. It is recommended that the incidence of granulation can be minimized by the application 2-3 sprays of NAA (300 ppm) and GA (50 ppm) during August to October.

Pruning practices for subtropical grapes: Pruning intensity has been standardized using bud dissection technique for early maturing grape varieties like 4-6 nodes (Perlette, Beauty Seedless, Cardinal, Flame Seedless, Pusa Urvashi, Pusa Navrang, Pusa Aditi), 8 node (Anab-e-Shahi), and 10-14 nodes (Pusa Seedless).

Hastening of bud burst and maturity in grape: Application of Dormex (49% aqueous solution of hydrogen cynamide) @ 1.5% a.i. on dormant buds in the beginning of January or thiourea (4.0-5.0%) immediately after pruning hastened the bud burst by about one month and ripening by 16,12 and 9 days in Pusa Seedless, Beauty Seedless and Perlette, respectively.

Berry elongation technologies in grape: For improving quality of grape, the application of GA_3 has been standardized. Application of GA_3 45 ppm at full bloom for Pusa Seedless; GA_3 45 ppm at full bloom for Beauty Seedless; 30 ppm at 50% bloom for Perlette are the most accepted recommendations by the north Indian grape growers. Application of GA_3 (60 ppm) at 50% bloom and three week later has been found better for Pusa Urvashi, Pusa Aditi and Pusa Trishar.

Technology for enhancing berry quality of grape: Trunk girdling at 15 cm height on trunk from ground about one month before ripening and followed by spray of 250-400 ppm ethrel at Veraison stage improved the berry quality in Pusa Urvashi. Water berry disorder in Perlette can be minimized by application of 20-30 kg/ha borax in soil or 0.5% sprays of boric acid. To improve berry quality, application of 0.2% $ZnSO_4$, 0.2% boric acid and 0.5% K_2SO_4 has been suggested.

Softwood grafting in mango: For faster multiplication of mango varieties Division has standardized soft wood grafting technique for faster and bulk multiplication of plants. In softwood grafting 45-60 day-old 8-12 cm long, healthy and vigorous scion sticks are suitable for successful grafting.

Epicotyle grafting in Mango: In Epicotyle grafting 15 to 20 days old seedlings are used as root stock for grafting. When the mango seedlings are 15 to 20 days old, (still in reddish, bronzy, coppery or greenish red colour and having stone kernel left with some food and nutrient supplement for getting quick joint in the grafts) these are uprooted carefully with stone and root undamaged.

***Strawberry yield improvement*:** GA_3 (75 ppm), given twice at 10 days interval after bud differentiation can increase the yield of strawberry by 80%. This treatment is the most effective if plants are covered with an alkathene sheet (400 gauge) immediately after first spray of GA_3.

***Phalsa yield improvement*:** Yield of phalsa can be increased by 30% with a single spray of 2,4-D (2.5 ppm) immediately after fruit set. This recommendation has given highly encouraging results at farmer's field.

Improved techniques of vegetative propagation

Mango-Bottom Heat Technique: Another simple technique, devised at the Indian Agricultural Research Institute, New Delhi has facilitated rooting of cuttings through bottom heat. The device employed consists of a double tin container, the outer one being 70 cm in height and 46 cm in diameter and an inner jacket having a height of 68 cm and a diameter of 44 cm.The space between inner jacket and cuter container is packed with glass wool for heat insulation. Another container which fits into the inner jacket (also constructed from the same material measuring 35 cm in height) and having two electric bulbs of 40 watt each at the bottom end- in between the vacant space in two containers, is fitted into the double container mentioned above. This container is then filled with the rooting media (I part sphagnum moss, I part sand: I part grit) and the non-leafy hardwood cuttings are planted in it. The temperature is maintained at about 30°C + 2°C through a thermostat which controls the heating bulbs. The cutting root in about a month and the success is as high as 97 per cent- Majumdar and Prasad (1985) further reported 85.8 per cent rooting at 35°C basal temperature when combined with wounding and 5000 ppm IBA treatment. When applied through sides, the low concentration of IBA (2500 ppm) increased rooting significantly (90.5%) as compared with higher concentration (5000 ppm). Rooting was the highest (95.6%) when lower concentration of IBA was applied through base and side.

Mango-Vaneer grafting: An improved method for mango propagation known as vaneer grafting was standardized in this Division. This method is highly economical, easy and gives very high degree of success (96%) for establishing in situ orchards. Besides, being detached method of grafting (unlike inarching), this technique can be utilized for preparing plants anywhere.

Mango-Top working: A technique known as top working, was developed in this Division. By this method, old declining orchards of mango can be rejuvenated with any superior variety by vaneer grafting.

Mango-Epicotyl/stone grafting: Epicotyl grafting – a highly successful method of mango propagation in Konkan region was standardized initially at this Division, although, success under north India conditions was limited owing to high temperature and low humidity.

Guava-Stooling: A new technique known as "stooling" was standardized for faster multiplication of guava. In this method, the plants are headed back during December and then allowed to develop multiple underneath shoots. These shoot are ringed, applied with IBA (2500 ppm) in Lanolin and earthed up as a mound to induce rooting. The rooted shoots are detached from mother plant and planted in nursery in April-May. It can be successfully practiced twice a year, as a result of which, one can get 30-40 stools/plant/year. This technique has attained the status of commercial propagation method in India and has replaced tedious and cumbersome method of inarching and air layering which were popular earlier.

Micropropagation

Papaya*:* Technique for *in vitro* multiplication of papaya through shoot tip culture has been standardized. *In vitro* propagation of grape with shoot tip culture and microcuttings have been perfected. A technique of *in vitro* zygotic embryo development and germination in fruitlets excised after 15-21 days of pollination was developed for increasing the breeding efficiency in mango which suffers from heavy fruit drop.

Grape: A protocol for mass multiplication of newly released grape hybrids employing two-node repetitive micro-cutting method was standardized. The protocol for bio-hardening of tissue culture raised plantlets has been developed, which improves the plant survival by 80-90% during hardening and subsequently during field transfer (95%). Genetic uniformity of plantlets was also confirmed using RAPD markers.

References

Awasthi O. P., Singh A. K. and Verma M. K. (2012). Orchard tourism an enterprise. In: Horticulture for Environment and Ecotourism (Eds. Jankiram et al). Published by Division of Floriculture and Landscaping, IARI, New Delhi. Pp.17-24.

Singh, A. K., Singh, S. K., Srivastava Manish, Prakash Jai and Verma M. K. (Eds.) 2013. Good Agricultural Practices GAP in Production of Horticultural Crops. Division of Fruits and Horticultural Technology, Indian Agricultural Research Institute, New Delhi 110 012, India, 144 + viii. TB-ICN: 114/2013.

Singh S. K., Singh A. K., Srivastava Manish and Verma M. K. 2013. Good agricultural practices for horticulture production in India vis-a-vis World. *In*: (eds. T. Jankiram, K. V. Prasad, K. P. Singh, K. Swaroop, D. V. S. Raju, Ritu Jain and Namita)). Division of Floriculture and Landscaping, Indian Agricultural Research Institute, New Delhi-110012. Pp-117-133.

Verma, M. K., Singh, S. K., Srivastava Manish and Prakash Jai. 2013. Good Agricultural Practices for Temperate Fruit Production. In: *Training Manual* "Good Agricultural Practices GAP in Production of Horticultural Crops". *Div*ision of Fruits and Horticultural Technology, Indian Agricultural Research Institute, New Delhi 110 012, India, Pp-28-44.

Improved Agriculture and Allied Practices for Farmers, pp 135-143
New India Publishing Agency, New Delhi, India

12

Diversification with Vegetable Crops for Nutritional Security and Higher Income

Harshawardhan Choudhary, P. K. Yadav and B. S.Tomar

Division of Vegetable Science, ICAR-Indian Agricultural Research Institute New Delhi-110012

Crop diversification has been well recognized as an effective strategy for achieving the objectives of food and nutritional security, income growth, poverty alleviation, employment generation, judicious use of land, water and other natural resources, sustaining agricultural development and environmental improvement. Vegetable crops being rich sources of various health building substances, especially vitamins and minerals, offer unique advantage for food and nutritional security, tackle malnutrition and dietary deficiency diseases. Besides, their high yield potential per unit area and time, diverse varietal wealth, labour intensiveness, high market price and prospects of processing, value addition and export, they play an important role in employment generation, and livelihood improvement. Since India is endowed with diverse agro-climatic conditions, almost all kinds of vegetable crops are grown in one or other corner of our country. More than 200 different vegetables are grown in India, of which, about 70 types are cultivated commercially. Vegetable production scenario in India has made a remarkable progress since independence from a meagre production of 17.5 million tonnes to 146 million tones, with a productivity of 17.3 t/ha. India is the second largest vegetable producer in the world after China. However, it needs to be doubled by 2020 in order to meet the nutritional requirement of burgeoning population in the country. With increasing purchasing power of people, the demand for vegetables has increased enormously, as a result the prices of vegetables have increased and it has been the dominant factor for high inflationary pressure in Indian economy during recent years. Inclusion of different vegetable crops with shorter maturity period in the prevailing cropping system will enhance the total system productivity, which may go a long way in improving the economic as well as physical health of small and marginal farmers.

Advantages of including vegetable in the cropping system

- Higher productivity and profitability per unit area and time.
- More number of crops can be grown on the same piece of land.
- Diversification minimizes weather and pests risk involved.
- Higher quantum of micronutrients, vitamins, minerals and other health protective phytochemicals would be available.
- More opportunities for employment generation due to labour intensive operations and related secondary activities like transportation and marketing.
- Better scope for value addition, processing and other agro-industries.
- Improvement in physical, chemical and biological health of soil.

Improved varieties of vegetable crops

Till now, Indian Agricultural Research Institute has developed 170 improved varieties including 26 F_1 hybrids in 43 different vegetable crops, which are popular among vegetable growers across the country. These varieties have formed a sound foundation for the establishment of private vegetable seed industry in India. Of these, fifteen varieties have resistance to different diseases; twelve have resistance to extreme temperature making them suitable for growing under changing climatic scenario. Twenty varieties have been specifically developed for high nutrition, suitability for processing, export and industrial uses such as nutraceuticals and extraction of food colorants. The important varieties (Table 1) and hybrids (Table 2) of different vegetable crops along with their production potential and estimated profit from their cultivation is given below, which may be chosen for inclusion in the prevailing cropping system as per the preference and specialization of individual farmer keeping in view the market demand and existing higher price for such vegetable in that locality during the availability of produce.

Table 1: High yielding varieties of different vegetable crops

Crop	Variety	Yield (q/ha)	Estimated profit (Rs./ha)
Solanaceous Vegetables			
Brinjal	Pusa Shymla	425	90,000
	Pusa Uttam	400	85,000
	Pusa Upkar	400	85,000
	Pusa Ankur	350	65,000
Tomato	Pusa Rohini	450	1,10,000
	Pusa Uphar	400	90,000
Chilli	Pusa Sadabahar	100	60,000
Cole Crops			
Cauliflower (Early)	Pusa Meghna	150	95,000
	Pusa Deepali	160	70,000
Cauliflower (Mid Early)	Pusa Sharad	250	50,000
Cauliflower (Mid Late)	Pusa Paushja	325	50,000
	Pusa Shukti	325	50,000
Cauliflower (Late)	PSB K-1	350	40,000
	PSB KT-25	350	55,000
Cabbage	Golden Acre	275	60,000
	Pusa Drum Head	400	65,000
Cucurbitaceous Vegetables			
Pumpkin	Pusa Vishwas	400	45,000
	Pusa Vikas	300	35,000
Bottle gourd	Pusa Samridhi	325	60,000
	Pusa Naveen	325	65,000
Bitter gourd	Pusa Vishesh	200	70,000
	Pusa Domausmi	325	65,000
Cucumber	Pusa Uday	160	50,000
Sponge gourd	Pusa Supriya	120	50,000
	Pusa Sneha	140	50,000
Ridge gourd	Pusa Nutan	185	60,000
Ash gourd	Pusa Ujwal	350	40,000
Muskmelon	Pusa Madhuras	200	50,000
Waterkmelon	Sugar Baby	300	60,000
Leguminous Vegetables			
Pea	Arkel	70	50,000
	Pusa Pragati	70	50,000
Root Crops			
Carrot	Pusa Rudhira (Red)	325	70,000
	Pusa Asita (Black)	275	55,000
	Pusa Vrishti (Red)	250	65,000
Onion	Pusa Madhvi	250	95,000
	Pusa Red	502	95,000
Leafy Vegetables			
Palak	Pusa Bharati	250	50,000
Veg. Mustard	Pusa Sag-1	350	40,000
Other Vegetables			
Okra	Pusa A-4	140	50000

Table 2: F_1 hybrids developed by IARI for higher yield

Crop	Variety	Days to first harvest	Av. Yield (q/ha)	Net Income (Rs./ha)
Brinjal	Pusa Hybrid-5	50-55	450	95,000
	Pusa Hybrid-6	55-60	450	95,000
	Pusa Hybrid-9	55-60	500	1,10,000
Tomato	Pusa Hybrid-1	70	450	90000
	Pusa Hybrid-2	80	550	1,10,000
	Pusa Hybrid-4	85	550	1,10,000
	Pusa Hybrid-8	80	650	1,30,000
Bittergourd	Pusa Hybrid-1	55-60	200	80,000
	Pusa Hybrid-2	55-60	250	1,10,000
Bottle gourd	Pusa Hybrid-3	50-55	450	95,000
Cauliflower	Pusa Hybrid-2	80	230	83,000
	Pusa Kartik Sankar	95	250	95,000
Cucumber	Pusa Sanyog	50	200	75,000
Pumpkin	Pusa Hybrid-1	60	450	55,000

Year round vegetable production through IARI varieties

Radish : Radish is one of the important salad vegetable crops required through out the year. A single variety can not be used for growing roots through out the year. Different cultivars developed at IARI having specific growing requirements (Table 3) can be grown under north Indian plains for sowing right from August and continued for the entire year so as to make fresh radish available to the consumers.

Table 3: Improved Radish cultivars for year-round production

S.No.	Cultivars	Period of sowing	Period of harvesting	Av. Yield (q/ha)
1.	Pusa Desi	Middle of August to Middle of October	Mid-September to Mid-December	175
2.	Pusa Mridula	First fortnight of September to Mid-November	2nd fortnight of October to 1st fortnight of January	180
3.	Japanese White	Mid-October to 2nd fortnight of December	Mid-December to 1st fortnight of March	250
4.	Pusa Himani	2nd fortnight of December to end of February	Mid-February to 3rd week of April	250
5.	Pusa Chetki	First week of April to mid-August	First fortnight of May to 2nd fortnight of September	170

Cauliflower

In earlier days, cauliflower was used to be grown only during winter season because only Snowball types were available during those days. But, with the evolution of heat tolerant, tropical types, popularly known as Indian cauliflower, it has become possible to grow cauliflower during the hotter part of the year also. Based on temperature requirement for growth and curd development, cauliflower varieties have been developed and now it has become possible to grow cauliflower for almost all the year-round (Table 4).

Table 4: Improved Cauliflower varieties and hybrids for different seasons of the year

Maturity group	Varieties and hybrids	Sowing time	Availability period	Av. Yield (q/ha)
Early I(a)	Pusa Meghna,	End of May	End of August-	110
	Pusa kartik Sankar		October	110
Early I(b)	Pusa Deepali	June	Oct.-Nov.	120
Mid Early II	Pusa Sharad,	End of July-	Nov.-Dec.	250
	Pusa Hybrid-2	August		200
Mid Late III	Pusa Paushja,	End of August-	Dec.-Jan.	325
	Pusa Shukti	Sept.		320
Late	Pusa Snowball K-1,Pusa	Sept.-Nov.	Jan.-March	350
	Snowball K-25			320

Carrot

Improved carrot varieties have been developed which can be grown round-the-year and the availability of carrot can be ensured right from May to June and November to January (Table 5)

Table 5: Improved Carrot varieties and hybrids for different seasons of the year

Variety	Sowing time	Availability	Yield (q/ha)
Pusa Vrishti	July to October	October to November	180-200
Pusa Rudhira, Pusa Asita	September to October	November to January	300-350
Pusa Yamdagini, Pusa Nayanjyoti	September to March	November to April	270-320
Pusa Vrishti	March to April	May to June	150-160

D. Vegetable Production out of normal season

i) *Kharif* Onion

Onion is, generally, grown as *rabi* crop in northern India, whereas in Maharashtra, Gujarat, Andhra Pradesh, Karnataka and Tamil Nadu, it is grown both in *kharif* and *rabi* seasons. Maharashtra, has the largest area under *kharif* onion and the crop is harvested during October to December. The *rabi* crop cannot be stored beyond October due to sprouting of bulbs in storage. Kharif crop raised in Maharashtra becomes available from October to April, before *rabi* crop comes into the market in north India. The prices of onion during this period remain high due to limited production and heavy transport charges.

In view of the importance of *kharif* onion crop, Division of Vegetable Science, IARI, New Delhi, took initiatives in exploring the possibilities of growing *kharif* onion crop in northern India and raised successful crop with N-53 variety. The technology generated by this Division, given here-under, has spilled to farmer's fields in Haryana, eastern U.P., Bihar, Rajasthan and M.P. The kharif onion can be produced both by raising seedlings and raising sets. However, raising of sets found to be more beneficial than raising the crop by seedling method.

Varieties: N-53, Agrifound Dark Red, Baswant-780 and Arka Kalyan

Seed Rate: 8-10 kg seeds per hectare

Transplanting: The seedlings are transplanted in the field in August at the spacing of 15 cm between rows and 10 cm within rows.

Harvesting and curing: The crop is harvested in about 5 months i.e., during December- January on discolouration of leaves and proper pigment development in bulbs after the attainment of good sizes.

Storability: Irrigation should be stopped 10 days before harvesting to check the growth and to enable the bulbs to become dry and compact.

Yield: The yield of kharif crop is 150-200 q/ha, which is less than rabi crop. With good management, yield upto 250 q/ha can be achieved.

Estimated net return: 1, 20,000/ha

ii) Growing Early Spring Crop of Tomato

Tomato is thermo-sensitive crop for its fruit setting. The maximum fruit set in this crop takes place at night temperature of 15-20°C. Most of the commercial varieties fail to set fruits below 13°C. Tomato is therefore, scarcely available during February-March in north Indian plains. The Indian Agricultural Research Institute, New Delhi has developed two varieties namely, Pusa Sheetal and Pusa Sadabahar, which are capable of setting fruit even if the night temperature touches a low level of 8°C. These cultivars are very much suitable for raising an additional early spring crop of tomato in north Indian plains providing fruits during lean period (February-March).

Estimated net return

Pusa Sadabahar: Rs 90,000/ha

Pusa Hybrid-1:Rs.1,30,000/ha

iii) Ratooning of Brinjal for Off-season Production

The summer crop of brinjal is planted in the end of February or in the first week of March which starts fruiting only in the beginning of May. This crop gives very poor return to the growers due to very late fruiting and also because of low yield in extremely hot months of summer season. An extra early and profitable production of this vegetable can be obtained in spring summer season by following ratooning technology. For this, the standing crop of *kharif* season is protected from frost and cold injury in December and January by giving 3-4 light irrigations in the field. When the danger of frost is over in the last week of January or first week of February, the plants of standing crop are pruned with the help of secateur. All the dried branches are removed and the green branches are pruned in such a way so that only 30-40 cm plant portion is left above the ground.. After that, weeding and spading is done and fertilizer di-ammonium phosphate @ 150 kg/ha is mixed properly in the soil. Irrigations are given from time to time as per the requirement of the crop.

The first picking of fruits can be made 30 to 40 days after pruning. Pickings should be done at regular intervals so that fruit setting may continue for longer period. Since temperature is high during spring-summer season (May and June) a special care should be taken to irrigate the crop frequently or as per the requirement. Extra early harvest of fruits available from middle of March gives higher profit to the growers. As per observations, 30 to 40 per cent extra yield can be obtained from ratoon crop as compared to newly planted brinjal in summer season.

Estimated net return: Rs 1, 10,000 per ha

Short duration vegetable varieties for intensive cropping system

Short duration vegetable varieties can produce higher yield per unit area, per unit time and enhance the system productivity in whole (Table 6).

Table 6: Important short duration vegetable varieties/ hybrids for enhanced productivity and higher income

Crops	Varieties	Days to first picking	Crop availability	Av.Yield (q/ha)	Estimated profit (Rs./ha)
Summer Squash	Pusa Alankar	45	Feb-March	400	80000
	Austalian Green	45-50	Feb-March	350	70000
Radish	Pusa Chetaki	40	Aug-Sept	250	50000
	Pusa Desi	50-55	Sept-Oct	300	45000
	Pusa Himani	50-55	Feb-March	325	40000
	Japanese White	55-60	Nov- Dec	375	35000
Tomato	Pusa Sadabahar	80	Dec-Jan	275	60000
Turnip	Pusa Sweti	40-45	Oct	250	45000
	Pusa Chandrima	50-55	Dec	350	45000
	Pusa Swarnima	50-55	Dec	350	45000
Pea	Arkel	55-60	Nov	70	50000
	Pusa Pragati	60-65	Nov	70	50000
Cabbage	Golden Acre	60-65	Nov	250	60000
Knol khol	White Vienna	40-45	Nov	250	40000

Crop intensification through vegetables

Vegetables crops are of shorter duration and they can be fit well in different cereal based cropping system prevalent in different regions of our country. Inclusion of vegetables will diversify the existing cropping system and increase the total profit manifolds (Table 7). Three or more vegetables can be grown on the same piece of land in a year, which will enhance the cropping intensity, productivity and thereby as profitability. Some important vegetable based cropping systems and their expected profitability have been presented in Table 8. Vegetable cultivation will also be beneficial in improving the physical, chemical and biological health of soil.

Table 7: Inclusion of vegetables in cereal based cropping system for higher profit

Cropping Sequence	Expected Net profit (Rs./ha)
Rice- Cauliflower- Cowpea	1,02,000-1,32,000
Rice-Pea-Bottlegourd	1,35,000-1,63,000
Rice- Cauliflower- Onion	1,34,000-1,69,000
Rice- Carrot-Wheat	1,10,000-1,40,000
Okra-Radish-Wheat	95,000- 1,20, 000

Similarly, different cropping systems by including the vegetables can be much profitable and must be adopted by the farmers for enhanced profitability (Table 8).

Table 8: Profitable cropping sequences of vegetables for different regions

Cropping Sequence	Expected Net profit (Rs./ha)
Northern Plains	
Early cauliflower- Peas - Tomato	2,00,000-2,45,000
Okra-Carrot-Cauliflower-Radish	1,53,000-1,92,000
Cucumber- Potato-Onion	1,24,000-1,52,000
Eastern Region	
Early Cauliflower-Spinach-Sweet Pepper	1,72,000- 2,00,000
Chilli-Broccoli-Radish-Bottle gourd	2,02,000- 2,31,000
North Western Hill Region	
French Bean-Knolkhol-Peas-Sweet Pepper	2,10,000- 2,39,000
Tomato-Cabbage-Turnip-Potato	2,05,000- 2,55,000

Conclusion

The ultimate objective of a farmer is to get maximum production of crop and earn higher profit from his piece of land per unit time. The demand for vegetable is growing continuously and to meet the ever increasing targets as per the demand, productivity needs to be raised further and/ or more area needs to be brought under vegetable cultivation. There is enough opportunities of introducing vegetables in prevailing cereal based cropping systems by introducing short duration vegetable crops between two cereal crops or by replacing less profitable or less environmental friendly cereal crop with high value vegetable crops. Vegetable crops will enhance the productivity and profitability and it will pave the way for nutritional security and healthy society.

Improved Agriculture and Allied Practices for Farmers, pp 145-159
New India Publishing Agency, New Delhi, India

13

Production Technology of Solanaceous Vegetables

Raj Kumar, R.K. Yadav, Sabina Islam, B.S. Tomar and H. Choudhary

Division of Vegetable Science ICAR-Indian Agricultural Research Institute New Delhi-110012

The solanaceous group of vegetables (tomato, eggplant, chili and bell peppers) generally take up large amounts of nutrients. The amount of nutrients they take up depends on the quantity of fruit and dry matter they produce, which in turn is influenced by a number of genetic and environmental variables. Tomato (*Solanum lycopersicum* L.) known as *Tamatar* in Hindi is one of the most popular and widely grown vegetables in the world ranking second in importance to potato in many countries. Tomato is widely grown in almost all parts of India and is adapted to wide range of soil and climate conditions. Tomato is consumed either as raw or cooked and is also very important crop for processing in the form of soups, juice, ketchup, puree, paste and powder. Tomato adds variety of colours and flavour to food.

Tomato is very well known for its nutritive value. Attractive red colour is due to lycopene pigment, pink and yellow colours are due to presence of anthocyanin and carotene pigments. Tomato is rich source of ascorbic acid, which varies from 15-31 mg per 100 g of fresh weight. Tomato also provides minerals like calcium, iron, magnesium, phosphorus and sodium. Tomato is also rich in medicinal value. The pulp and juice is digestible, promoter of gastric secretion and blood purifier. It is also considered to be intestinal antiseptic. It is one of the richest vegetables, which keeps our stomach and intestine in good condition.

Improved varieties

Pusa Ruby, Pusa Early Dwarf, Pusa-120, Pusa Gaurav, Pusa Sheetal, Pusa Sadabahar and Pusa Rohini.

F_1 hybrids

Pusa hybrid –1, Pusa hybrid–2, Pusa hybrid-4, Pusa hybrid-8 and Pusa Divya

Crop production

Climate

Tomato is a warm season crop. It is highly susceptible to frost. Environmental factors such as temperature, light intensity and humidity effect fruit set, fruit development, yield and quality in tomato. Fruit set in tomato is poor when temperature is either relatively low or high. The ideal night temperature for fruit set is 15-25^{0}C. Now varieties are available which can set fruit at a wide range of night temperature.

Soil

It can be grown on many type of soil but a well-drained sandy loam soil is ideal for growing good tomato crop. Tomato prefers a pH of 6-7.

Seed rate : 250-300 g/ha for OP varieties and 150 g for hybrids

Sowing

Seeds are sown on raised nursery beds of 60 cm width in rows at a distance of 2-3 cm. Seedlings are protected against heavy rains and extreme weather conditions. Seedlings become ready for transplanting in 4-5 weeks time. When seedlings attain pencil thickness and 15 cm length they are ready for transplanting. The transplanting should be done in the evening on the sides of flat ridges of 60 cm width, followed by irrigation.

The crop can be sown two times a year in North Indian plains. First crop can be sown in July-August and transplanted in late August-September. Second crop is sown in October-November and transplanted in November-December. The crop should be transplanted only in frost-free areas or protected against frost by maintaining desirable moisture. In hilly areas crop is sown in March-April.

Spacing: 60 x 45 cm.

Manure and fertilizers

About 25-30 metric tonnes per hectare of well rotten farmyard manure should be added at the time of preparation of land. In addition 60 kg phosphorus and 60 kg potash is added at the time of transplanting. 150 kg of nitrogen is applied in split doses, half at the time of transplanting and other half in the form of two top dressings, first at 30 days after transplanting and second 60 days after transplanting. Foliar spray of 1% urea is beneficial.

Irrigation

It is necessary to maintain even moisture supply, as overwatering is harmful. Water is essential at the time of flowering and fruiting. Adequate moisture also helps in better colour development during summer months. Irrigation is applied at 8-10 days interval in winter season where as during summer months, irrigation is applied at 4 days interval depending upon weather conditions.

Interculture

Weeds are often a limiting factor in tomato production as they share light, water, nutrients and space, harbor insect pests and diseases. Frequent shallow cultivation should be given as often as necessary to control weeds. Stomp @ 1.5 liters a.i./ha should be applied before transplanting for controlling weeds.

Staking

Staking is an important operation for tomato. Staking should be done 20-25 days after transplanting. The plants should be loosely tied on horizontal stakes. The timely staked plants produce more and better quality fruits.

Harvesting

Harvesting depends upon purpose for which they are harvested and distance over which they are to be transported. Tomato is harvested at mature green stage for long distance transportation. For short distance transportation fruits are harvested at pink stage and for processing fully ripe red colour tomatoes should be harvested. If the harvested tomatoes are cooled rapidly at 13°C the total shelf life increases.

Average yield

Yield depends upon variety, climatic factors and culture practices. On an average, a normal crop produces 300 q/ha and hybrids can yield upto 450-500 q/ha of tomato fruits.

Plant protection

Disorders

Blossom end rot

Water soaked spots appear at the point of attachment which enlarges rapidly. The disease occurs due to reduced soil moisture especially at fruit development stage. This disease also occurs due to calcium deficiency. Any cultural practices, which conserve soil moisture and spraying of calcium ($CaCl_2$ @ 0.5%), help in controlling disease.

Cracking

Radial cracking is more damaging as compared to concentric cracking. Genetical and environmental factors are responsible for cracking. Use resistant varieties and harvest the fruits before full ripening.

Sun scald

Exposed portions of fruits get blistered and water soaked due to extreme heat of scorching sunshine. This trouble is common plants that have lost considerable foliage. Grow cultivar with heavy foliage.

Insect-pests

Tomato Fruit Borer

Moths are brown to yellowish brown in colour. Caterpillars are green in colour, they attack during October-March. They rollover the leaves and find its way to the fruits where it cut holes.

Control: Collect and destroy egg masses on affected fruits. In case of severe infestation spray Thiodan 1ml/ litre, Emamectin benzoate or spinosad (4 ml / 10 litre water), of water. Spraying should be repeated at fortnight intervals. Keep pheromone traps @20 traps/ha.

One row of marigold should be planted after every 10-15 row of tomato.

Whitefly

They are minute insects of white colour and suck the sap of plant. It transmits leaf curl disease. Affected leaves get shriveled and gradually curl up.

Control: Spraying flonicamid@ 3 ml/10 litre water or imidachloprid (3ml/ 10litre water), fipronil (2ml/litre) in the early stage of crop growth can control whitefly and jassids. Grow bajara or maize as border crop.

Root knot nematode

Three species, *Meloidogyne incognita*, *Meloidogyne arenaria* and *Meloidogyne javanica* are common in tomato. Symptom consists of foliage discoloration to pale then normal unthrifty development, dwarf and wilting.

Control: It can be controlled by use of resistant varieties namely Pusa 120 or Pusa Hybrid-2. Use of Dichloropropane mixture (DD) is recommended.

Diseases

Tomato crop is affected by many diseases. Some of the most important diseases, which are found in India, are presented below.

Damping off

A very destructive disease in nursery beds especially during rainy season. The seedlings are attacked at ground level and as a result they topple over.

Control: Seed treatment with Thiram or Captan @ 3 g/kg of seed. Sow nursery in a raised bed with proper drainage. Drench the nursery bed with Captaf or Bavistin @ 2 g of fungicide per litre of water at 10-15 days interval.

Leaf curl virus

This disease is more severe in monsoon crop. It is characterized by curling of leaves, reduction in leaf size, excessive branching and stunted plant growth. The virus spreads through a vector, white fly (*Bemisia tabaci*). Spraying imidachloprid (4ml/10litre water) or fipronil (2ml/litre) in the early stage of crop growth can control whitefly.

Control: The disease can be checked by control of vectors whitefly by spraying the crop with Rogor or Thiodan at fortnight interval.

Early blight

This disease is caused by *Alternaria solani*. The fungus is soil-borne. Circular, angular, dark brown spots appear on leaves, stem and fruits. Concentric circles of 2-5 mm in diameter are formed at advanced stage. Fruits are also affected showing sunken large dark brown leathery spots on green fruits.

Control: Spray of Dithane M-45 @ 2 g/litre of water is helpful in controlling the disease.

Late blight

This disease is caused by *Phytophthora infestans*. Disease occurs on the foliage at any stage of growth. Brown to purple black lesions occurs in the leaflet, petiole, fruit and stem. At later stage plants die. The disease appears on low temperature and high humidity.

Control: Use of resistant variety is the effective control. Application of Dithane M-45 or Ridomil (2g/Litre) is effective at early stage of plant growth.

Fusarium wilt

The disease starts with yellowing of lower leaves, leading to wilting and death of leaves on few or all branches. In advanced stage, whole plant wilts and die.

Control: Crop rotation and use of resistant varieties is effective.

Bacterial wilt

Most serious disease in tropical as well as temperate regions. Lower leaves may drop before wilting, plants are stunted with yellow leaves. Pith of root is dark brown to black in colour and water soaked.

Control: Use of resistant varieties, crop rotation and application of oil cakes, neem cakes are the measures for controlling the disease. Drenching of root zone of infected plant by Carbendazim @ 0.1% solution.

Brinjal

Brinjal or eggplant (*Solanum melongena* L.), known as *Baigan* in Hindi is one of the most common, popular and principal vegetable crop grown in India. It can be grown in almost all parts of India except higher altitudes round the year. The unripe fruit is used as cooked vegetable, for the preparation of various dishes such as roasted, fried, stuffed, pickled and cooked forms. Brinjal fruits are fairly good source of calcium, phosphorus, iron and vitamins A, B and C. It has some Ayurvedic medicinal properties and white brinjal is said to be good for diabetic patients, it has also been recommended as an excellent remedy for those suffering from liver complaint.

Improved varieties

Pusa Purple Long, Pusa Kranti, Pusa Uttam, Pusa Upkar, Pusa Bindu, Pusa Ankur and Pusa Shymala.

Hybrids

Pusa Hybrid-5, Pusa Hybrid-6, Pusa Hybrid-9 and Pusa Hybrid-20.

Crop production

Climate

Brinjal is a warm season crop and requires long warm season during growth and fruit maturation. The temperature 22 to 30°C is most favorable for good growth and yield.

Soil

It can be grown in all types of soils but loam or clay loams are well suited for high yield. Since brinjal is a long duration crop, deep cultivation and well-drained soil is preferred.

Sowing time

There are three main seasons of sowing brinjal.

1. Autumn-winter crop: Seeds for this crop are sown in June and seedlings are transplanted in July.
2. Spring-summer crop: Seeds are sown in the first fortnight of November and transplanted during the end of January when the danger of frost is over.
3. Rainy-season crop: Seeds are sown in February-March and transplanted during March-April.

Seed rate: 350-500g/ha.

Sowing and transplanting

The seeds are sown in raised nursery beds about one cm deep in rows 5cm apart. The topsoil of the nursery is mixed with fine, well-decomposed farmyard manure or leaf mold. 4-5 weeks old seedlings are ready for transplanting.

Spacing : Spacing are generally recommended depending upon the size and spread of the plant.

Spreading varieties: 75x60 cm

Non spreading varieties: 60x60 cm

Small varieties: 60x45 cm

Manure and fertilizers

30-35 tonnes of FYM, 50 kg urea, 370 kg of single *super phosphate* and 100 kg muriate of potash/ha should be mixed in soil at the time of field preparation. Another 100 kg of urea in 2 top dressing at one month intervals after 3 to 4 weeks of transplanting.

Irrigation

Irrigation is given after transplanting the crop and before top dressing of urea if there is no rain. Timely irrigation is quite essential for good growth, flowering, fruit setting and development of fruits. During winter, soil should remain moist by light irrigation to protect the crop from winter injury.

Harvesting

Fruits are harvested when they have developed a good colour and marketable size and still immature and tender.

Average yield

The average yield of fruits varies (variety to variety) from 300 to 600 q/ha.

Plant protection

Insect-pests

Shoot and fruit borer

These are the worst enemies of this crop. Its larvae bore in to the young shoot and fruits.

Control: Cut out the infested shoots, half inch below the bore point and bury them deep in to the soil.

Control: Spray the crop with Malathion @ 2 ml/litre of water or Spinosad(4ml/ 10litre of water). Repeat the spray two times at fortnight interval. 20 Pheromone traps per ha should be used.

Jassids

Small green insects, suck the sap of the leaves. The leaves become cup shaped.

Control: Spraying of Demecron @1.5 ml/litre at 10 days interval.

Diseases

Damping off

It is a very destructive disease in nursery bed especially during rainy season. The seedlings are attacked at ground level.

Control: Treat the seeds with Thiram or Captaf @ 2g per kg of seed. Drenching the nursery bed with Captaf @ 2 g/litre of water.

Little leaf

The diseased plants produce small yellow leaves and do not bear fruits.

Control: Uproot the diseased plants from the field as soon as they are observed. Spray systemic insecticide to control leaf hopper.

Phomopsis blight

Dark brown spots are formed on the leaves and fruits. This disease is a major problem in seed crop.

Control: Treat the seed with Thiram. Spray the crop with Dithane M-45 @ 2g/litre of water at 7 to 10 days interval.

Chilli

Chilli (*Capsicum annuum* and *C. frutescens*) commonly known as *Mirch* in Hindi is indispensable in the individual kitchen. Chilli is one of the most valuable; cash crop produced almost everywhere from kitchen garden to truck farming. Its unripe fruits are mostly pungent used as table vegetable in salads and ripe fruits as a spice condiment. Fruits are nutritious, colourful and aromatic giving individuality to food dishes with different recipes. Its pungency factor capsaicin is an alkaloid, used in medicine.

Improved varieties

Pusa Jwala, Pusa Sadabahar, Bhagyalakshmi (G-4), Sindhur (CA-960), Pant C-1 and Punjab Lal

Crop production

Climate

The chilli is a crop of tropical and sub-tropical region requiring a warm humid climate. It can be grown in hills most successfully in frost-free summers without infection of leaf-curl complex. Ideal temperature range for good plant growth and fruit-development is 15-35°C. Chilli is mainly grown during monsoon or *kharif* season (June to October) as a rainfed crop or with supplemental irrigation.

Soil

Practically it can be grown in all types of soils, but sandy loam is preferable. But the soil should be with well drainage facility. Excessive moisture can cause defoliation and rotting of plants.

Seed rate

1 to 1.5 kg seed is required for one hectare or depending upon the variety.

Sowing/planting

Treated chilli seed with Thiram or Captan @ 2 g/kg seed is sown in the raised nursery beds in lines 5-7 cm apart. The nursery beds too, should be treated with Captaf (0.3% @ 5 litre solution/m^2) for raising healthy seedlings. Seedlings are ready for transplanting at 4-6 true leaf stage. Before planting seedlings are first hardened by gradually withholding the water 6-9 days before transplanting and exposing them to direct sunlight. Nursery is thoroughly irrigated 12 to 14 hours before transplanting. Transplanting may be done in the evening in shallow ridges.

Spacing: Row to row - 45 to 70 cm

Plant to plant- 30 to 45 cm

Manure and fertilizers

Apply 15 to 20 tonnes FYM in one hectare at the time of ploughing. At the time of final land preparation apply 40:40:40 kg NPK/ha and prepare light ridges. Then 30 days after transplanting again apply 30:40:40 kg NPK/ha as top dressing. After first fruit picking apply final dose of 30 kg N/ha as top dressing.

Irrigation

Number of irrigation may be varied depending upon soil condition. However, in light soil, frequency of irrigation is more than in heavy soil. Excess water should be drained out from the field to avoid wilting problems.

Interculture

Hoeing and weeding should be done at 7 to 15 days interval so as to keep the field free from weeds and to facilitate soil aeration and proper root development. Earthing up should be done after application of both the doses of fertilizers.

Harvesting

Although this crop is mainly grown for ripe fruits but picking at green stage also fetches good price and one harvest of green chillies should be done to stimulate further flush of flowering and fruit set. The biologically riped fruits are picked at interval of 1-2 weeks and harvesting continues. The number pickings vary from 5-8, depending upon, season, variety and cultural practices. The ripe chillies are dried under the sun.

Average yield

Average yield 90-110 q/ha fresh chillies and 9-10 q/ha dry chillies.

Plant protection

Insect-pests

Aphids

Small tiny insects sucks the plant sap from leaves and growing tips, flower buds etc.

Control: Spray the crop with Malathion @ 1-2 ml/litre of water at 15-20 days interval.

Thrips

The insect feed on the sap of tender leaves causing curling of leaf. Buds and flowers are also damaged.

Control: Spray Dimethoate @ 0.025 per cent at fortnightly interval or imidachloprid (4ml/10litre water) or fipronil (2ml/litre) in the early stage of crop growth can control whitefly.

Mites

This tiny spider like insect feed on leaves are found large in numbers on the underside of leaves covered with fine webs.

Control: Spraying with Dicofal @ 2 ml/litre or Spiromecifen @ 0.5 ml/ litre of water controls the mites in chillies.

Diseases

Damping off

The seedlings are attacked at or just below the soil level and they gradually die.

Control: Drenching of nursery bed with Captaf @ 2g/litre of water and treating the seeds with Captan (2g/kg of seed) checks the disease.

Anthracnose

Severe on ripe fruits also called as ripe fruit rot. The spots are usually circular and sunken with black margins.

Control: Seed treatment with Captan (2g/kg of seed) and spray of Dithane M-45 @ 1.5-2 g/1itre of water at 10-15 days intervals.

Leaf curl

Most serious disease causing curling of leaves accompanied by puckering and blistering of interveinal areas. In advance stage whole plant appears bushy with stunted growth.

Control: Removal of infected plants and application of Rogor or Metasystox @ 1.5 ml/litre of water at regular 10-15 days interval before harvesting.

Chilli mosaic

The virus causes cleaning of veins of younger leaves followed by irregular pattern of light yellow and green patches of whole leaf and stunted growth of infected plants at later stages.

Control: Application of Rogor or Metasystox @ 2ml/ litre, of water or imidachloprid (4ml/10litre water) or fipronil(2ml/litre), at regular intervals to reduce the vector or use of resistant varieties like Pant C-1, Pusa Sadabahar etc.

Capsicum

Capsicum (*Capsicum annuum* L.) commonly known as sweet pepper or bell pepper and *Simla Mirch* in Hindi. It is an important cash crop cultivated for blocky, round, conical, thick fleshed, non-pungent fruits. The fruits are used for salad, stuffing or cooked as vegetable. It is one of the best known luxury vegetable appreciated world wide for its flavour, aroma and an excellent source of pro-vitamin A and vitamin C. Its fruits are important constituents of many special recipes.

Improved varieties/hybrids

Variety: California Wonder, Yolo Wonder, Arka Mohini, Arka Gaurav and Arka Basant

Hybrid :Bharat and Pusa Deepti

Crop production

Climate

It requires long and warm growing season, which is frost-free for 4 to 5 months. The crop is cultivated mainly in tropical and sub-tropical regions during cooler season and in temperate regions during summers. The ideal temperature for growth is 21°C at night and 27°C during the day. Flowers will drop and fruit develop very slowly if temperature deviate beyond this ideal range.

Soil

It requires well drained, warm, fertilie sandy loam or clay. The pH of 5.6-6.8 is ideal.

Seed rate

600 g/ha for open pollinated varieties and 350-400g/ha for F_1 hybrids.

Sowing and planting

Treated nursery beds either with formalin solution (1 part-formaldehyde and 48 parts water) or Captaf (0.3% @ 5 litre solution/m^2) for raising healthy seedlings. Treated seed with Thiram or Captan (2g/kg seed) is sown 2.5 mm deep in rows 5-7 cm apart in raised nursery beds. Seedbeds are then covered with grass and watering is done by Rose can. On germination mulch grass is removed. Seedlings become ready for transplanting at 4-6 true leaf stage. Keep the nursery beds moist but not too wet. Seedlings before harvesting are first hardened by gradually withholding the water for 6-9 days. These are then fit to be exposed to direct sunlight. Thoroughly water the seedlings 12-14 hours before transplanting to the field. Healthy seedlings may be transplanted in the evening hours.

Spacing: Hills 45 x 45 cm row to row and plant to plant.

Plains 60 x 40 cm row to row and plant to plant

Manure and fertilizers

Apply 20-25 tonnes of well rotten farmyard manure while preparing the field. A fertilizer dose having 120,100,90 kg/ha of N, P and K should be applied. Phosphorus and potassic fertilizers are applied before transplanting as basal dose, while nitrogenous fertilizer in 3 equal split doses. First dose should be applied with P and K, whereas second and third as top dressing 2-4 and 6-8 weeks after transplanting.

Irrigation

The crop should be irrigated at 7-15 days intervals depending upon weather conditions. There is no need of irrigation at onset of rainy seasons. The drainage of excessive water should be properly done to prevent the crop losses from disease complex.

Interculture

Hoeing and weeding should be done regularly to keep the crop weed free as well as facilitate better aeration in the root zone. Earthing up after first hoeing (10-15 days after planting) is beneficial. Earthing up should also be done after both the top dressings with nitrogen fertilizer.

Harvesting

When the fruits attain marketable maturity they are harvested graded and packing is done properly to avoid damage during transit. Ripe fruits too, is used in salads or cooked is more delicious. Harvesting should be done continuously to earn high prices.

To increase the yield, it is always beneficial to remove the buds from the first dichotomous branching, followed by pinching two or more buds if possible from the nodes of the first stage branching.

Average yield: 200-250 q/ha.

Plant protection

Physiological disorder

Sun scald

When large fruits are exposed directly to sunlight, it forms a white papery patch, which is further infected by fungus on the exposed region.

Control: Fruits should be covered after harvesting and in the plants 35% shading may be useful.

Insect-pests

Aphids

These small insects suck the plant sap from leaves and growing tips in both nursery seedlings and transplanted crop.

Control: Spray the crop with imidachloprid @ 4ml/ 10 litre of water.

Thrips

These also feed upon tender leaves, flower petals and buds.

Control: Spray of imidachloprid @ 4ml/ 10 litre of water.

Hairy caterpiller

Damage young as well as developed fruit and its pedicel.

Control: Rogor or Metasystox @ 1 ml/litre of water if sprayed in 15-20 days interval checks the damage.

Diseases

Damping off

Withering of emerging seedlings from the hypocotyl region.

Control: Spray/drench with Captan 2g/litre of water checks the damage.

Phytopthora blight

On the outcome of disease plants starts withering showing a complex symptom.

Control: Spray of Copper-oxychloride @ 2.5g/litre of water of 15-20 days interval. Drainage of excess water from the fields.

Leaf curl

This is very serious disease causing curling of leaves that starts yellowing, shrinking into small leaves and buds.

Control: Removal of infected plants and control of insects by regularly spraying with insecticides.

Section 3: Crop Protection

Improved Agriculture and Allied Practices for Farmers, pp 161-173
New India Publishing Agency, New Delhi, India

14

Weed Management in Crops and Cropping Systems

T.K. Das, Mamta Bisht and Neelmani Rathi

Division of Agronomy, ICAR-Indian Agricultural Research Institute New Delhi-110012

Weeds have been existing on the earth ever since the man started domesticating /cultivating plants and animals around 10,000 B.C. and recognized as a problem since then. The battle against weeds is a never ending one and often the costliest agronomic input for successful crop production. Weeds are most under estimated crop pests in tropical agriculture although they cause higher reductions in crop yields than other pests and diseases. Of the total annual loss of agricultural produce from various pests in India, weeds roughly account for 37%, insects for 29%, diseases for 22% and other pests for 12%. Weeds decrease quantity and quality of the produces [food, fibre, oil, forage/ fodder, animal products (meat and milk)] and cause health hazards for humans and animals. Therefore, weed control is very important for a successful crop production.

1. Weed Flora During Rainy (Kharif) Season

1.1 Annual Grass Weeds

Echinochloa colona L., *Echinochloa crusgalli* (L.) Beauv., *Echinochloa glabrescens, Amischophacelus* (*Cyanotis*) *axillaris/cuculata, Leptochloa chinensis* (L.) Nees, *Acrachne racemosa* Heyne ex Rhoem, *Panicum sp, Paspalum distichum* L., *Commelina benghalensis/ nudiflora, Eleusine indica, Brachiaria platyphylla,Setaria glauca/verticillata,Dactyloctenium aegyptium, Dicanthium annulatum, Digitaria sanguinalis/adscendens, Ischaemum rugosum, Leersia hexandra, Oryza sativa* var *fatua* (wild red rice).

1.2 Annual Broad-leaved Weeds

Trianthema portulacastrum/monogyna, Digera arvensis, Amaranthus viridis/ retroflexus, Parthenium hysterophorus,Physalis minima, Phyllanthus niruri, Tribulus terrestris, Ammania baccifera L., *Ageratum conyzoides, Bidens pilosa, Celosia argentea, Corchorus aestuans (acutangulus), Centella asiatica, Cleome* (*Gynandropsis*) *gynandra/viscosa*, *Datura stramonium, Eclipta alba* (Linn) Hask, *Euphorbia hirta, Galinsoga parviflora, Ludwigia* (*Jussiaca*) *parviflora/ perennis/ octovalvis, Marsilea quadrifolia/minuta, Monochoria vaginalis* (Burm.f.) Presl., *Scoparia dulcis, Solanum nigrum, Sphenoclea zeylanica* Gaertn., *Striga asiatica (lutea), Xanthium strumarium, Caesulia axillaries* Roxb.

1.3 Annual Sedges

Cyperus iria/difformis/compressus/compactus, Fimbristylis miliacea (L.) Vahal., *Fimbristylis tenera/ dichotoma, Scirpus supinus* var *lateriflorus, Eleocharis atropurpurea.*

1.4 Perrenial Weeds

Grasses: *Cynodon dactylon*

Sedges : *Cyperus rotundus / esculentus, Scirpus supinus / maritimus, Scripus tuberosus* (Desf.), *Eleocharis dulcis, Fimbristylis littoralis / barbata*

Broad-leaved weeds : *Ipomoea repens, Ipomoea aquatica* Forsk.

2. Critical Period of Weed Competition

Weeds do not cause harm to crops equally all through the growing period. There are certain stages in crop growth cycle when weeds are more damaging to crop growth and yield. Usually early season weed competition is most detrimental to crop and, therefore, early season weed control is indispensable, although weeds present at later stages of crop growth cause yield loss and other inconveniences. The critical period of weed competition (Table 1) may be defined as "the short time span in the life cycle of a crop, when weed causes maximum reduction in its yield or in other words, when weed control measure if adopted may fetch near maximal or maximum acceptable crop yield." It is, therefore, simply the specific duration of weed-free situation of a crop resulting into near maximal yield, which is sufficiently close or equal to that obtained by the season-long weed-free situation. A "thumb rule" is that the first one-fourth ($1/4^{th}$) to one-third ($1/3^{rd}$) period of the total growing duration of a crop, irrespective of growth stages, weed species and environmental (climatic and soil) conditions may be assumed as "the critical period for weed competition."

Table 1: The critical period of weed competition in crops

S.No.	Crops	Critical period (DAS)*
1.	Rice (upland) direct-seeded	15-45 DAS
2.	Rice (lowland transplanted)	30-60 DAT
3.	Rice (lowland direct-seeded)	40-60 DAS
4.	Dwarf wheat, barley, pea	30-45 DAS
5.	Tall wheat	35-50 DAS
6.	Maize	30-60 DAS
7.	Sorghum	15-45 DAS
8.	Pearl millet	30-45 DAS
9.	Soybean	First 60 days
10.	Groundnut	42-56 DAS
11.	Chickpea, lentil	30-60 DAS
12.	Green gram	15-30 DAS
13.	Black gram	15-30 DAS
14.	Cowpea	First 30 days
15.	Cotton (rainfed)	20-60 DAS
16.	Jute, mesta	15-45 DAS
17.	Sugarcane	30-60 DAS
18.	Potato	First 4 weeks
19.	Rapeseed	10-40 DAS

*DAS, days after sowing

3. Weed Management Options

Before recommending as well as adopting any weed control measure, a little knowledge about weed characteristics, morphology, life cycle/ontogeny, nature of competition, degree and duration of association with crops is pre-requisite for suggesting efficient weed control measures. Some common characteristics of weed species, which are clearly visible and easily understandable by farmers/users are to be exploited for making their classes/ groups for possible recommendations. For example, based on life cycle/ ontogeny, weeds are annuals, biennials and perennials. Knowing weed species annuals, biennials or perennials which are dominant in the field would be of immense use towards recommending measures for their control. Perennial weeds are generally more difficult to control than annuals.

To control perennial weeds, annual weed killers are less useful. Deep tillage sometimes does not reach to the deep roots of perennials. Systemic herbicide could be an option in this situation, but is season-specific. It also sometimes cannot reach to the roots and vegetative structures lying deeper in soil.

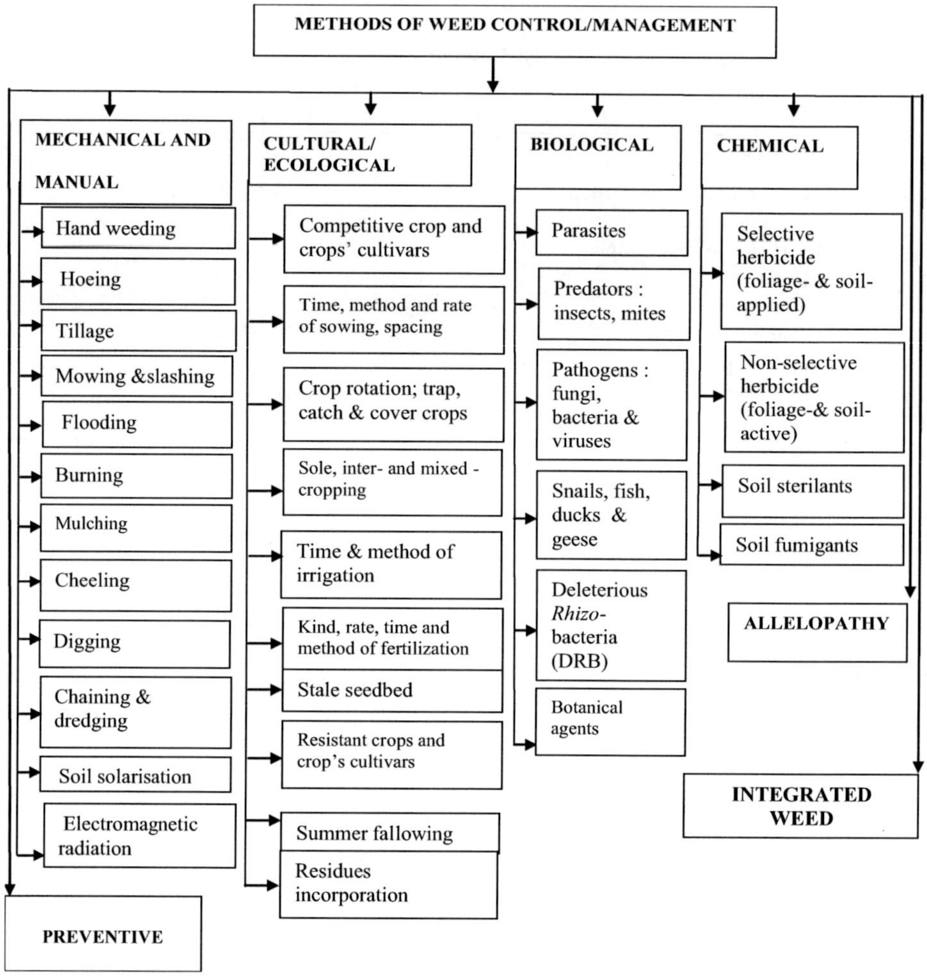

Fig. 1: Methods of weed control

Many methods of weed control are available (Fig. 1), but adoption of more than one method in a cohesive manner is always advisable. Integrated approach manages weeds better for longer periods. Integrated weed management although considered as a method of weed management, is basically a concept of weed management, which put together two or more methods, e.g. preventive, physical, cultural, biological, chemical, allelopathy in mutual and compatible way to work better and more efficiently towards weed management.

4. Herbicide Recommendation for *Kharif* Crops

Table 2 Herbicide recommendation for weed control in *kharif* crops

Crops	Herbicides	Dose (kg/ha)	Time of application	Conditions/remarks
Rice Nursery	Pendimethalin	1.0-1.5	Pre-em (1-2 DAS)	Broad-spectrum control of weeds except sedges and *Digera arvensis*; In dry nursery, do not spray on dry soil; irrigation should follow immediately if moisture less; moisture is a pre-requisite for its application, but not standing water; In wet nursery, drain out water and spray on saturated soil; irrigate after 1-2 days
	Pyrazosulfuron-ethyl	0.015-0.020	Pre-em (1-2 DAS) or early post-em (10-15 DAS)	Broad-spectrum control of weeds except sedges and *Polygonum*; do not spray on dry soil; irrigation should follow immediately if moisture less; In wet nursery, drain out water and spray on saturated soil; irrigate after 1-2 days
	Butachlor	1.0-1.5	Pre-em (1-2 DAS)	Broad-spectrum control of weeds except sedges; *Echinochloa colona* and *Ischaemum rugosum* reported not being controlled or resistant; do not spray on dry soil; irrigation should follow immediately if moisture less; In wet nursery, drain out water and spray on saturated soil; irrigate after 1-2 days
	Pretilachlor (S)	0.75	Pre-em (3-5 DAS)	Broad-spectrum control of weeds except sedges; do not spray on dry soil; irrigation should follow immediately if moisture less; In wet nursery, drain out water and spray on saturated soil; irrigate after 1-2 days
Direct-seeded Upland Rice	Pendimethalin	1.0-1.5	Pre-em (1-2 DAS)	Broad-spectrum control of weeds except sedges and *Digera arvensis*; moisture in soil is a pre-requisite for its application; irrigation should follow immediately if moisture less
	Pyrazosulfuron-ethyl	0.015-0.020	Pre-em (1-2 DAS) or early post-em (10-15 DAS)	Broad-spectrum control of weeds except sedges; moisture in soil is a pre-requisite for its application; irrigation should follow immediately if moisture less
	Butachlor	1.0-1.5	Pre-em (1-2 DAS)	- do -
	Pretilachlor (S)	0.75	Pre-em (3-5 DAS)	- do -
	Bispyribac-Na	0.020-0.025	Post-em (20-30 DAS)	Broad-spectrum control of grass, sedge and broad-leaved weeds; moisture in soil is a pre-requisite for its application; irrigation should follow immediately if moisture less; Do not spray if rains expected within 6 hours; Do not mix with sulphur- or copper-containing pesticides
	2,4-D (ester, Na/K or Amines)	0.75-1.0	Post-em (25-30 DAS)	Good control of broad-leaved weeds and few sedges; Apply if field dominated by broad-leaved weeds
	Cyhalofop-butyl	0.08-0.09	Post-em (20-30 DAS)	Good control of grassy weeds; Apply if field dominated by grassy weeds

Direct-Seeded Puddled and Transplanted Rice	Pendimethalin	1.0-1.5	Pre-em (3-5 DAT)	Broad-spectrum control of weeds; Drain out water from the field before application, apply herbicide in saturated soil and refill the field with water after 2-3 days
	Butachlor	1.0-1.5	Pre-em (3-5 DAT)	- do -
	Pretilachlor (S)	0.75	Pre-em (3-5 DAT)	- do -
	Pyrazosulfuron-ethyl	0.015-0.020	Pre-em (3-5 DAT) or carly post-em (15-20 DAT)	- do -
	Bispyribac-Na	0.020-0.025	Post-em (20-30 DAT)	Broad-spectrum control of grass, sedge and broad-leaved weeds; Do not spray if rains expected within 6 hours; Do not mix with sulphur- or copper-containing pesticides
	Oxadiargyl	0.07	Pre-em (3-5 DAT) or early post-em (15-20 DAT)	Broad-spectrum control of grass, sedge and broad-leaved weeds; Do not spray if rains expected within 6 hours
	Ethoxysulfuron	15	Pre-em (3-5 DAT) or early post-em (15-20 DAT)	Broad-spectrum control of grass, sedge and broad-leaved weeds
	2,4-D (ester, Na/K or Amines)	0.75-1.0	Post-em (25-30 DAS)	Good control of broad-leaved weeds and few sedges; Apply if field dominated by broad-leaved weeds; Drain out water from the field before application, apply herbicide in saturated soil and refill the field after 2-3 days
	Cyhalofop-butyl	0.08-0.09	Post-em (20-30 DAS)	Good control of grassy weeds; Apply if field dominated by grassy weeds
Maize	Atrazine	1.0-1.5	Pre-em (1-2 DAS) or early post-em (15-20 DAS)	Broad-spectrum control of weeds but basically a broad-leaved killer; Does not have effective control against *Acrachne racemosa*, *Commelina beghalensis* and sedges; Moist soil is a pre-requisite for its action
	Pendimethalin	1.0-1.5	Pre-em (1-2 DAS)	Broad-spectrum control, but basically a grass killer herbicide; Does not control sedges, *Digera arvensis, Parthenium hysterophorus* effectively; Moist soil is a pre-requisite for its action
	Metolachlor	1.0-1.5	Pre-em (1-2 DAS)	Broad-spectrum, but controls mainly grasses; does not control *Rottboellia cochinchinensis*, but controls *Cyperus esculentus*; Moist soil is a pre-requisite for its action
	Atrazine + pendimethalin (Tank-mix)	0.75+0.75	Pre-em (1-2 DAS)	This tank-mix application controls a broad-spectrum of weeds and can render maize field almost free of weeds except sedges
	Atrazine + metolachlor (Tank-mix)	1.0+1.0	Pre-em (1-2 DAS)	- do -
	2,4-D	0.75	Post-em (30-35 DAS)	Good control of broad-leaved weeds and few sedges; Apply if field dominated by broad-leaved weeds
	Bentazon	1.0-2.0	Post-em (15-20 DAS)	Control of broad-leaved weeds effectively, but *Cyperus esculentus* moderately

Sorghum/ Pearl Millet	Atrazine	1.0	Pre-em (1-2 DAS)	Broad-spectrum control of weeds but basically a broad-leaved killer; Does not have effective control against *Acrachne racemosa, Commelina beghalensis* and perennial sedges
	Pendimethalin	0.75-1.0	Pre-em (1-2 DAS)	Broad-spectrum control, but basically a grass killer herbicide; Does not control sedges, *Digera arvensis, Parthenium hysterophorus* effectively; it is less selective to sorghum and pearl millet than maize
	Atrazine + pendimethalin	0.5+0.5	Pre-em (1-2 DAS)	The tank-mix application controls a broad-spectrum of weeds and render maize field almost free of weeds except perennial sedges
	2,4-D	0.75	Post-em (30-35 DAS)	Good control of broad-leaved weeds and few sedges; Apply if field dominated by broad-leaved weeds
Pigeon pea/ Black gram /Green gram	Pendimethalin	1.0	Pre-em (1-2 DAS)	Broad-spectrum control, but basically a grass killer herbicide; Does not control sedges, *Digera arvensis, Parthenium hysterophorus* effectively
	Fluchloralin	1.0	Preplant incorporatio n (PPI)	Control broad-spectrum of weeds; After final land preparation and before sowing of a crop, spray herbicide over soil followed by slight stirring of surface soil for incorporation of the herbicide into soil to reduce vaporization
Soybean	Pendimethalin	0.75-1.0	Pre-em (1-2 DAS)	Broad-spectrum control, but basically a grass killer herbicide; Does not control sedges, *Digera arvensis, Parthenium hysterophorus* effectively
	Metribuzin	0.400	Pre-em (1-2 DAS)	Broad-spectrum control of weeds including *Cyperus rotundus* to a moderate extent
	Metolachlor	0.75-1.0	Pre-em (1-2 DAS)	Broad-spectrum, but does not control *Rottboellia cochinchinensis*, but controls *Cyperus esculentus*; soil moisture is a pre-requisite for its desired activity
	Pendimethalin + imazethapyr (Tank-mix)	0.75 + 0.075	Pre-em (1-2 DAS)	Broad-spectrum control of weeds including sedges, e.g., *Cyperus rotundus*
	Pendimethalin followed by (fb) imazethapyr	0.75 & 0.075 (resp.)	Pendi as pre-em (1-2 DAS) fb Imazetha post-em (15-20 DAS)	Broad-spectrum control of weeds including sedges, e.g., *Cyperus rotundus*
	Pendimethalin followed by chlorimuron-ethyl	0.75 & 0.009 (resp.)	Pendi as pre-em (1-2 DAS) fb Chlori post-em (15-20 DAS)	Broad-spectrum control of weeds including sedges, e.g., *Cyperus rotundus*

Groundnut/ Sunflower	Pendimethalin	1.0	Pre-em (1-2 DAS)	Broad-spectrum control, but basically a grass killer herbicide; Does not control sedges, *Digera arvensis, Parthenium hysterophorus* effectively
	Fluchloralin	1.0		Control broad-spectrum of weeds; After final land preparation and before sowing of a crop, spray herbicide over soil followed by slight stirring of surface soil for incorporation of the herbicide into soil to reduce vaporization
	Oxadiazon (for sunflower)	0.5-1.0		Moist soil is a pre-requisite for its activity
Sugarcane	Atrazine	2.0	Pre-em (1-2 DAT)	Broad-spectrum control of weeds but basically a broad-leaved killer; Does not have effective control against *Acrachne racemosa, Commelina beghalensis* and perennial sedges
	Atrazine + Alachlor (Tank-mix)	1.0+2.0		The tank-mix application controls a broad spectrum of weeds and provides a weed-free condition
	Metribuzin	1.0-1.5		Broad-spectrum control of weeds
	2,4-D	1.0	Post-em (35-40 DAT)	Only for broad-leaved weed control where there is post-transplanting broad-leaved weed problem
Cotton	Pendimethalin	1.0	Pre-em (1-5 DAS)	Broad-spectrum control, but basically a grass killer herbicide; Does not control sedges, *Digera arvensis, Parthenium hysterophorus* effectively; Moist soil is a pre-requisite
	Diuron	0.5-0.75		Pre-emergence; not to apply in coarse sandy soil; crop damage imminent if seeds are treated with systemic insecticide
	Metolachlor	0.75-1.0		Pre-emergence at 1-2 DAS; broad-spectrum, but does not control *Rottboellia cochinchinensis*, however, *Cyperus esculentus* is controlled; moist soil required for its activity; effect usually does not persist more than 30 days.
	Trifluralin	0.75-1.0	Preplant incorporation (PPI)	Control broad-spectrum of weeds; After final land preparation and before sowing of a crop, spray herbicide over soil followed by slight stirring of surface soil for incorporation of the herbicide into soil to reduce vapourization
Other	Quizalofopethyl	0.050	Post-em (25-30 DAS)	For grassy weed control
	Pyrithiobac-Na	0.062	Post-em (25-30 DAS)	For broad-leaved weed control

5. Weed Flora During Winter (Rabi) Season

5.1 Annual Grass Weeds

Avena fatua/ludoviciana/sterilis, Agrostemma githago, Asphodelus tenuifolius, Bromus tectorum/ pectinatus, Cyanotis barbata, Eragrostis tenuifolia, Lolium temulentum/ multiflorum/ rigidum, Phalaris minor/ paradoxa, Poa annua, Polypogon monspeliensis, Setaria pallidefusca, Snowdenia polystachya.

5.2 Annual Broad-leaved Weeds

Chenopodium album, Chenopodium murale, Melilotus indica/alba/ parviflora, Anagallis arvensis, Argemone maxicana, Cannabis sativa, Carthamus oxyacantha, Cleome(~Gynandropsis) gynandra, Coronopus didymus, Daucus carota, Erucastrum arabicum, Fumaria parviflora, Gnaphalium luteo-album/indicum, Launaea asplenifolia, Malva parviflora L., *Medicago denticulata/polymorpha, Oenothera laciniata, Polygonum plebejum/ aviculare/ nepalense, Rumex dentatus/retroflexus, Sisybrium irio, Spergula arvensis* (narrow-leaved dicotyledonous weed), *Trigonella polycerata, Vicia hirsuta/faba/ sativa*, *Sonchus oleraceous*

5.3 Perrenial Weeds

Grasses : *Agropyron repens, Cynodon dactylon*

Sedges : *Cyperus rotundus/esculentus*

Broad-leaved weeds: *Convolvulus arvensis, Cirsium arvense, Launaea cornuta, Rumex crispus/ obtusifolius /acetosella* (simple perennial), *Sonchus arvensis* (simple perennial), *Plantago lanceolata* (simple perennial).

Chenopodium murale, Cannabis sativa used to grow earlier on the field bunds, have encroached to wheat and other winter crops in recent years, and are plenty in number. Recently two other species of *Chenopodium* (not identified) are growing in huge number in winter crops at IARI, New Delhi. *Parthenium hysterophorus* L. has also encroached to wheat and winter crops in some places of North and North-western India.

6. Herbicide Recommendation for *Rabi* Crops

Table 3: Herbicide recommendation for weed control in *rabi* crops

Crops	Herbicides	Dose (kg/ha)	Time of application	Conditions/remarks
Wheat	Sulfosulfuron	0.03	Post-em (30-35 DAS)	Broad-spectrum weed control; can control resistant *Phalaris minor*; it can be followed by metsulfuron-methyl (0.006 kg/ha), isoproturon (0.75 kg/ha), or carfentrazone (0.0.03 kg/ha) after 5 days if broad-leaved weed control not at desired level; Don't apply tank-mix of these herbicides, which causes slight phytotoxicity to wheat. Slight phytotoxicity observed on the succeeding sorghum, cotton
	Clodinafop-propargyl followed by metsulfuron-methyl	0.06 & 0.006 (resp.)	Post-em (30-35 DAS)	Ensures broad-spectrum weed control; clodinafop controls resistant *Phalaris minor*; however, recently biotypes of *Phalaris minor* have shown cross-resistance to it in different parts of Punjab and Haryana, and, therefore, its use may be restricted to areas where cross-resistance has not cropped up; Don't apply tank-mix of these herbicides, which causes slight phytotoxicity to wheat.
	Clodinafop-propargyl followed by carfentrazone-ethyl	0.06 & 0.03 (resp.)	Post-em (30-35 DAS)	- do-
	Clodinafop-propargyl followed by 2,4-D	0.06 & 0.50 (resp.)	Post-em (30-35 DAS)	-do- ; 2,4-D should not be used in sensitive varieties of wheat, e.g., Arjun (HD 2009), Kundan
	Isoproturon	0.75-1.0	Post-em (30-35 DAS)	Highly economical herbicide; controls both grass and broad-leaved weeds; however, in Punjab, Haryana and Western U.P., resistance in *Phalaris minor* has cropped up against it, should not be advocated in those areas
	Metsufuron-methyl	0.004-0.008	Post-em (30-35 DAS)	At 30-35 DAS; exclusively a broad-leaved killer; could be a substitute of 2,4-D, which causes ear malformation in some wheat varieties, e.g. Arjun (HD 2009), Kundan.
	2,4-D	0.4-0.5	Post-em(30-35 DAS)	At 30-35 DAS; exclusively a broad-leaved killer with some sedge killing action; should not be used in sensitive varieties of wheat, e.g. Arjun (HD 2009), Kundan.
Chick Pea/ Lentil/ Pea	Fluchloralin	1.0	Pre-plant incorporation (PPI)	Control broad-spectrum of weeds; After final land preparation and before sowing of a crop, spray herbicide over soil followed by slight stirring of surface soil for incorporation of the herbicide into soil to reduce vapourization
	Pendimethalin	1.0	Pre-em (1-2 DAS)	Broad-spectrum control, but basically a grass killer; Does not control sedges, *Parthenium hysterophorus* effectively
Rapeseed and Mustard	Fluchloralin	1.0	Pre-plant incorporation (PPI)	Broad-spectrum control, but basically a grass killer; Sprayed into soil before sowing of a crop followed by slight stirring of surface soil for incorporation into soil to reduce vapourization
	Pendimethalin	1.0	Pre-em (1-2 DAS)	Broad-spectrum control, but basically a grass killer; Does not control sedges, *Parthenium hysterophorus* effectively
	Isoproturon	0.75-1.0	"	Broad-spectrum weed control; Does not control sedges
	Oxadiazon	0.50	"	Broad-spectrum weed control

7. Herbicide Recommendation for Vegetable Crops

Table 4: Herbicides recommended for vegetable crops

Crops	Herbicides	Dose (kg/ha)	Time of application	Conditions/remarks
Potato	Pendimethalin	1.0	Pre-em (1-2 DAP)	Broad-spectrum control, but basically a grass killer; Does not control sedges, *Parthenium hysterophorus* effectively; moisture in soil is essential for activity
	Metribuzin	0.5		Broad-spectrum weed control; controls *Parthenium hysterophorus* and *Cyperus rotundus* fairly
	Paraquat	0.5	Post-em at 2-5% sprouting	Delay in spraying beyond 5% sprouting may cause phytotoxicity to potato plants and reduces its plant stand
	Isoproturon	0.75-1.0	Pre-em (1-2 DAP)	Broad-spectrum control
Tomato/ Brinjal	Pendimethalin	1.0	Pre-em	Apply before transplanting followed by irrigation or post-transplanting pre-emergence; one hand weeding at 35-40 DAT is supplementary to herbicide
	Fluchloralin	1.0	PPI	Sprayed into soil before sowing followed by irrigation for incorporation into soil to reduce vapourization; one hand weeding at 35-40 DAT is highly useful
	Metribuzin (in tomato only)	0.5	Pre-em	Avoid contact with tomato plants; Apply before transplanting followed by irrigation; one hand weeding at 35-40 DAT is highly useful
Cabbage/ Cauliflower/ Knolkhol	Pendimethalin	1.0	Pre-em	Apply before transplanting followed by irrigation or post-transplanting pre-emergence at 2-3 DAT; one hand weeding at 35-40 DAT is supplementary to herbicide
	Fluchloralin	1.0		--- do ---
	Trifluralin	1.0		--- do ---
Onion/Garlic	Pendimethalin	1.0	Pre-em (1-2 DAS)	Apply before transplanting followed by irrigation or post-transplanting pre-emergence; one hand weeding at 35-40 DAT is supplementary to herbicide
	Fluchloralin	1.0	PPI/Pre-em	--- do ---
	Oxyfluorfen	0.25	Pre-em (1-2 DAP)	--- do ---
	Bentazon	1.0-1.5	Post-em (25-30 DAP)	Applied as post-emergence to control broad-leaved weeds and sedges
Beet/Radish/ Carrot	Pendimethalin	1.0	Pre-em (1-2 DAP)	Broad-spectrum control, but basically a grass killer; Does not control sedges, *Parthenium hysterophorus* effectively; moisture in soil is essential for activity; one hand weeding at 35-40 DAT is supplementary to herbicide

8. Weed Management in Fruit and Plantation Crops

i) **Live mulch:** Intercropping with fodders and food legumes, namely *Trifolium alexandrinum* (Egyptian clover/berseem), *Medicago sativa* (Lucerne/alfalfa), *Desmodium sp* and French bean serves dual purposes; controls weeds as well as gives some extra yield from the intercrops. Some low-growing vegetables, e.g. tomato, potato, cabbage, cauliflower may also be cultivated if found compatible with the crops.

ii) **Dead mulch**: plant residues may be embedded in between rows of the tree crops.

iii) **Soil solarization**: Soil solarization using transparent polyethene film (50-100 m thickness) may be employed during hot summer months. It will kill most annual weed seeds at the upper surface layer of soil and reduce weed competition.

iv) **Tillage and mechanical or hand weeding**: Tillage during summer season and mechanical or hand weeding at the early part of wet rainy season, when huge population of mainly annual weeds have emerged, are useful

v) **Chemical control**

In the fields of established fruit and plantation crops, post-emergence directed spray/application of some non-selective herbicides on the actively growing weeds may be advocated as follows :

a) Glyphosate @ 1.0-1.5 kg/ha as directed spray can be applied at later stages on the actively growing perennial weeds. Use clean water, low drift nozzle and do not mix with other herbicides. Do not apply if rain expected within 8 hours.

b) Paraquat @ 1.0-1.5 kg/ha as directed spray also controls broad-spectrum of weeds. Bright sunshine is a pre-requisite for its greater efficacy.

c) Glufosinate-AM @ 0.35-1.0 kg/ha can also be applied as directed spray to control broad-spectrum of weeds.

d) In the transplanted new/young plantations, non-selective herbicides, namely glyphosate, paraquat, glufosinate-AM could safely be used if the individual sapling/ seedling of the fruit crops is covered by earthen pots or by some other means. By doing so, the herbicide droplets or drift won't come in contact/touch with the seedlings/saplings while spraying. This will give a very effective control of annual and perennial weeds as well as ensure crop safety.

Conclusion

A handful of weed control options are available now-a-days. They, however, individually are not a fool-proof strategy/technology. They have usefulness, as well as limitations. Therefore, a suitable integration of the relevant options for weed control available to the growers is necessary for achieving higher crop yields through improved weed management. Practices like summer tillage, stale seedbed, mulches/crop residues, herbicide along with other cultural/non-monetary options (appropriate variety; time, method & rate of sowing; row-spacing; proper fertilization, etc) may be integrated/combined in a compatible and mutually exclusive way to harness a better, efficient and longer weed management in crops and cropping systems.

References

Das T K. 2008. *Weed Science: Basics and Applications*. Jain Brothers Publishers, New Delhi, India, 901p.

Improved Agriculture and Allied Practices for Farmers, pp 175-182
New India Publishing Agency, New Delhi, India

15

Nematode Pest of Crops and their Management

H.K. Sharma and Pankaj

Division of Nematology, ICAR-Indian Agricultural Research Institute New Delhi -110012

Plant parasitic nematodes are microscopic, thread like, soil-born organism, found in every conceivable environment. Severity of nematodes as plant pathogen is more a problem in tropical than temperate climate. These organisms are basically root feeders causing injury to the plants, which adversely affect the function of the root to absorb and translocate nutrients and water efficiently. This result into nutrient deficiency like symptoms, and as such nematode problems go unnoticed. Such symptoms are not corrected even after a dose of fertilizers.

Based on feeding habits, they are classified as migratory ectoparasitic, sedentary semi-endoparasitic or endoparasitic nematodes. They feed silently but constantly without killing the host producing debilitating symptoms and in the process reproducing to great numbers at the end of the crop season. Normally, nematodes complete their life cycles in a span of 20-25 days at temperature of 25-30 °C. They have a great egg laying capacity, which varies from 40-1000 eggs.

The most important nematode is root-knot nematode, *Meloidogyne* sp. distributed worldwide and can infest more than 2000 species of plants. Cereal cyst nematode, *Heterodera avenae* - an important nematode pest of wheat, barley and oat, citrus nematode, *Tylenchulus semipenetrans* associated with citrus dieback and slow decline, burrowing nematode, *Radopholus similis* causing spreading decline in plantation crops, the reniform nematode, *Rotylenchulus reniformis* in vegetables and cotton and the rice root nematode, *Hirschmanniella* spp in lowland rice are few other damaging nematodes.

Various types of nematode feeding habits produce various types of symptoms both in aerial and below ground parts of plant. Aerial symptoms include yellowing, wilting, crinkling and curling of leaf, deformity of buds and flowers and seed gall formation. Below ground symptoms include production of galls, branching of root systems, stubbiness of the roots or deformation of roots.

The average worldwide annual yield loss due to nematodes is estimated at 10% but sometimes there may be total loss of the crop under heavy infestation conditions. Thus, economic loss accrued due to them warrants remedial and control measure to be taken so that such damages could be contained and production and productivity of different crops is increased. The nematodes can be controlled and managed by major four methods:

A number of management options based on physical, chemical, biological, and cultural methods have been found effective, to different degrees, in reducing nematode population densities and improving crop performance. None of the methods, except on very small areas or quantities of plant material, have been found sufficient or economic to completely control, or to maintain the population densities at levels below the economic tolerance level. Therefore, reliance has to be placed on a judicious combination of different control options suitable for location specific conditions.

The main aim of a control measure is to reduce the pest population density to the minimum possible levels. But the various limitations of cost, environment and safety do not permit attainment of such a goal in case of most of the nematode pests. Therefore, a suitable management approach is adopted with an objective to maintain nematode population densities at levels below the economic injury threshold or to reduce their levels such as to derive maximum profits out of the management cost incurred.

Cultural Control

Crop rotation: Rotation of crops on seasonal basis is the age old practice of cultivation of crops. It is well known fact that continuous growing of crop in a particular field lead to gradual reduction of productivity and a stage come when cultivation is no more profitable. From the scientific knowledge it has been confirmed that some of the pest problems could be better managed by if proper cropping system is adopted.

a) **Root- knot disease of vegetable and pulse crops**: Rotate the field at least for 1 year with wheat barley/ mustard/ toria/ African Marigold in rabi and rice/ bajra/sorghum during *kharif* season.

b) **Molya disease of wheat** : Replace wheat/barley with mustard, gram, fenugreek or crops in *rabi* at least for 2-3 years in infested fields, helps in reducing the cyst population by 40-60%.

c) **Intercropping** : It has been found that root-knot nematode problem could be managed by mix cropping or intercropping with *Tagetes* spp. either as 1:1 or 1:2 ratio of crop/tagetes. It is helpful in reducing the population by 20%.

Deep Summer Ploughing : Deep ploughing during peak summer months (May-June) is very helpful to control harmful soil born pest and diseases. Two-three deep summer ploughings at 7-10 days interval during May-June months drastically reduces *H. avenae* population. During peak summer turning soil exposes the pest and disease organism to high temperature. Where flooding can be practiced, alternating periods of 2-3 weeks flooding, drying and again flooding are more effective in controlling nematodes. This method reduces the harmful pest population by more than 60%. Besides, it also encourages beneficial microflora and helps to improve the soil fertility.

Organic amendment : Incorporation of organic matter into soil is very good for soil health and is a normal cultivation practices and it is a fact that there is correlation of organic carbon with yield of crop. By organic amendment C:N ratio of soil improves which further help in availability of nutrients and consequently improve crop growth and yield. Application of deoiled cakes of mustard, groundnut, mahua and Karanj (@ 1 t/ha), 15-20 days prior to sowing of the crop is highly effective. It is also recommended to incorporate the crop residue in soil, which not only improve the soil physiochemical properties but also helps to sustain the crop productivity by encouraging beneficial microflora and make the soil as soil suppressant of pest and diseases.

Solarization : Prepare the area for nursery bed for sowing, give light irrigation and cover with polythene sheets (40-100 μm thickness). Care must be taken that the edges of the polythene sheet are properly sealed by covering them well with soil. Leave it for preferably 3 weeks. By this time most of the plant parasitic nematodes in the upper soil layers are killed. The nursery raised in such beds is very healthy, and vigorous with well developed root system, which enables the seedlings to stand better when transplanted in the main field.

Use of resistant varieties : Resistant cultivars affect the multiplication of nematodes. This is easy, cheap and environmentally friendly. Grow resistant varieties to minimise the yield loss due to nematodes as under ;

1. Tomato : VFN-8, SL-120, PNR-7, Ronita, Hisar Lalit, Patriot against root-knot nematode.

2. Chilli : Malagachi yellow, Julandhari, Pusa Jwala against root-knot nematode.

3. Cowpea : C-152, 82-113, GAU-1 against root-knot nematode.
4. Barley : Rajkiran, C-164 against cereal cyst nematode, *Heterodera avenae*
5. Wheat : Raj-MR-1, against cereal cyst nematode, *Heterodera avenae*
6. Brinjal : Blackbeauty, Vijay, Jaunpuri Long against *Meloidogyne* spp.

It is also advised to grow resistant cultivars for two years only otherwise virulent population may develop which could be more devastating.

Adjustment of sowing or planting time may help in avoiding nematode damage in regions with well defined winters, such as in the north Indian plains. Most tropical plant parasitic nematode species (except ear-cockle, potato cyst and cereal cyst nematodes) are slow in invasion and reproduction during winter than in summer. Late planting of susceptible *rabi* crops therefore, helps the crop to escape early nematode damage. On the same principle early sowing of spring crops in the temperate regions can escape nematode damage to the young crop.

Good crop husbandry practices including tillage, nutrient and moisture management, in general, improve plant tolerance to nematode attack. Proper availability of micronutrients also helps in overcoming some of the nematode damage, e.g. *Heterodera avenae* caused more damage to wheat in soils deficient in Mn. The integration of crop management as well as compatible cultural options might be viewed as the basic foundation for successful nematode management and crop production.

Clean and healthy cultivation: It is always advised to use certified seed free of any disease causing organism. Earcockle disease of wheat is seed born caused by *Anguina tritici*.

Control of ear-cockle and tundu disease

The control of this nematode can be obtained if the seed material is free from contaminated wheat seed galls. This is very easily achieved by :

i) Sieving (0.2 mm) the wheat seed material to eliminate gall, the one which passes through it should be discarded.
ii) Winnowing the seed in air to remove light seed galls.
iii) Dipping the wheat seed material in salt solution (500 g/10 litres of water), removing the floating seed galls, washing the settled grain twice with

fresh water and followed by drying in shade. With the adoption of this technique it is possible to eliminate this problem.

Thus, it is better to destroy all the crop residue left over after crop harvest, it would be helpful to deprive disease organism of the food.

Physical Control

Slash and burn farming: By burning the crop residue for a period of 10 min. can raise the temperature of upper layer (10 cm depth) of soil to 45°C. This is not only helpful to reduce the pathogen and also but also kill aerial parasitic worm like *Anguina tritici*, this is also helpful in reducing stem & bulb as well as foliar nematodes on rice. Important physical methods include cleaning of seed (other planting materials like tubers, bulbs, rhizomes, rooted cuttings etc. in dormant state), by winnowing, washing, filtration, flotation, paring etc., as appropriate.

Hot water treatment: The use of heat in the form of insolation, heat or hot water treatments to free the planting materials and soil, of nematodes provides some degree of control and prevention of spread of nematodes to uninfested areas. Banana rhizome, bulbs of narcissus and Tuberose infected with burrowing, root-knot and stem nematode could be disinfested by Hot Water Treatment at about 50°C for 10 min and at 55°C for 3-4 min. It has also been found that efficiency of hot water treatment could be improved with addition of formalin. The efficiency could also be enhanced by first giving cold water treatment for 2-3 wks, keep them for 37 °C for one day and then finally give hot water treatment of root system at 48°C for 30 min. Efficiency of treatment may also be achieved by peeling of hard surface of tissue material.

Biological Control

Besides, *Pasteuria penetrans*, a bacterium, several fungi like *Dactylella cynopaga*, *Dactylaria eudermata, Arthrobotrys* spp., *Cytopage* spp. and *Stylopage* spp. etc. have been reported as nematophagous. Also some of the endoparasite fungi like *Paceilomyecs lilacinus*, *Verticillium chlamydosporium*, *Trichoderma viride* and *Aspergillus niger* have shown potentialities against various nematodes and there is a scope of commercialization. Application of *P. penetrans* in nursery bed in combination with carbofuran @ 1.5 kg a.i./ha was as good as carbofuran applied @ 3 kg a.i./ha in respect to plant growth parameters and nematode reduction.

Chemical Control

Chemical nematicides have been found to give quick and demonstrable control of nematodes; these facts have been responsible for the recognition of nematodes as economically important pests and hence, for the establishment of the science of nematology. Nematodes being soil born are difficult pathogen to be controlled, could be control by fumigants and non-fumigants chemicals. Fumigants are required in large quantity and are broad spectrum, and are not preferred. In comparison, systemic or granular nematicides are required in a small quantity and are specific in action which are less toxic to beneficial microflora and microfauna. However, fumigants are regarded as true nematicides as they kill nematodes while non fumigants affect the behavior of nematode primarily and are thus considered more as nematostatic rather than nematicides. Despite of development of chemical industry there is paucity of chemical nematocides. To this date there are not many registered nematicides except carbofuran/ carbosulfan. That is because of the fact that these chemicals are required in large quantities to manage soil nemic population which is harmful for beneficial microflora. Also it contaminates groundwater so are environmentally unsafe. Hence their usage are restricted. Moreover, it has application difficulties like instruments and also efficacy is dependent on number of factors like organic matter, structure and texture of soil. To cope with negative aspect of chemicals scientific communities has devised methods of application like seed treatment, spot treatment, row application, nursery bed treatment, chiesel application, spray, split dose etc. which are judicious uses of chemicals without having harm to environment. Indiscriminate use of chemicals be avoided as it can cause pest resurgence. The most important nematicide recommended include Carbofuran 3G @ 1-2 kg ai/ha, Triazophos 40 EC.@ 1-2 l/ha, carbosulfan 25 ST as Seed treatment @ 1-2% w/w are being recommended for transplanted and direct seeded crop plants:

Transplanted crops

Nursery bed treatments: Prepare the nursery seed beds thoroughly. Apply carbofuran @ 0.3 g a.i./m^2 (approximately 10G of Furadan 3G/m^2). Mix it in the upper 5 cm soil and sow the seed. The nursery so raised is healthier and establishes better under field condition with minimized nematode infestation and enhanced yields.

Bare root dip treatment: The nursery seedling raised in untreated / treated bed may be given bare root dip in triazophos (40 EC) solution @1000ppm (2.5ml/ l water) for six hours during day time prior to its transplant in the main field.

Direct seeded crops

Seed Treatment

The seeds of directly seeded vegetable and pulse crops (okra, cucurbits, cowpea and beans) may be treated with carbosulfan (Marshal 25 ST) @ 3%w/w

(120g/kg seed) using need based gum as sticker. The plants raised from such treated seeds are healthier, carry reduced nematode infection and give more yield.

Integrated nematode management

Beside, use of 2-3 above mentioned method of management reduces the pest population below threshold limits and also would be socially feasible and environmentally compatible with the hallmark of high benefit cost ratio.

Summer : Two or three deep summer ploughing with a soil turning harrow at fortnightly intervals, preferably with a light irrigation between two ploughings; soil solarization of nursery-beds or pit soil using clear thin polyethylene mulch for 3 - 6 weeks before sowing; application of granular systemic nematicides to nursery-beds; growing of non-host or antagonistic commercial crop, and ploughing back non-commercial crop residues before monsoon.

Monsoon : Green manuring with non-host or trap crop or application of organic soil amendments/ manure.

***Kharif*:** Growing non-host or resistant commercial crops; nematicidal treatment of nursery-beds and seed; spot application of systemic nematicides in high value crops; uprooting and burning of roots of host crops and weeds after harvest.

***Rabi* :** Delaying sowing/ planting to mid-November (when soil temperature falls to 15-18°C); restricting growing of susceptible or tolerant crops to *rabi* season with a suitable period of rotation; removal and burning of roots; ploughing back non-commercial but disease-free crop residues.

Improved Agriculture and Allied Practices for Farmers, pp 183-192
New India Publishing Agency, New Delhi, India
© 2019, Editors: *J.P.S. Dabas, Ambrish Kumar Sharma, Nishi Sharma B.K. Singh and Pratibha Joshi*

16

Integrated Pest Management: An Ecofriendly Approach to Pest Suppression

Subhash Chander and R.K. Palta

Division of Entomology, ICAR-Indian Agricultural Research Institute New Delhi-110012

Insect pests, diseases and weeds are major biotic bottlenecks in the production of crops, inflicting on an average about 30% yield loss. Monetary value of these losses has been estimated to be exceeding Rs 1,00, 000 crores. There is thus ample scope to enhance our food production by curtailing losses due to pests to certain extent. Only option to boost our production remains in growing improved crop cultivars with better pest management options. Mainly synthetic pesticides have been used to combat pest menace in agriculture and public health. Pesticides have, of course, played a commendable role in increasing our food production and protecting us against disease vectors. However sole reliance on pesticides has created several problems such as development of resistant pests, pest outbreaks, mortality of useful organisms, adverse effect on human health and environmental degradation. Adverse effects of pesticides prompted scientists to look for safer and environment friendly methods of pest control and consequently concept of pest management came in to being.

Pest management is known as an ecological approach to tackle pest problems, because here emphasis is on containing pests in such a way that other biotic components of the system such as natural enemies, human beings and wildlife etc are not harmed and environment is preserved in general. Pest management works on the premise that all pest population levels are not injurious to crops and crops can always compensate for certain injury due to pests. Moreover, some pest population is always required for the survival of natural enemies of the pests. This has given rise to the concept of economic injury level (EIL), which helps to avoid unwarranted application of pesticides. The EIL based pesticide application ensures favourable cost-benefit ratio to farmers. Pest management also underlines that pest control tactics should be socially acceptable and within the reach of farmers. It is a holistic approach to pest problems in which we aim to protect our commodities against all the pests and ensure production of healthy crops.

Purpose of IPM

- To ensure pest suppression and safety of our environment
- To discourage indiscriminate use of pesticides and conserve natural enemy fauna
- To evolve pest management strategies based on host plant resistance, cultural practices, mechanical methods, natural enemies and selective pesticides and botanicals.

In IPM different methods of pest suppression such as resistant varieties, cultural methods, physical methods, natural enemies and pesticides are integrated to contain pest population without jeopardizing other components of the environment. Any pest management programme should be preferably based upon use of resistant cultivars and natural enemies and other control methods should be blended with them for tackling pest menace. The utility of different pest management components has been described in the following discussion.

A. Resistant cultivars

Preference should be given to the cultivation of resistant cultivars beacuase their use greatly reduces need for other methods of pest control. The resistant cultivars enhance the efficiency of some natural enemies against the pests, for example there is more predation of spiders against plant hoppers on moderately resistant varieties than susceptible varieties. Likewise, on moderately resistant cultivars, the need for pesticide application is greatly reduced. Similarly these are also compatible with cultural methods of pest control. However continuous use of resistant varieties has given rise to the problem of biotype development in some insects. Biotypes of brown plant hopper BPH and gall midge have developed due to cultivation of monogenic resistant varieties of rice. Resistant cultivars of various crops against insect-pests are presented in table 1.

B. Cultural methods

Routine agronomic practices can be utilized for minimizing pest infestation by slight modification in timing or method of their application. These function as preventive method. The field operations right from field preparation to harvesting or post harvesting can reduce population of one pest species or other.

a) *Clean cultivation*: Destruction of crop residues helps to prevent carryover of pest to next crop, for example destruction of cucurbit vines and stubbles of rapeseed mustard after crop season kills population of pumpkin beetle and painted bug, respectively. Similarly disposal of stubbles in rice, maize, sorghum reduces carry over of stem borers.

b) *Tillage:* Deep ploughing of fields exposes the soil inhabiting stages of pests such as army worm, cut worms, borers and white grubs, termites and mole cricket to vagaries of nature and thus helps to reduce pest infestation in crops. The raking of bunds of rice fields destroys egg pods of grasshopper. Similarly raking the soil at the base of mango tree destroys eggs of mango mealy bug.

c) *Sowing time:* The change in sowing/planting time is aimed for disturbing synchrony between host and pest populations. In rice, early planting (June-July) results in low populations of gall midge, leaf folder and armyworm while late planting (August) reduces incidence of grasshopper, gundhi bug and BPH. Early sowing of cucurbits escapes the attack of pumpkin beetle. However change in planting time should be adopted based on relative importance of different pests in the area. As far as possible, crops should be planted around the same time in a locality as staggered planting increases pest incidence. If there is incidence of pests in nursery then it should be preferably tackled in nursery itself as it is easier to contain a pest problem at small scale.

d) *Intercropping:* Intercropping helps in reducing incidence of certain pests by making microclimate less favourable for them. It also hinders the free movement of pests among the plants of same species. Intercropping of cabbage with tomato reduces infestation of diamond back moth. Intercropping of chickpea with wheat decreases incidence of pod borer while intercropping of mustard with wheat causes decline in aphid population on rapeseed-mustard.

e) *Croprotation:* The crops sharing the same pests should not be grown in a rotation. It has been found that with rice-wheat rotation, the incidence of pests of rice and wheat has increased. Similarly multiple cropping of rice results in outbreaks of pests.

f) *Trap crops:* Trap crops provide protection by preventing the pests from reaching the main crop. The trap crop must be more attractive to the pests and less economical than the main crop. Mustard as trap crop with cabbage reduces the incidence of diamond back moth, leaf webber, web worm and aphids. Marigold planted with tomato as a trap crop is highly effective against fruit borer.

g) *Nutrient and water management:* Fertilizers should be applied in a balanced manner. Excess use of nitrogen intensifies incidence of several pests such as plant hoppers and leaf folder in rice. Most of the times farmers only apply nitrogen and not phosphorus and potash. Potash induces resistance in plants against pests. Likewise excess use of water should also be avoided. It has been found that continuous flooding of rice fields is as good as alternate wetting and drying as far as yield is concerned. Drying of field at some intervals is helpful in reducing incidence of plant hoppers in rice.

h) *Harvesting:* Crops should be harvested at the optimum time. These should be harvested in such a manner that pests inside the stubbles get destroyed. The harvesting of rice at ground level destroys larvae of stem borer, thereby reducing pest survival for the next season.

C. Mechanical and physical methods

The removal of infested plant parts and conspicuous pest stages manually or mechanically can reduce pest incidence to a certain extent. Dead hearts, whiteheads, silver shoots, aphid infested shoots, bored brinjal and bhindi fruits should be picked and destroyed. Similarly egg masses, hairy caterpillars and beetles can also be collected and killed by dipping in kerosene. Light traps can be used for attracting and killing moths of stem borers, leaf folders and case worms etc. Although mechanical methods are labour intensive but these prove very effective if practiced at the initiation of pest incidence.

D. Biological control

There is rich diversity of natural enemies viz. predators, parasitoids and pathogens in fields and these help to reduce pest incidence if not harmed by broad spectrum pesticides. The egg parasitoids *Trichogramma* spp. are very effective against stem borers, fruit borers and bollworms. Likewise there are several effective larval and pupal parasites. Predators such as coccinellids, syrphids, spiders, dragon flies, ground beetles and rove beetles etc also devour pest populations in the field. The vertebrate predators like fish, ducks and frogs also consume pests in rice ecosystems. Likewise bacteria (Bt), fungi and viruses (NPV) also kill pests. Artificial releases of natural enemies are also carried out to combat crop pests (Table 2).

Farmers must differentiate between pests and useful insects and should conserve the useful ones. Efforts should be made to enhance natural enemies by providing artificial foods and shelters. The application of broad spectrum pesticides poses greatest hazard to the natural enemies of pests. The preference should be given to selective pesticides such as neem, Bt, NPV, diflubenzuron, propargite, cartap etc which do not harm non target organisms.

In the absence of selective pesticides, even non selective pesticides can be applied to achieve same purpose through suitable formulations. The granular formulations of pesticides pose less risk to natural enemies of pests.

E. Chemical Control

Pesticides have definite role to play in pest management. However their application should be need-based and in a judicious manner. The pesticide consumption can be greatly reduced and these can be made safer to environment by adopting improved techniques and proper timing of application *e. g.* plant hoppers are controlled more effectively if spray is directed at bases of rice plants. Regular monitoring of pests should be carried out and pesticides should only be applied if population has reached economic threshold level (ETL). The ETL based pesticide use causes substantial reduction in pesticide requirement. In pest infestation begins in nursery then it should be nipped there itself because it is easier to manage pests at smaller scale and this will also not pollute the environment. The perusal of various control methods reveals their effectiveness in general against pests. Depending upon the pest problem, the relevant and mutually compatible methods can be selected and integrated to achieve pest suppression. Following is the example of integration of different control methods for managing rice pests.

Integrated management of rice insect pests

1. Cultivation of resistant cultivars like Vajram, Chaitanya, Nagarjuna, Pratibha, Manasarovar and other high yielding varieties/ cultures like Badava Mahsuri and IET 7251.
2. Deep ploughing and raking of bunds to kill soil dwelling pest stages and grasshopper egg pods
3. Altering planting time, preponing or postponing depending upon pests in the area.
4. Clipping of seedling tips at the time of transplanting to destroy stem borer and hispa eggs
5. Formation of alleyways in the field for proper air circulation
6. Synchronous planting for avoiding food availabilty to pests for long period
7. Balanced use of fertilizers with emphasis on recommended N dose.
8. Use of trichocards against stem borer and leaf folder @ 50000-10000 eggs/ha.

9. Installation of stem borer sex pheromone traps. Scirpolure @ 5/ha for monitoring and 20-25/ha for mass trapping.
10. Field sanitation- removal of excess nurseries and weeds from field
11. Water management- alternate wetting and drying rather than continuous flooding.
12. Conservation of natural enemies through need-based and selective use of pesticides.

In this way pest management packages can be formulated against key pests of other crops also. The adoption of IPM packages will help to reduce dependence on pesticides and pests can be managed in an eco-friendly and sustainable manner.

Contrary to the belief, pesticide is not a bad word in pest management. Pesticides have to play definite role in pest suppression. However, their use has to be judicious and need-based. The need for pesticides should be determined through regular monitoring for economic threshold level of pests.

Table 1: Resistant cultivars/ sources of resistance against insect-pests in different crops

Crop	Pest	Cultivars
Rice	Stem borer	Deepti, Rama, Ratna, Sasyasree, TKM-6, Vikas, IR 20, IR 26
	Gall midge	Karna, Lalat, Sarasa, Ahaya, Ruchi, Kshira, Dhanyalaxmi, Pothan, Mahaveer, Divya, IR36, Kakatiya, Kunti, Phalguna, Shakti, Surekha
	Brown planthopper	Bhadra, Chandana, CO 42, Jyoti, Kanaka, Pratibha, Vajram, Chaitanya, Nagarjuna, Sonasali, Manasarovar, Krishnaveni, Kakatiya, Aruna, Makom, Remya, HKR 120
	Green leafhopper	IR 20, Vani, Vikramarya
	Leaf folder	ADT-34, Dular, Giza-171, Basmati Dwarf, Pattambi-29, IR-28, IR-50, Tainan-3, Pusa- 112073
Multiple resistant		
	Gall midge, BPH, WBPH	Suraksha, Shaktiman
	Gall midge, BPH, GLH	Daya, Lalat
	Gall midge, BPH, WBPH, GLH	Kshira
Wheat	Shoot fly	UPD 8, UP 85042, HD 2307, PDW 215, PDW 206, PBW 225, Raj 3214, HUW 206, WH 841, WH 283, HI 8381
	Brown wheat mite	C 306, A-9-30-1, DT 18, HDR 132, HPW 42, WH 589, Hyb 65, ISW 38, UP 2377, UP 2378, UP 2379
Barley	Aphids	DL117, DL200
Maize	Stem borer and shoot fly	Deccan hybrid, Ganga-5, Kundan, ChandaComposite-217
Sorghum	Shoot fly	SPV 1015, SPV 1413, SPV 1450, SPV 1467, SPH 1174, SPH 1077, SPH 1166, Selection 3, P 311, M 35-1, Swati, CSV 14R
	Stem borer	SPV 1155, SPV 1359, SPV 1466, SPH 1162, SPH 1078, SPH 1079, SPH 1165
	Midge	ICSV 197, ICSV 88032
	Aphids	SPV 1450, SPV 1359, SPV 1155, SPV 1380, SPV 1413, SPV 1452, SPV 1453, SPV 1462, SPV 1465, Selection 3, SPH 1162, SPH 1174, SPH 1020, SPH 1026, SPH 1079
	Multiple resistance	SPH 1161, SPV 1413, SPV 1450

Chickpea	Gram pod borer	PDE-2, ICC 738, ICC 506, ICCV 7, ICC 6663, PDE-5, N 37, GL 1002
Cowpea	Aphids	TVu 57, TVu 408, TVu 410, TVu 1937, TVu 3273, P2, R-1476, R-1473
Lentil	Pod borers (blue butterfly, gram pod borer)	DPL 7, DPL 16, DPL 26, DPL 34, PL 81-1, PL 88-54, PL 406, L 19-12, L 4136, L 4150, LL 56, LH 88-8, DHL 90-13
Pigeonpea	Gram pod borer	ICPL (Abhaya), ICPL 87088, ICPL 84060, ICPL-1, ICPL-2, ICPL 87-1, ICP 10466-E3, IPCL-6, JAT-10, UPAS-120, ICP-903E, ICP-1903E
	Pod fly	ICPL 82064-E15-E, PDA-881E, PDA-882E,PDA-883E, ICP 7946, PDA-89-2E, ICP 8102-5-6, ICP 7946, MA2, PDE-89-23
Soybean	Stem fly, grey weevil & YMV (whitefly)	PK 1069, PK 1107, MM-39-2(DS-93-39-2)
	YMV transmitted by whitefly	PK 1180, PK 1189, PK 1092, PK 1060, DS 016, Pusa 9702, DSK-18, SL-443, SL-444
Groundnut	Tabacco caterpillar	NCA-17840, NFG-79, EC-21898
Sugarcane	Top borer	CoJ 65, Co 859, C0 1158, Co 7224
	Internode borer	Co 975, Co 62175, Co 6806, CoC 77-1
	Scale insect	CoC 671, Co 62174, CoC 6907, CoS 611, Co1132
	White grub	Co 6304, CoS 510
Cotton	Jassid	Bikaneri Nerma
	Whitefly	Surekha, PS-141, Kanchan
	Boll worms (*Earias* & *Helicoverpa*)	A-1318, Bes-9-7, FBRN-3-7
Okra	Leafhopper	AE-30, Punjab-7, IIHR-4, AC-3375, Reshmi, Sangli-2, Okra Red, Selecton 6, Selection 8
	Shoot and fruit	AE-79, AE-22, AE-71, Narnaul special, borer Perkins Long Green
Tomato	Fruit borer	Parker, Bonus, VFN-8, Rick x Solan Gola (Lal, 1985)
Brinjal	Shoot and fruit	Sm 202, Sm 17-4, Pant Samrat, Pusa Purple borer Cluster
	Leafhopper	BI-7, 435-20
Cabbage	Aphid	Red Rock Mammoth, Red Drum Head, Glory, Early Queen, Express Mail, Rainy Princess, All Season, Red Pickling
Cauliflower	Head borer	ES-97, ES-96, Kartiki, KW-5, KW-8, KW-10, Kuwari, Kathmandu Local, Early Patna, EMS-30, PSK-16
Chilli	Aphid, jassid, mosaic, leaf curl	Pusa Jawala

Table 2: Biological suppression of crop pests with released predators, parasitoids/sprayed pathogens

Pest	Natural enemy	Release/application
Rice yellow stem borer & leaffolde	*Trichogramma japonicum* & *T. chilonis*	Inundative release @50,000- 100000/ha/week for 5 weeks
Sugarcane pyrilla	*Epiricania melanoleuca*	Periodic release @8000-10,000 cocoons or 800,000-1000,000 eggs/ha
Maize stem borer	*T. chilonis*	Inundative releases @ 100,000/ha/3 days - 5 times (Singh, 2002)
Tobacco caterpillar	*Telenomus remus* (egg parasitoid)	3 fortnightly releases @ 40,000/ha
Cotton bollworms	*T. chilonis* (Bio C1, Bio C6)*Chrysopa carnea*	@ 1,50,000/ha/7-10 days during egg laying period@ 2 larvae/plant during peak egg hatching
Apple San Jose scale	*Encarsia perniciosi* or *Aphytis* sp. *Chilocorus infernalis*	@ 2000/infested tree@20/infested tree
Apple wooly aphid	*Aphelinus mali*	-
Codling moth	*T. embrophagum*	@2000/tree during active egg laying period
Mealybugs & scale insects	*Cryptolaemus montrouzieri*	@10-50 beetles/plant
Cottony cushion scale	*Rodolia cardinalis*	-
Anar butterfly *Deudorix isocrates*	*T. chilonis*	Inundative release @50,000/ha/week
Coconut back headed caterpillar Opisina arenosella	*Goniozus nephantidis*	Relaesed at fort nightly interval during peak availability of pest
Subabul psyllid *Heteropsulla cubana*	Coccinellid, *Curinus coeruleus*	Inoculative release
Tomato fruit borer *Helicoverpa armigera*	*T. chilonis*/*T. pretiosum*	Inundative release @50000/ha/week
Spider mites	*Amblyseius* spp.	-
Tomato fruit borer	Ha NPV (nuclear polyhedrosis virus)	@250 larval equivalents(LE)/ha
Tobacco caterpillar *Spodoptera litura*	Sl NPV	@250 LE/ha
Sugarcane shoot borer	Granulosis virus (GV)	-

Chilo infuscatellus		
Cotton bollwormsCoffee berry borer	*Beauveria bassiana*	-
Sugarcane pyrilla	*Epiricania pyrillae*	
Rhiniceros beetle	*Metarhizium anisopliae*	-
Tomato fruit borer	*Nomuraea rileyi*	-
Tobacco caterpillar		
Castor semilooper		
Lepidopteran (moths and butterflies) pests	*Bacillus thuringiensis* var. Kurstaki	-

Section 4: Natural Resource Management

Improved Agriculture and Allied Practices for Farmers, pp 193-200
New India Publishing Agency, New Delhi, India

17

Soil Testing and Its Importance in Agriculture

V.K. Sharma, Kapila Chobhe, Mandira Barman and Abir Dey

Division of Soil Science and Agricultural Chemistry, ICAR-Indian Agricultural Research Institute, New Delhi-110012

Scientific agriculture is the only solution for increasing productivity of various crops and the input resources most effectively. Fertilizer, the most efficient and often the most expensive input can be applied to crops on the basis of generalized state level fertilizer recommendations. However, state level fertilizer recommendations do not take into consideration fertility status of the field which is neither uniform nor constant and which varies from field to field. Sometimes more fertilizers are added leading to unnecessarily higher investment, while under another situation, inadequate quantities of fertilizers may be added, resulting in lower returns. Inefficient and imbalanced uses of fertilizers affect both soil health and agricultural productivity.

Therefore, the best and most effective method for fertilizer recommendation is soil testing. It is a mode of assessing the nutrient supplying capacity of a soil.

Soil testing helps the farmers in the following ways

- To evaluate the fertility status (organic carbon - for nitrogen, phosphorus, potassium, zinc, iron, copper and manganese etc.) of the soil and to sort out the deficient areas from the non-deficient ones, for efficient use of fertilizers in soil.
- To know about the physical (Texture) and chemical (pH and EC – Total soluble salts) characteristics of the soil in respect of soil reclamation and management practices for improving salt affected soils.
- To assess the suitability/unsuitability of soil (pH, EC and calcium carbonate) for growing fruit crops in addition to fertility evaluation.
- Time to time monitoring of fertility status of soil through preparation of fertility status maps of the village/block/sub-division/district/state *w.r.t.*

different nutrients for review of the policy makers, administrators, field functionaries and other stakeholders for making strategies, dissemination of knowledge and timely arrangements of respective input (fertilizer) in advance for achieving maximum fertilizer efficiency, profitability, yield sustainability and quality.

Due to variability in soils, it seems impossible to device an entirely satisfactory method for soil sampling, hence, procedure depends on the purpose for which the sample is taken. Samples collected improperly should be discarded, because their analysis will lead to erroneous interpretation.

Collection and processing of soil samples for analysis

The chemical analysis becomes meaningless if there is an error in collection and preparation of soil samples. Hence, the collection and preparation of soil samples should be done with utmost perfection. Moreover, proper collection of soil sample (s) is an important as the reliable methods for their analysis. Unless a sample is not truly representative the analytical results are of little or no value. The steps involved are:

1. Collection of soil samples

The soil sample collected should be representative of the area sampled. A field can be treated as a single sampling unit if it is uniform in all respects. Variation in slope, texture, colour, crop(s) growth and management levels should be taken into account for soil sampling. Separate sets of composite samples need to be collected from areas differing in these characteristics. Recently fertilized plots, bunds, channels, marshy tracts, area near trees, farm ways, buildings, wells, compost piles or other non-representative locations must be avoided during sampling. When crops are grown in rows, samples can be taken in between the rows. Soil samples should be taken in zig zag pattern or randomly.

2. Depth of sampling

The plant roots penetration is very important for deciding the sampling depth. The following parameters may be kept in consideration:

a) For field crops (cereals, vegetables and others seasonal crops) a sampling depth of 0-15 cm (plough layer) is desired.

b) For deep-rooted crop like sugarcane or under dry farming conditions, samples from different depths may be taken.

c) For horticultural crops, sampling is done at depths of 0-15, 15-30, 30-60, 60-90, 90-120, 120-150 and 150-180 cm.

d) In saline-alkali soils, salt crust (visible or suspected) should be sampled separately and sampling depth be recorded.

e) Sampling should be done every year if the field is under intensive cultivation. If one crop per year is grown, sampling once in three years is sufficient. Soil sampling should be done at the same time in each year.

3. Sampling tools

Soil sampling can be done with the help of following tools:

a) Tube auger

b) Screw type auger

c) Post-hole auger

d) Spade or Khurpi.

For sampling soft and moist soil, a tube auger, spade or khurpi is quite satisfactory. A screw type auger is more convenient on hard or dry soil, while the post-hole auger is useful for sampling in excessively wet areas viz. rice fields. Tube Auger is convenient for sampling from lower depths. If a spade or khurpi is used, a V-shaped cut may be first made up to the plough layer and a uniform 2 cm thick slice is taken out.

4. Time of Sampling

For most field crops, soil should be sampled every three years. Soil samples may be taken after crop harvest, when the field is vacant and at least a fortnight before the crop sowing.

5. Methods for taking composite Soil Samples

a) For Fertility Evaluation

A composite soil sample gives a mean analytical value representative of the soil sample from a field/area. Collection of composite samples by mixing 5 soil cores – which are taken from plough layer (up to 15 cm depth) with the help of Khurpi and is the vital nutrient supporting zone to the plant(s) from different parts of an area is an acceptable method of sampling. Combine the soil collected from each spot and mix thoroughly and take about 250 to 300 g soil as a representative of that field. Put it in cloth bag. Label the bag clearly with name of the farmer, father's name, killa No., complete postal address and then transport the sample to the concerned soil testing laboratory, you desire.

b) For Soil Reclamation

For ascertaining the extent of salinity/Alkalinity or even both in a field, soil samples should be collected (from different soil depths 0-15 cm, 15-30 cm, 30-60 cm and 60-90 cm depth) up to 90 cm depth in one acre area with the help of post hole auger, which is affected by this problem. Store the depth wise soil sample in separate cloth bag. Label the soil samples; write the other information as stated above in fertility evaluation. If two profiles have been dug in one acre, then site A and site B must also be mentioned, along with killa No. on the cloth bags. Likewise, deposit these soil samples in soil testing laboratory for analysis, after proper tying together all the four soil samples from one pit/profile.

c) For Garden Plantation

Dig one or preferably two pits of 180 cm depth each, per acre, from which soil is to be got tested for orchard plantation. Procedure for digging the soil profile is exactly the same as stated for soil reclamation, except the soil depth, In case of orchard plantation, the depth of the pits is up to 180 cm and the various depth intervals are 0-15 cm, 15-30 cm, 30-60 cm, 60-90 cm, 90-120 cm, 120-150 cm and 150-180 cm, thus seven soil samples of 250-300 g each, to be collected per pit as per method stated before, labeled with required information, tied together and sent to the soil testing laboratory for desired analysis.

6. Labelling of samples

For identification, label the soil samples. A label of thick paper with identification mark along with the details of the sample should be put inside the sample bag and another label carrying same details tied outside the bag. In addition to location, field number, name of cultivator and relevant information about slope, drainage, previous cropping history, irrigation, fertilizer, manure used etc. must be recorded.

7. Preparation of composite sample

For making composite sample, mix the collected soil samples thoroughly by hand on a clean thick paper or cloth or polythene sheet. Reduce the bulk sample to about 500 g by quartering process in which entire soil mass is spread, divided into four quarters, two opposite samples are discarded and the remaining two are remixed. Repeat this process until about 500 g soil is left.

8. Processing of soil samples for analysis

Processing of soil samples involves several steps in sequence as follows: a) Drying, b) Grinding, c) Sieving.

a) Drying

The soil samples should be dried in shade at room temperature.

b) Grinding

Crush the soil clods lightly and grind with the help of wooden pestle and mortar. Care is taken so that primary sand and gravel particles are not crushed.

c) Sieving

Sieve the entire quantity of soil through 2 mm stainless steel sieve. Remove plant residues, gravel and organic material as much as possible by retaining them on the sieve. For specific type of analysis (e.g. organic C) grind the soil further and pass it through the 0.2 to 0.5 mm sieves. Remix the whole quantity of sieved soil before a sample is weighed for analysis.

9. Precautions in collection and processing of soil samples

In order to avoid the contamination, special care is required in the collection and processing of soil samples. Following precautions should be taken to minimize error:

- Collect the soil sample when the soil is in workable condition i.e. at field capacity.
- Avoid taking samples near bunds, water channels and trees in the field.
- Sampling should not be taken from the fertilized field(s).
- Do not take samples from the location where the cow dungs, crop residues have been heaped after harvest.
- Keep away the sample from chemicals, fertilizers or manure.
- For sampling of micronutrient analysis, always use auger made up of stainless steel instead of rusted iron khurpi or spade.
- Do not use bags previously used for storing fertilizers or any chemical.
- Store soil samples in clean cloth or polythene bags.
- Use glass or polythene jars for storage of soil samples for a longer duration.

Interpretation of results on the basis of soil test: After analysis of soil samples, keep the result of soil samples in low, medium and high categories of nutrient rating as given below:

Table 1: Rating of available nutrients for different categories after soil analysis

S.No.	Nutrient	Nutrient rating		
		Low	Medium	High
1.	Organic Carbon (%)	<0.50	0.50-0.75	>0.75
2.	Available nitrogen ($KMnO_4$ Oxidizable N) ($kgha^{-1}$)	<280	280-560	>560
3.	Available Phosphorus (Olsen P) ($kgha^{-1}$)	<10	10-25	>25
4.	Available Potash (K) in NH_4OAC ($kgha^{-1}$)	<120	120-280	>280
5.	Available Sulphur (S) ($kgha^{-1}$)	<20	20-30	>30

Table 2: Critical limits of different micronutrient elements

S.No.	Nutrient	Soil Rating	Critical limit (ppm)
1.	DTPA extractable Zinc (Zn)	Deficient	<0.6
2.	DTPA extractable Iron (Fe)	Deficient	<4.5
3.	DTPA extractable Manganese (Mn)	Deficient	<2.0
4.	DTPA extractable Copper (Cu)	Deficient	<0.2

Fertilizer recommendations on the basis of soil test

In this approach medium soil fertility is equated with general recommended dose. In case of low and very low, and high and very high fertility categories the fertilizer doses are increased or decreased by 25 to 50 per cent of the general recommended dose as per the situation. At present most of the fertilizer recommendations issued from soil testing laboratories in India are based on this approach. The rating were developed in 1965 for old varieties of crops. Unfortunately since then these ratings are the same irrespective of types of soils and varieties of crops.

10. Remedial Measures

Correction of nutrient deficiencies involves the use of right choice of fertilizer after knowing the quantity of fertilizer required along with appropriate mode and/or method as well as optimum time of application. The appropriate methods for efficiently correction of deficiencies of different nutrients in soil and plants are illustrated as under:

Table 3: Remedial measures of essential plant nutrients

Nutrient	Suggested Remedial Measures	
	Soil	Plant
Nitrogen	Apply recommended doses of nitrogenous fertilizers e.g. Urea, CAN etc.	Foliar application of 2.0% urea
Phosphorus	Apply recommended phosphatic fertilizer such as DAP, SSP or nitro phosphate	Nil
Potassium	Apply recommended doses of potassic fertilizers e.g. Murate of Potash , sulphate of Potash	Foliar application of Murate of potash @1.0% solution
Sulphur	Apply gypsum @ 200kg ha^{-1} followed by mixing and light irrigation	Foliar application of potassium sulphate solution @ 0.5%
Zinc	25 kg Zinc Sulphate ha^{-1} 0.5% with 0.25% unslaked lime	Foliar application of $ZnSO_4$ @
Iron	50-100 kg Ferrous sulphate ha^{-1}	Foliar application of $FeSO_4$ solution @ 2.0%
Manganese	40-50 kg Manganese sulphate ha^{-1}	Foliar application of $MnSO_4$ @ 2.0% solution
Copper	10 kg Copper sulphate ha^{-1}	Foliar application of $CuSO_4$ solution @ 0.1% +0.05% Lime
Boron	1.0 to 1.5 kg Boron ha^{-1}	Foliar application of borax @0.2% solution
Molybdenum	1 kg Sodium Molybdate ha^{-1} molybdate	Apply 0.1% solution of Sodium

Improved Agriculture and Allied Practices for Farmers, pp 201-207
New India Publishing Agency, New Delhi, India

18

Use of Biofertilizers for Improving Soil Health for Higher Crop Production

Y.V. Singh

Centre for Conservation & Utilization of Blue Green Algae, ICAR-Indian Agricultural Research Institute, New Delhi-110012

What is biofertilizer?

Biofertilizers are ready to use live formulates of such beneficial microorganisms which on application to seed, root or soil mobilize the availability of nutrients by their biological activity in particular, and help build up the micro-flora and in turn the soil health in general.

Why should we use biofertilizers?

With the introduction of green revolution technologies the modern agriculture is getting more and more dependent upon the steady supply of synthetic inputs (mainly fertilizers), which are products of fossil fuel (coal+ petroleum). Adverse effects are being noticed due to the excessive and imbalanced use of these synthetic inputs. This situation has lead to identifying harmless inputs like biofertilizers. Use of such natural products like biofertilizers in crop cultivation will help in safeguarding the soil health and also the quality of crop products.

Advantages of biofertilizers

- Provide plant nutrients at very low cost.
- Biological Nitrogen made available to plants.
- Replace chemical nitrogen, phosphorus potassium and zinc by 25%.
- Solubilise phosphate and sulphate and increase nutrient uptake efficiency of plants.
- Enhance plant growth by release of vitamins, auxins and harmones.
- 10-30% crop yield increases with their use.

- Improve physical, chemical and biological properties of soil.
- Have no harmful effect on soil fertility and plant growth.
- Required in small quantity.
- Provide residual effects for subsequent crops.
- Hasten seed germination, flowering and maturity in crops.
- Pollution free and eco-friendly.
- Helps in recyling/decomposition of organic waste.
- Provide protection against drought and some soil borne diseases.

Types of available Biofertilizers

The use of biofertilizers is quite important while practicing the concepts of integrated plant nutrient management and organic farming. Some of the commonly used biofertilizers are as follows.

For Nitrogen

Rhizobium for legume crops

Azotobacter/Azospirillum for non legume crops

Acetobacter for sugarcane only

BGA and Azolla for low land paddy

For Phosphorous

Phosphatika or PSB for all crops to be applied with Rhizobium, Azotobacter, Azospirillum and Acetobacter

For potassium

Potassium solubilizing bacteria (KSB)

For Zinc

Zinc solubilizing bacteria (ZnSB)

For enriched compost

Cellulolytic fungal culture

Phosphotika and Azotobacter culture

---

1. **Rhizobium:** It induces better root nodulation and stem nodulation (*Azorhizobium*) in inoculated plants and thereby brings down the

requirement of nitrogen fertilizer for the cultivation of pulses, oil seeds and legume green manures. Commercially it is available as carrier based inoculums. It can be applied through seed treatment.

2. **Azotobacter:** Suitable only for upland crops like cereals, oil seeds, vegetables, tapioca, plantation and orchard crops. It is available as carrier as well as liquid based inoculum. It fixes N about 15-20 kg/ha under ideal upland conditions and thereby reduces the requirement of nitrogen fertilizers by 10-20 per cent.Methods of application are seed treatment, seedling root dip and direct soil application.

3. **Azospirillum:** It is suitable for both upland and wetland conditions and is available as carrier as well as liquid based inoculum. It fixes about 20-25 kg N per ha under ideal conditions thereby effecting a reduction of 25 per cent in the quantity of N fertilizers required.Treatment with *Azospirillum* also induces better root formation in inoculated plants. Hence this biofertilizer is also recommended for root induction in polybag-raised seedlings of plantation and orchard crops and also for vegetable crops. The isolates of *Azospirillum* brasilense strains AZR 15 and AZR 37 are highly effective for rice, vegetables and nursery plants. The strains AZ 1 and AZ 2 are effective in vegetable and nursery plants.

4. **Blue green algae (BGA):** Mainly recommended for wetland rice cultivation. It is available as carrier-based inoculum and it fixes N about 25-30 kg/ha under ideal conditions. It is applied through direct broadcasting in the rice fields @ 10 kg/ha (soil based) or @ 1.5 kg/ha (multani mitti based) 2-3 days after transplanting the seedlings.

5. ***Azolla:*** It is suitable for wetland rice cultivation. The required quantity of *Azolla* will have to be raised in the farmers' field itself.It fixes about 25 to 30 kg N / ha. Incorporate fresh *Azolla* @ 10 t/ha before transplanting or apply @1.0 t/ha 3-4 days after transplanting of rice.

6. **Phosphate solubilising bacteria (PSB):** Recommended for all the crops raised in neutral and slightly alkaline soils. Available as carrier-based inoculum. Enables the efficient utilization of cheaper sources of phosphatic fertilizers such as rock phosphate by crop plants in neutral and alkaline soils.

7. **Vesicular / arbuscular mycorrhiza (VAM/ AM):** Vesicular arbuscular mycorrhiza is mostly recommended for upland especially for raising container and tissue culture plantlets and transplanted crops. It mainly improves the uptake of available P by inoculated plants. There is also an enhanced absorption of water and other nutrients such as N and K and

certain micronutrients. Mycorrhiza inoculation can improve the survival and establishment of tissue culture plantlets under field conditions. Also induces better resistance against certain soil borne plant pathogens. It is commercially available as granular inoculum consisting of infected roots and soil with mycorrhizal spores. It is given as soil application.

VAM fungus Glomus microcarpum var. microcarpum is suitable for tropical tuber crops. The inoculation can be done by placing inoculum (3-5 g/set) beneath the set before planting. The rate of spore load in the inoculum should be to the tune of 50 to 400 spores per 100 g soil medium

Table 1: Contribution of different biofertilizers in crop production

S. No.	Biofertilizer	Nutrient Fixation	% yield increase	Crop
A.	**Nitrogen fixing**			
1.	*Rhizobium* sp.	50-200 Kg N/ha	15-30%	Leguminous crops
2.	*Azotobactor*	20-30 Kg N/ha	5-20%	Cereals, oilseeds, vegetables
3.	*Acetobactor*	10-25 Kg N/ha	10%	sugarcane
4.	*Azospirillum*	10-20 Kg N/ha	5-20%	Cereal crops
5.	Blue GreenAlgae (BGA)	10-40 Kg N/ha	5-20%	rice
6.	*Azolla*	20-30 Kg N/ha	25-30%	rice
B.	**Phosphate supplying**			
1.	Phosphate solubilizing Micro-organism (PSM)	20-50 Kg P_2O_5/ha		All crops
2.	Phosphate Mobilizing Microbes-Mycorrhizae-VAM	10-30Kg P_2O5/ha		All crops

Doses of Biofertilizers

1. Seedlings treatment

   Carrier based : 200 gram for seedlings required per acre.

   Liquid : 50 ml for seed required for one acer.

   Soil Treatment : 4-5 kg per hectare.

Application Techniques of Biofertilizers

1. Seed treatment

Five hundred grams of commercially available inoculum will be required for treatment of seeds for one-hectare area. For this, thick slurry of the carrier-based inoculum is initially prepared by mixing 500 g of the inoculum in 1.25 litre of water. The stickiness of the biofertilizer on seed surface can be significantly improved by using 10% jagerry solution or 5% sugar solution supplemented with 40% boiled and cooled gum arabic solution or rice-gruel water. The required quantity of seed material is then gently mixed with this slurry by taking care not to damage the seed coat. The treated seeds are spread evenly over a gunny bag and dried in shade and sown immediately in moist soil.Under no circumstances, the treated seeds are exposed to direct sunlight for a longer period of time since the UV rays of solar radiation will reduce significantly the population of inoculated bacteria on seed surface.

Recommended dose of liquid biofertilizer may be diluted with water as per seed quantity and mixed with rod. Than seed may be dried in shade for 30-40 minutes before sowing.

2. Seedling treatment

This method of application is mainly recommended for transplanted crops. In this procedure, the roots of seedlings to be transplanted are dipped in loose water slurry of the biofertilizer (500 g in 2.5 litre of water) for 20 minutes, prior to transplanting.

3. Soil application

Soil application is generally recommended for all types of biofertilizers except Rhizobium, Bradyrhizobium and Azorhizobium. The method is to apply the biofertilizer after mixing with dried FYM, compost or vermicompost @ 1:25. For crops of six-month duration, the recommended dose is 1-2 kg/ha. This can be increased to 2-4 kg/ha for crops of more than six-month duration. For perennial crops, 10 to 25 g of the biofertilizer is to be applied in the root zone during the first year and 25 to 50 g during subsequent years. This can be done at the time of sowing, transplanting or during intercultivation.

Factors influencing the efficient use of bio-fertilizers

- Use adequate quantity of organic manure (as per the recommendation for each crop) along with biofertiliser application. This is essential to ensure better survival, growth and activity of the introduced microbial inoculums in acidic soils.
- Liming is essential if the soil pH is below 6.0. In moderately acidic soils, the application of lime @ 250 kg/ha is recommended along with biofertilizer treatment.
- Irrigation is essential during summer months after biofertilizer application to ensure the survival of the introduced microbial inoculum in the soil.

- Since N biofertilisers can supplement only a part of the nitrogen requirement of the inoculated plant, low dose of nitrogen and full doses of phosphorus and potassium as per the recommendation may be applied. This is essential to ensure better plant growth and yield. Similarly, in the case of P biofertilisers, the full doses of nitrogen and potassium should be applied. However, there should be a gap of at least one week between biofertilizer and chemical fertilizer applications.
- Use only biofertilisers, which are manufactured as per the quality parameters prescribed by the Bureau of Indian Standards. In the case of bacterial biofertilisers, the prescribed standard is that in the final product, the population of the desired bacterium should not be less than ten million per gram of the carrier material and there should not be any contamination with other microorganisms when examined at 1:100000 dilution. Further, it should have a shelf life of at least six months.
- The commercially available biofertiliser should always be used before the expiry date marked on the culture packet.
- Topdressing with super phosphate 25 kg/ha 10 days after inoculation of BGA will enhance its growth under field conditions.
- Since the occurrence of green algae in rice field can affect the normal growth and proliferation of BGA, the population of green algae should be controlled initially by applying copper sulphate @ 4 kg/ha.
- In moderately acidic soils of pH around 6.5, root nodulation by Rhizobium and Bradyrhizobium can be improved by pelleting with finely powdered calcium carbonate. (See recommendation under cowpea)
- Application of P_2O_5 @ 1 kg/ha is recommended once in 4 days in P_2O_5 deficient soils to ensure good growth of *Azolla*. The development of a reddish purple colour in Azolla is a typical symptom of P_2O_5 deficiency.
- Since a floating population of *Azolla* can release its bound nutrients only during decay in the soil, it is essential to incorporate *Azolla* in the soil prior to the transplanting of rice seedlings.

Precaution During Use of Biofertizers

- Use recommended biofertilizer for particular crop.
- Check expiry date before use.
- Do not expose treated seeds to direct sunlight.
- Soil/Transplant treated seeds/seedling within 2-3 hours.

- Do not mix chemical fertilizers insecticides or pesticides directly with biofertilizers.
- Use double dose of biofertilizers, if seeds are treated with insecticides or pesticides or fungicides.
- Follow instruction given on packets.

Improved Agriculture and Allied Practices for Farmers, pp 209-221
New India Publishing Agency, New Delhi, India

19

Effective Utilization of Water for Enhanced Crop Productivity

Neelam Patel

Centre for Protected Cultivation Technology, ICAR-Indian Agricultural Research Institute, New Delhi-110012

Indian agriculture, accounting for 25% of the nation's Gross Domestic Product (GDP), 15% of exports and 60% of the employment continues to be the mainstay of economy. Having achieved laudable success in agricultural production in the last 50 years (50 Million ton to 210 Million ton) India has transformed herself from a food deficit country to a food surplus country. Still there are many challenges, which Indian agriculture is facing, in a fast changing technical and socio-economic scenario. Relating to the natural resources and production base, water has emerged as the most crucial factor for sustaining the agricultural sector. India accounts for 16 percent of the world's human population and nearly 30% of the cattle with only 2.4% of the land and 4% of the water resources.Even if the full irrigation potential is exploited, about 50% of the country's cultivated area will still remain unirrigated, particularly with the current level of irrigation efficiency. The share of water for agriculture would reduce further with increasing demand from other sectors. But the water demand for agricultural purposes is estimated to increase in order to produce more cereals, horticultural produce and raw material for a fast expanding food industry. Efficient management of water is, therefore, key to future growth of Indian agriculture. The requirement of water by all sectors in 2025 is estimated to be 105 M ha m (Water Resources Development in India, 2010). Share of water for agriculture is expected to get reduced from the present level of 85 % to 69 % by 2025. On the other hand, the actual agricultural water demand is estimated to increase from 470 Billion Cubic Meters (BCM) in 1985 to 740 BCM in 2025. During the same period, the demand for non-agricultural use of water will multiply fourfold from 70 BCM to 280 BCM. The water use efficiency (WUE) in Indian agriculture, at about 30-40%, is one of the lowest in the world, against 55% in China.

The vulnerability of Indian agriculture is bound to be severe if the present trend of water use and management efficiency is not changed. The International Water Management Institute forecasts that by 2025, 33% of India's population will live under absolute water scarcity condition. Rainfed lands are not only low in productivity and sustainability, but also more prone to risks, as compared to irrigated areas. Rainfed areas are also the location where, proportionately a greater concentration of poor and hungry persons lives. This can be obviated to some extent by expanding irrigated areas through improving water management and water use patterns. Presently, the problem facing the country is not the development of water resources but the management of the developed water resources in a sustainable manner. By adopting efficient water management practices, the bulk of India's agricultural lands could be brought under irrigation. Micro irrigation is one such practice.

Development of micro irrigation in India

Micro irrigation is based on the fundamental concepts of irrigating only the root zone of the crop rather than the entire land surface and maintaining the water content of the root zone at near optimum levels.It is one of the latest methods of irrigation, which is becoming increasingly popular in areas with water scarcity and salt problems. Economic considerations usually limit the use of drip irrigation system to orchards and vegetables in water scarcity areas. The concept of drip irrigation goes back at least to the ancient Egyptians who placed porous pots in the soil and filled them with water.Use of subsurface clay pipes for irrigation in Germany led to doubling of crop yields in 1860 and perhaps that was the beginning of the concept of applying irrigation water directly to the root zone. Nehemiah Clark, in 1873, obtained the first known US patent for a water-emitting device as a dripper (a simple hole).

Almost at the same time experiments of irrigating with porous pipes began. Drip irrigation in principle evolved from subsurface irrigation when in 1920, the drainage tubes (with open joints) of subsurface irrigation were replaced experimentally by porous pipes for the purpose of irrigation. Another version of porous pipe was the perforated pipe, which allowed the exit of water through holes or slits in its wall and the flow caused by the pressure inside the pipes. The perforated pipe was used in Germany, which made the concept feasible, and various experiments were then centered on the development of drip system using perforated pipe made of various materials. During early 1940, Symcha Blass, an Israeli engineer observed that a large tree near a leaking faucet exhibited a more vigorous growth than the other tree in the area, which were not reached by the water from the faucet. This example of leaking faucet led him to the concept of an irrigation system that would apply water in small amounts.

In earlier times, in several parts of India, a system similar to drip was used for irrigating 'tulsi' plant. An earthen pitcher used to be hanged above the plant with the help of ropes. The pitcher used to have a small hole through which water used to discharge in drops over the plant. In India, micro irrigation was originally practiced through some indigenous methods like, micro irrigation using bamboo pipes in Meghalaya, perforated clay pipes and pitchers in Maharashtra and Rajasthan. In Meghalaya an ingenious system of tapping of stream and spring water by using bamboo pipes to irrigate plantations is widely prevalent. The tribal farmers of Khasi and Jaintia hills use this 200-year-old system to irrigate the betel leaf or black pepper crops planted in mixed orchards. Bamboo pipes are used to divert perennial springs on the hilltops to the lower reaches by gravity. The channel sections, made of bamboo, divert and convey water to the plot site where it is distributed without leakage into branches, again made and laid out with different forms of bamboo pipes.

During the initial phase of drip irrigation, viz., during 1971, the area under drip irrigation totaled 56,000 hectares worldwide. In 1998, the area under micro irrigation was around 2.8 million ha. In India, the area under drip irrigation had grown from a negligible level in 1975 to around 18.97 lakhs ha in 2009-10 (Anonymous, 2010). In India, Maharashtra, Tamil Nadu and Karnataka are some of the states which have taken a lion's share in drip irrigated area constituting 39.7 per cent, 28 per cent and 24.3 per cent respectively of total area under micro irrigation. In India, the farmers are conversant with micro irrigation techniques since 1970s (Sivanappan et. al., 1987). However, micro irrigation systems were initially not popular due to poor quality of components and lack of services from the manufacturers/ traders. It took more than ten years before Indian business started showing interest in the systems, after getting convinced that the system has potential in the country. By the time researches in different ICAR Institutions and Agricultural Universities had provided adequate data to prove that the systems not only save water but also help in improving crop yields and the quality of the produce. The positive results of the experimentations by the grape growing farmers in Maharashtra proved the much needed catalyst in increasing the popularity of the micro irrigation system by assuring the farming community of its benefits including its cost effectiveness. The Task Force estimated potential for Drip irrigation of 27 Mha and for Sprinkler irrigation as 42.5 Mha as given in the following table (Table 1). Based on the recommendations of the Task Force, Government of India had initiated a Centrally sponsored micro irrigation scheme in the country with 50% financial support to farmers from Central as well as State Government budgets in the ratio of 4:1. National Committee on Plasticulture Applications in Horticulture was made its coordinating agency. The results of the efforts of the scheme may be seen in Table 2.

Table 1: Theoretical potential of micro irrigation in India

Crop	Area (Million Ha)		
	Drip	Sprinkler	Total
Cereals	-	27.6	27.6
Pulses	-	7.6	7.6
Oil seeds	3.8	1.1	4.9
Cotton	7.0	1.8	8.8
Vegetables	3.6	2.4	6.0
Spices and condiments	1.4	1.0	2.4
Flowers, Medicinal, aromatic plants	-	1.0	1.0
Sugarcane	4.3	-	4.3
Fruits	3.9	-	3.9
Coconut & Plantation crops, Oil palm	3.0	-	3.0
Total	27.0	42.5	69.5

Table 2: The state-wise total area under Drip/ Sprinkler Irrigation till 31.03.2010

S.N	State	Drip, ha	Sprinkler, ha	Total
1.	Andhra Pradesh	505205	256911	762116
2.	Arunachal Pradesh	613	0	613
3.	Assam	116	129	245
4.	Bihar	301	435	737
5.	Chhattisgarh	6360	95740	102100
6.	Goa	793	582	1375
7.	Gujarat	226773	180572	407345
8.	Haryana	11351	533740	545090
9.	HP	116	581	697
10.	Jharkhand	208	742	950
11.	Karnataka	209471	385579	595050
12.	Kerala	15885	3540	19425
13.	Madhya Pradesh	51712	143233	194946
14.	Maharashtra	604440	295382	899822
15.	Manipur	30	0	30
16.	Mizoram	72	106	178
17.	Nagaland	0	3962	3962
18.	Orissa	11046	33015	44061
19.	Punjab	17925	11414	29339
20.	Rajasthan	30047	866592	896639
21.	Sikkim	23460	11339	34799
22.	Tamil Nadu	153437	27834	181271
23.	UP	12636	13310	25945
24.	Uttranchal	38	6	44
25.	West Bengal	247	150196	150443
26.	Others	15000	30000	45000
Grand Total		1897280	3044940	4942220

Source: Data compiled from progress reports from state government

Drip irrigation can be used for most crops, such as:

Orchard crops	:	Grapes, Banana, Pomegranate, Orange, Citrus, Tamarind, Mango, Fig, Lemon, Custard Apple, Sapota, Guava, Pineapple, Coconut, Cashew nut, Papaya, Aonla, Litchi, Watermelon, Muskmelon etc.
Vegetables	:	Tomato, Chilly, Capsicum, Cabbage, Cauliflower, Onion, Okra, Brinjal, Bitter gourd, Bottle gourd, Ridge gourd, Cucumber, Peas, Spinach, Pumpkin etc.
Cash Crops	:	Sugarcane, Cotton. Arecanut, Strawberry etc..
Flowers	:	Rose, Carnation, Gerbera, Anthurium, Orchids, Jasmine, Lily, Mogra, Tulip, Dahilia, Marigold etc.
Plantation	:	Tea, Rubber, Coffee, Coconut etc.
Spices	:	Turmeric, Cloves, Mint etc.
Oil seed	:	Sunflower, Oil palm, Groundnut etc
Forest crops	:	Teakwood, Bamboo etc.

The Government of India (GOI) plans to cover 17 Mha under drip and sprinkler systems by the end of XI plan. The Micro Irrigation Scheme has been launched by the GOI with this objective. Micro irrigation system was found to result in 30 to 70 % water savings in various orchard crops and vegetables from along with 10 to 60 % increases in yield as compared to conventional methods of irrigation. Mulching with drip further enhanced the crop yield to the tune of 10-20 % and controlled weeds up to 30-90%.

Table 3: Area Covered under Drip/Sprinkler irrigation under CSS on Micro Irrigation

Financial Year	Drip Irrigation(ha)	SprinklerIrrigation (ha)	Total (ha)
2006-07	155000	144000	299000
2007-08	209000	241000	450000
2008-09	250000	324000	574000
2009-10	1897280	3044940	4942220
Total	2511280	3753940	6265220

Source: Progress reports from state governments

Impact of micro irrigation

A review of selected literature on micro irrigation technology is strongly suggestive of economic viability of this technology for many crops of India. What is no less striking is that this economic viability exists even without taking into account the Government subsidy available on micro irrigation under various development programs. The cost for installing the system works out Rs. 15,000 to 25,000/ ha for widely spaced crops like coconut, mango etc to Rs. 60,000 to 90,000/ ha for closely spaced crops like cotton, sugarcane, vegetables etc. The cost depends upon the crop, spacing, water requirements, source of water supply etc. The economics of micro irrigation has been calculated and payback period was worked after interviewing more than 50 farmers for different crops (Table 4). It is observed that the payback period is about one to 2 years for most of the crops and the benefit cost ratio varies from 2 to 5 (Sivanappan, 2009). Micro irrigation led agriculture should be viewed as one of the eco-technological approaches to attain sustained and enhanced agriculture production and productivity. Through micro irrigation, the green revolution could be transformed into an evergreen revolution to ensure the sustainability, profitability and equity. Since micro irrigation greatly enhances water, fertilizer and energy use efficiency and promotes precision agriculture, the green revolution could be achieved without the burden of environmental degradation.

Micro irrigation research in India

Different Institutes of Indian Council of Agricultural Research, State Agricultural Universities and other Water Management Research Institutes investigated different aspects of micro irrigation technology which are summarized in brief in the following paragraphs.

i) Studies on water saving and yield enhancement

Many studies across the country compared different micro irrigation methods with conventional methods of irrigation like border, check basin, furrow or surge irrigation and established beyond doubt the superiority of micro irrigation methods in different aspects of water management (Table 4) (Annonymous, 2001).

Table 4: Increase in yield and water savings in micro irrigation as compared to surface irrigation

Crop	Increase in yields (%)	Water saving (%)	Crop	Increase in yields (%)	Water saving (%)
Tomato	25-50	40-60	Chilli	10-40	60-70
Sugarcane	50-60	30-50	Bottle Gourd	20-40	40-50
Okra	25-40	20-30	Cauliflower	60-80	30-40
Potato	20-30	40-50	Pomegranate	20-40	50-60
Cabbage	30-40	50-60	Cabbage	50-60	50-60
Brinjal	20-30	40-60	French Bean	55-65	30-40

ii) Crop Geometry

Micro irrigation methods encourage researchers to investigate different planting arrangements for maximizing crop yields. Also the cost of the micro irrigation system depends on the crop geometry. Some of the significant findings of the research efforts in terms of optimal plant to plant and row to row spacing of different crops for getting the maximum yields and least cost of the system are presented in Table 5 (Anonymous, 2001).

Table 5: Effect of crop geometry on the yield and water saving under micro irrigation

Crop	Location	Recommended Row to Row and Plant to Plant Spacing, cm x cm	Increase in Yield (%)	Water Saving (%)
Okra	Delhi	30 x 15	66	71
Tomato	Delhi	40 x 40	81	54
Banana	Coimbatore	200 x 400	70	-
Okra	Coimbatore	30 x 40	30	-
Banana	Kharagpur	200 x 400	50	50
Pea	Solan	20 x 11.50	80	-
Tomato	Solan	20 x 42.1	42	-
Cabbage	Hyderabad	20 x 72	50	50
Banana	Navsari	120 x 200	28	40
Chilli	Coimbatore	20 x 72	50	50
Sugarcane	Navsari	60 x 120	17	46

iii) Fertigation Studies

The introduction fertigation opened up new possibilities for controlling water and nutrient supplies to crops and maintaining the desired concentration and distribution of ions and water in the soil. Most research is focused on comparing the conventional fertilization method with fertigation, especially the NPK use efficiency and their impact on the quantity and quality of produce. Adoption of fertigation by farmers in several crops registered significant fertilizer savings (20 to 60%) in different crops (Table 6). New fertilizers for fertigation include slow and controlled release fertilizers which ensure continuous plant nutrition over months besides matching nutrient release rate with plant needs, labour saving and reduced leaching.

Table 6: Effect of fertigation on yield of different crops

Crop	Saving in Fertilizer	Increase in yield	Crop	Saving in Fertilizer	Increase in yield
	(%)	(%)		(%)	(%)
Tomato	40	18	Castor	60	32
Okra	40	18	Cotton	-	20
Onion	40	16	Potato	40	-
Broccoli	40	10	Tomato	50	28
Banana	20	11	Tomato	40	33
Banglore blue grapes	20	-	Sugarcane	50	-
Sapota	40	-	Gerbera	-	37
Banana	40	-	Tomato	20	25

Source: Anonymous, 2001

Advances in Micro Irrigation

Initially, micro irrigation was used primarily for high-value crops such as fruits, vegetables, nuts, and sugarcane. As system reliability and longevity improved, its application was extended to lower-valued agronomic crops (viz., cotton, tobacco, peanuts, corn, castor, chickpea, soybean, sunflower etc.) and spices & condiments (ginger, turmeric, cardamom etc), primarily because the system could be used for multiple crops & years owing to varied planting patterns (mainly paired row pattern), reducing the annual system cost with enhanced rate of return on investment.

i) Emitter types and design

In many countries still the dominant emitter type used by the farmers in several field & horticultural crops is either point source (individual button drippers) or line-source (integral drip lines and drip tapes). In the USA, Israel, Australia, Spain, South Africa, India, Mexico, China, France, Germany, Thailand, Philippines, Malaysia etc there is an increasing tendency to use line-source emitters primarily for sugarcane, corn, tobacco, mulberry, vegetables, flowers etc. Subsurface drip with pressure compensating dripper and flap outlet is gaining momentum in several crops viz., turf grass, cotton, sugarcane, corn, potatoes, vegetables, grapevines etc in the USA, South Africa, Israel, Australia, Turkey, Argentina, France etc.

ii) Low-flow irrigation

Soil water potential and water content in the vicinity of active roots generally controls the rate of water and nutrient uptake by plants. In drip irrigation, frequency and emitter discharge determine the variation in soil water potential, and consequently, root distribution and plant water uptake patterns. Drip irrigation systems generally consist of emitters that have discharge rates varying from 2 to 8 Lph. In semiarid summer climates crop water use is generally 6 to 8 mm/day, and water is supplied two or three times a week. Therefore the duration of water application is much shorter than the time over which plants take up water. Even if the water is supplied on a daily basis, a water application rate of 2 Lph delivers the consumptive needs of plants in a small fraction of the time over which plants photosynthesize and transpire. This means that even for water applications exactly equal to plant water needs, part of the water may not be used by the plant and would most likely drain below the root zone. Lowering emitter discharge rates to as close as possible to plant water uptake rates may reduce these percolation losses. Recently, low pressure drip systems have been developed that provide emitter discharge rates lower than 1.0 Lph (Battalwar, 2009).

The low pressure drip irrigation system (LPS) is a systematic development of low cost drip system which performs as drip except that the water is applied on the soil surface or below the soil surface through discrete emitters with low discharge rates (0.6 – 1.0 LPH) and narrow spacings. The low pressure capability of LPS (0.3 – 0.5 bars) provides an effective low energy and economical upgrade for flood/furrow irrigation. LPS is specifically designed to enable growers to use existing infrastructure like leveled fields, water sources and pumps, minimize front end investment, provide fast return on investment, reduce energy cost on pumping and pressurizing, move and reuse equipment easily, provide low system maintenance and management.

iii) Automation

There is an increasing trend to shift from manual to automatic operations. Irrigation can be almost totally automated by using micro irrigation systems. This requires linking the irrigation water lines to a remote control mechanism that uses small computers. Israel first introduced these computerized controls into its farm management systems in the early 1970s. In either small or large micro irrigation systems and greenhouses, the computer can control water flow, detect leaks, shut off faulty lines, adjust water application for wind speed, air temperature, and soil moisture content; filter backwash; apply fertilizer on schedule; monitor plant growth & water relations. An added advantage of the computerized systems is their ability to locate malfunctions and alert the operators to make necessary repairs before the faulty element causes too much damage or water loss. Accurate and precise application rates, increased yields (2 – 10%) savings in water

& energy (10 – 30%), reduced labour cost, minimized return flow, aquifer & stream pollution; improved pest and disease control are among the major advantages that are offered by adopting automation techniques in micro irrigation (Battalwar, 2009).

iv) Use of waste water through micro irrigation

Use of wastewater for irrigation of industrial crops, fruits, managed landscapes, golf & race courses has been actively practiced in other countries for many years, especially in Israel & USA. However, wastewaters often contain microbial and chemical contaminants that may affect public health and environmental integrity. Wastewater pre-treatment strategies and advanced irrigation systems may limit contaminant exposure to crops and humans. Subsurface drip irrigation (SDI) shows promise for safely delivering reclaimed wastewater. The closed system of SDI subsurface pipes and emitters minimizes the exposure of soil surfaces, above ground plant parts, and groundwater to reclaimed wastewater. Experiments in several countries revealed that subsurface drip irrigation prevented the virus movement onto crop leaf surfaces. However, beneficial and safe use of reclaimed wastewater for subsurface drip irrigation will depend on management strategies that focus on irrigation pre-treatment, virus monitoring & inactivation strategies, field and crop selection, appropriate drip design & component (emitter type & discharge rate; filter type) selection, assessment of water quality, monitoring and periodic leaching of salts.

v) Linking of major and medium irrigation projects with micro irrigation

Though in some foreign countries like Australia etc. this type of micro irrigation linkage with surface irrigation is reported yet in India it is only practiced in few states (though the scope and needs are very high). But, some attempts were made in Punjab to store canal water in pond and then its application through drip, Haryana to link it with sprinkler irrigation system. Recently the Government of AP has initiated a move linking a major irrigation project in the district of Nalgonda with sprinkler irrigation.

In Gujarat, the government is giving a very serious thought of bringing the whole Sardar Sarovar (Narmada Project) project of 18 lakhs hectares under drip and sprinkler (Raman, 2009). The water availability, soil irrigabilty and the proposed cropping pattern go very much in favour of this. A techno economic feasibility study on pilot scale has been initiated in this regard in the state. It has been given to understand that the Government of Tamil Nadu is also contemplating of introduction of MIS in surface irrigated systems.

Though the linking of major, medium and minor irrigation projects with MIS is technically feasible and profitable, yet, modalities for the sharing of the cost involved for the additional storage structures, pumping and networking have to be worked out. Further crop/ cropping pattern selection and diversification will be the major factors in deciding its economic viability (Srivastava, 2009). Presently water is applied once in every 7 to 15 days in surface / gravity irrigation depending on the soil and crop. Hence, moisture / water stress will be noticed just before irrigation and the growth of the crop is affected. Furthermore, it is difficult to constantly give the required quantity of water to the root zone using surface irrigation so yields are often less than optimum. Water productivity can be maximized by sifting from surface irrigation to drip irrigation (Table 7) (Sivanappan, 2009).

Table 7: Water productivity gains in percentage from shifting to drip irrigation from surface irrigation in India

Crop	Change in yield	Change in water use	Change in water productivity
Banana	+52	-45	+173
Cabbage	+2	-60	+150
Cotton	+27	-53	+169
Cotton	+25	-60	+255
Grapes	+23	-48	+134
Potato	+46	-0	+46
Sugar cane	+6	-60	+163
Sugar cane	+20	-30	+70
Sugar cane	+29	-47	+91
Sugar cane	+33	-65	+205
Sweet Potato	+39	-60	+243
Tomato	+5	-27	+49
Tomato	+50	-39	+14

Source: Adapted from data in Indian National Committee on Irrigation and Drainage, Drip Irrigation in India, Compiled by R.K.Sivanappan (New Delhi 1994).

Constraints in large scale adoption

In-spite of the technical and economic feasibility, its large scale and adoption is not so encouraging in India. The following factors responsible for slow application of drip were identified in an all India survey of farmers:

a) Low level of awareness (16 percent) and inadequate knowledge about its application, utility, method of operation and maintenance.

b) High initial cost, non-availability of basic finance.

c) Operating problems such as:

- Non availability of spares/micro-tubes
- High cost of component
- Leakage due to damage of pipes/drippers
- Blockage/clogging of drippers/micro-tubes
- Improper after sales service
- lack of technical knowhow like the quantity of water to different crops under different situations etc.

Forward Path for Promotion of Micro Irrigation in the Country

Viewing micro irrigation technology as an integral part of comprehensive water management strategy with the focus on enhancing water productive efficiency (more revenue per drop of water) and equity & equality must be the guiding principle promotion of micro irrigation on a sustainable basis. This requires a include holistic approach including capacity building & training of stakeholders, institutional support, standardization & quality control, credit support from international financial institutions like World Bank, ADB, JBIC etc, tax & fiscal incentives, strengthening of R & D for improved technological support, improving the availability of water soluble fertilizers at affordable prices etc.

Steps can be taken immediately for the promotion of micro irrigation system-

- Water to be treated as a national resource
- Integration of Macro & Micro Irrigation technologies
- Expenditure for laying Micro Irrigation Systems should be treated as an investment rather than an expense.
- Adequate Policy measures for financial assistance as well as availability of funds at soft interest rates.
- Reduction of system cost through cost economical designs
- Development of Package of Practices for various crops in different agro-climatic zones.
- To create technology awareness among various stake holders through audio, video aids & media.

References

Anonymous. 2001. Annual progress report, National Committee on Plasticulture Applications in Horticulture, Department of Ag. and Co-operation, Min. of Ag., GOI, New Delhi, p. 98.

Anonymous. 2004. National Committee on Plasticulture Applications in Horticulture, Review meeting of North Indian PFDC, October, 30. Department of Agricultural and Co-operation, Ministry of Ag, GOI, New Delhi.

Anonymous. 2010. National Committee on Plasticulture Applications in Horticulture, NCPAH, Ministry of Agriculture, GoI.

Battalwar, R.V. 2009. Advances made in the field of micro irrigation. Micro irrigation, Proceedings of winter school on Micro irrigation sponsored by ICAR., pp 19-26.

Raman, S. 2009. Linking of major and medium irrigation projects with micro irrigation. Micro irrigation, Proceedings of winter school on Micro irrigation sponsored by ICAR pp 99-103.

Singh, H.P. 2001. Emerging scenario of micro irrigation in India. Proceedings of International Conference on Micro and Sprinkler Irrigation Systems, 8-10 Feb. 2000. CBIP Publication. 282, pp 18-30.

Sivanappan, R.K., Padmakumari, O. and Kumar, V. 1987. Drip irrigation. Keerti Publishing House (P) Ltd., 278-281.

Sivannapan, R.K. 2009. Development of micro irrigation technology in India. Micro irrigation, ISBN 978-81-88708-48-2, pp 1-7. Proceedings of winter school on Micro irrigation sponsored by ICAR.

Srivastava, R.C. 2009. Canal supply cased pressurized irrigation in flow based minor irrigation. Micro irrigation, ISBN 978-81-88708-48-2, pp 90-98. Proceedings of winter school on Micro irrigation sponsored by ICAR.

Water Resources Development in India. 2010. Edited by C.D. Thatte, A.C. Gupta, M.L. Baweja, and Central Water Commission. (2010). Min. of Water Resources, GoI, 173 p.

Improved Agriculture and Allied Practices for Farmers, pp 223-236
New India Publishing Agency, New Delhi, India

20

Enhancing Water Holding Capacity and Organic Carbon Content of Soil Through Biogas Slurry

J.P.S. Dabas

Centre for Agricultural Technology Assessment and Transfer (CATAT)
ICAR- Indian Agricultural Research Institute, New Delhi-110012

As early as in the 18th century, the anaerobic process of decomposing organic waste was known. During 19th century, it became clear that anaerobic bacteria are involved in the decomposition process (DST, 1981). Since that time, the applications of anaerobic digestion have grown steadily, in both its microbiological and chemical aspects. The process of anaerobic digestion possesses a double advantage i.e. the process in addition to providing good quality manure also supplies a large volume of combustible gas. Anaerobic digestion provides some exciting possibilities and solutions to such concerns as alternative energy production, handling of human, animal, municipal and Agro-industrial wastes safely, controlling environmental pollution and expanding food supplies. The technology of anaerobic digestion (Kumar *et al.,* 2015) has not yet realized its full potential for manure and energy production even in developed and industrialized countries. In developing countries expansion of this technology has been hindered because of the need for better economic initiatives, organized supervision, proper training and education.

During twentieth century, the method of anaerobic fermentation or digestion was successfully applied by sewage chemists. A considerable amount of research work was carried out by sewage chemists on different aspects of the digestion process and sludge digestion plant, were installed. The gas produced was used for burning as domestic fuel, operating machinery and for tractors. In this context, the studies on anaerobic digestion of cattle dung and vegetable waste has been important. The work is particularly important ones in the Indian context as long as gas production and plant design efforts are concerned. For research on anaerobic digestion energy part (methane generation) received more attention since the early decades of this century when research into bio digestion had gradually begun to take shape. Fertilizer aspect received occasional attention a couple of decades later. The engineering and micro biological aspects of gas production remain a dominant preoccupation even up to 1990s. Emphasis on the energy aspect still dominates in many parts of the world today.

However, the environmentalism of the more recent years has somehow turned to be conducive for the appreciation of slurry manure as well as biogas technology as a whole.

By the end of 1950s, many researchers had already reported that manure production by anaerobic digestion was more efficient in nitrogen conservation than aerobic methods. Aanaerobicaly digested manure containing 1.7 per cent nitrogen in the dry matter which showed better effect on crops than farm yard manure. Biogas slurry helps to bind soil particles together into aggregates. This improves the physical properties of the soil making it easier for root to develop and penetrate. Tillage becomes easier and soil becomes well drained. The binding effect also reduces wind and water erosion. With organic matter the water retention capacity of soil is also considerably increased. Organic matter plays an important role because of its beneficial effects in supplying plant nutrients, enhancing the caution exchange capacity, improving soil aggregation, stabilizing its humus content and increasing its water holding capacity. Organic soil amendments support biological activities and also control root pathogens. Biogas slurry has proved to be a high quality organic manure compared to FYM. Digested slurry will have more nutrients, because in FYM, the nutrients are lost by volatilization (especially nitrogen) due to exposure to sun (heat) as well as by leaching. When fresh cow dung dries, approximately 30 to 50 per cent of the nitrogen escapes within 10 days. While nitrogen escaping from digested slurry within the same period amounts to only 10 to 15 per cent. Therefore, the value of slurry as fertilizer, if used directly in the field as it comes out of the plant, is higher than when it is used after being stored and dried. Cattle dung contains about one percent total nitrogen. Ammonia is less likely to leach away and hence more apt to become fixed to exchange particles like clay and humus. Experiments have shown that compared to fresh dung, the ammonical nitrogen in the digested slurry increases by 260 per cent whereas it decreases by 17.5 per cent in the FMY. The slurry thus has more free ammonia than available in composted manure. As a result of anaerobic fermentation, about 30 to 40 per cent of organic carbon present in the dung is decomposed as carbon dioxide and methane.

Animal waste causes environmental pollution when applied to land without appropriate controls and management (Balsari *et al.,* 2005), whereas agronomic utilization of biogas slurry represents the best solution for its reuse. It is obvious that biogas generation is technically feasible through the anaerobic digestion of a variety of organic materials. In developing countries animal and human fetal matter are dominant inputs for the production of biogas. During the anaerobic fermentation process 25 to 30 per cent of the organic matter from fecal matter is converted into biogas while the rest becomes available as residual manure. Such manure is generally considered to be rich in major plant nutrients as well as micro-nutrients such as zinc, iron, manganese and copper which are generally in short supply in many soils. The nutrient constitutions of biogas slurry vary from site to site. The slurry can with easily be brought to places that need organic fertilizers. The most important benefit is that the slurry is a very effective fertilizer that can improve the growth of the crops (Ahmad *et al.* 2009). The factors affecting this constitution range from the type of fodder being eaten by the livestock to the type of organic biomass fed to the digester.

India has huge number of livestock population near about 512.05 million heads in 2012 (Dhikshit *et al.,* 2010). Biogas production is practiced on a high proportion of farms, but there is a lack of experience to operate these systems efficiently (Groot and Bogdanski, 2013). Bio-digesters and the bio-digester market will tremendously improve living conditions in poor households thanks to their enabling independent production of clean energy and fertilizer. Biogas installation and use has many positive effects. The biogas production of 25 kg of dung per day replaces 5 kg of firewood, 1.5 kg of charcoal and 0.6 liters of kerosene per day. Throughout the world, a countless number of designs of biogas plant have been developed under specific climate and socio-economic conditions. The performance of a biogas plant is dependent on the local conditions in terms of climate, soil conditions, the substrate for digestion and building material availability. The design must respond to these conditions. In areas with generally low temperatures, insulations and heating devices may be important, if bedrocks occur frequently, the design must avoid deep excavation work. The amount and type of substrate to be digested have a bearing on size and design of the digester and the inlet and out let construction. The choice of design will also be based on the building materials which are available reliably and at reasonable cost.

Types of Biogas plants

Considering the ***feeding method,*** three different forms of biogas plants can be distinguished

Batch plants

These are filled and then emptied completely after a fixed retention time. Each design and each fermentation material is suitable for batch filling, but batch plants require high labor input. As a major disadvantage, their gas-output is not steady.

Continuous plants

These are fed and emptied continuously. They empty automatically through the overflow whenever new material is filled in. Therefore, the substrate must be fluid and homogeneous. Continuous plants are suitable for rural households as the necessary work fits well into the daily routine. Gas production is constant, and higher than in batch plants. Today, nearly all biogas plants are operating on a continuous mode.

Semi batch plants

If straw and dung are to be digested together, a biogas plant can be operated on a semibatch basis. The slowly digested straw-type material is fed in about twice a year as a batch load. The dung is added and removed regularly.

Considering the ***construction,*** two main types of simple biogas plants can be distinguished.

1. Fixed-dome plants
2. Floating drum plants.

Fixed-dome plants

The cost of a fixed-dome biogas plant is relatively low. It is simple as no moving parts exist. There are also no rusting steel parts and hence a long life of the plant (20 years or more) can be expected. The plant is constructed underground, protecting it from physical damage and saving space.While the underground digester is protected from low temperatures at night and during cold seasons, sunshine and warm seasons take longer to heat up the digester. No day/night fluctuations of temperature in the digester positively influence the bacteriological processes. The construction of fixed dome plants is labor-intensive, thus creating local employment. Fixed-dome plants are not easy to build. They should only be built where construction can be supervised by experienced biogas technicians. Otherwise plants may not be gas-tight (porosity and cracks).

Types of fixed-dome plants

Chinese fixed-dome plant is the arche type of all fixed dome plants. Several million have been constructed in China. The digester consists of a cylinder with round bottom and top.

Janata model was the first fixed-dome design in India, It is not constructed anymore. The mode of construction lead to cracks in the gasholder, - very few of these plants had been gas-tight.

Deenbandhu the successor of the Janata plant in India, with improved design, was more crack-proof and consumed less building material than the Janata plant, with a hemisphere digester.

Camartec model has a simplified structure of a hemispherical dome shell based on a rigid foundation ring only and a calculated joint of fraction, the so-called weak / strong ring. It was developed in the late 80s in Tanzania.

Influent collecting tank

Fresh substrate is usually gathered in an influent collecting tank prior to being fed into the digester. Depending on the type of system, the tank should hold one to two days' substrate. An influent collecting tank can also be used to homogenize the various substrates and to set up the required consistency, e.g. by adding water to dilute the mixture of vegetable solids (straw, grass, etc.), or by adding more solids in order to increase the biomass. The fibrous material is raked off the surface, if necessary, and any stones or sand settling at the bottom are cleaned out after the slurry is admitted to the digester. The desired degree of homogenization and solids content can be achieved with the aid of an agitator, pump or chopper. A rock or wooden plug can be used to close off the inlet pipe during the mixing process.

Inlet and outlet

The inlet (feed) and outlet (discharge) pipes lead straight into the digester at a steep angle. For liquid substrate, the pipe diameter should be 10-15 cm, while fibrous substrate requires a diameter of 20-30 cm. The inlet and the outlet pipe mostly consist of plastic or concrete. Both the inlet and the outlet pipe must be freely accessible and straight, so that a rod can be pushed through to eliminate obstructions and agitate the digester contents. The pipes should penetrate the digester wall at a point below the lowest slurry level (i.e. not through the gas storage).The points of penetration should be sealed and reinforced with mortar. The inlet pipe ends higher in the digester than the outlet pipe in order to promote more uniform flow of the substrate. In a fixed-dome plant, the inlet pipe defines the bottom line of the gas-holder, acting like a security valve to release over-pressure. In a floating-drum plant, the end of the outlet pipe determines the digester's (constant) slurry level. Inlet and outlet pipe must be placed in connection with brick-laying. It is not advisable to break holes into the spherical shell afterwards, this would weaken the masonry structure.

Digester- No matter which design is chosen, the digester (fermentation tank) must meet the following requirements:

- **Water/gas tightness** - water tightness in order to prevent seepage and the resultant threat to soil and groundwater quality; gas tightness in order to ensure proper containment of the entire biogas yield and to prevent air entering into the digester (which could result in the formation of an explosive mixture).
- **Insulation** - if and to which extent depends on the required process temperature, the local climate. Heat loss should be minimized if outside temperatures are low, warming up of the digester should be facilitated when outside temperatures are high in day time.
- **Minimum surface area** - keeps cost of construction to a minimum and reduces heat losses through the vessel walls. A spherical structure has the best ratio of volume and surface area. For practical construction, a hemispherical construction with a conical floor is close to the optimum.
- **Structural stability** - sufficient to withstand all static and dynamic loads, durable and resistant to corrosion.

Digesters can be made from any of the following materials:

- **Steel vessels-** Steel vessels are inherently gas-tight, have good tensile strength, and are relatively easy to construct (by welding). In many cases, a discarded steel vessel of appropriate shape and size can be salvaged for use as a biogas digester. Susceptibility to corrosion both outside (atmospheric humidity) and inside (aggressive media) can be a severe problem. As a rule, some type of anti-corrosive coating must be applied and checked at regular intervals.
- **Concrete vessels-** Concrete vessels have gained widespread acceptance in recent years. The requisite gas tightness necessitates careful construction and the use of gas-tight coatings, linings and/or seal strips in order to prevent gas leakage. Most common are stress cracks at the joints of the top and the sides. The prime advantage of concrete vessels is their practically unlimited useful life and their relatively inexpensive construction. This is especially true for large digesters in industrialized countries.
- **Masonary-** Masonary is the most frequent construction method for small scale digesters. Only well-burnt clay bricks, high quality, pre-cast concrete blocks or stone blocks should be used in the construction of digesters.

Cement-plastered/rendered masonry is a suitable and inexpensive approach for building an underground biogas digester, whereby a dome like shape is recommended. For domes larger than 20 m^3 digester volume, steel reinforcement is advisable. Masons who are to build masonry digesters have to undergo specific training and, initially, require close supervision.

Fixed-dome gasholders

A fixed-dome gas-holder can be either the upper part of a hemispherical digester (CAMARTEC design) or a conical top of a cylindrical digester (e.g. Chinese fixed dome plant). In a fixed-dome plant the gas collecting in the upper part of the dome displaces a corresponding volume of digested slurry. The following aspects must be considered with regard to design and operation:

- An overflow into and out of the compensation tank must be provided to avoid overfilling of the plant.
- The gas outlet must be located about 10 cm higher than the overflow level to avoid plugging up of the gas pipe.
- A gas pressure of 1 m WC or more can develop inside the gas space. Consequently, the plant must be covered sufficiently with soil to provide an adequate counter pressure.

Special care must be taken to properly close the man hole, which may require weighing down the lid with 100 kg or more. The safest method is to secure the lid with clamps.

Gas pipe, valves and accessories

Biogas piping

At least 60% of all non functional biogas units are attributable to defective gas piping. Utmost care has to be taken, therefore, for proper installation. For the sake of standardization, it is advisable to select a single size for all pipes, valves and accessories.

The requirements for biogas piping, valves and accessories are essentially the same as for other gas installations. However, biogas is 100% saturated with water vapor and contains hydrogen sulfide. Consequently, no piping, valves or accessories that contain any amounts of ferrous metals may be used for biogas piping, because they would be destroyed by corrosion within a short time. The gas lines may consist of standard galvanized steel pipes. Also suitable and inexpensive is plastic tubing made of rigid PVC or rigid PE. Flexible gas pipes laid in the open must be UV-resistant.

Steel pipes

Galvanized steel water supply pipes are used most frequently, because the entire piping system (gas pipe, valves and accessories) can be made of universally applicable U.S. Customary system components, i.e. with all dimensions in inches. Pipes with nominal dimensions of 1/2" or 3/4" are adequate for small-to-midsize plants of simple design and pipe lengths of less than 30 m. For larger plants, longer gas pipes or low system pressure, a detailed pressure-loss (pipe-sizing) calculation must be performed. When installing a gas pipe, special attention must be paid to gas-tight, friction-type joints line drainage, i.e. with a water trap at the lowest point of the sloping pipe in order to empty water accumulation protection against mechanical impact.

Cattle dung and manure

Cattle dung is the most suitable material for biogas plants because of the methane producing bacteria already contained in the stomach of ruminants. The specific gas production, however, is lower and the proportion of methane is around 65% because of prefermentation in the stomach. Its homogenous consistency is favorable for use in continuous plants as long as it is mixed with equal quantities of water. Fresh cattle dung is usually collected and carried to the system in buckets or baskets. Upon arrival it is hand-mixed with about an equal amount of water before being fed into the digester. Straw and leftover fodder or hay is removed by hand in order to prevent clogging and reduce scum formation. Since most simple cow-sheds have dirt floors, the urine is usually not collected. When it is, it usually runs along the manure gutter and into a pail standing in a recess at the end of the gutter. The pail is emptied into the mixing pit, thereby replacing some of the mixing water in preparation for charging the digester. Urine can considerably increase the gas production. A cemented stable floor, directly attached to the mixing pit, is the best solution to make optimum use of dung and urine and to save time for charging the digester. Liquid cattle manure, a mixture of dung and urine, requires no extra water. However, the simple animal housing found on most farms in developing countries normally does not allow the collection of all animal excreta. Hence, most of the urine with its valuable plant nutrients is lost. Factors that influence nitrogen availability from slurry are its inorganic N content, digestion process (aerobic or anaerobic), C:N ratio, pH, the method and timing of application, and soil type and properties (Warman and Termeer, 2005).

Operation and Use of Biogas plant

The day-to-day operation of a biogas unit requires a high level of discipline and routine to maintain a high gas production and to ensure a long life-span of the biogas unit. Many problems in the performance of biogas plants occur due to user mistakes or operational neglect. Often, these problems can be reduced by

- Less complicated designs that are adapted to the substrate, the climatic conditions and the technical competence of the user,
- High-quality and user-friendly appliances,
- Design and lay-out of the biogas for convenient routine work,
- Proper training and easy access to advice on operation problems.

During design selection, planning, construction, handing over and follow-up, the biogas extension program should emphasize further on a reduction of the users' workload for operating the biogas unit and using the gas and the slurry. In particular during work peaks for farm work, it is important that the biogas unit relieves the user from work rather than adding to the workload. As a general rule, the farming family should have less work with a biogas unit than without it, while enjoying the additional benefits in terms of a clean fuel and high quality fertilizer.

Feeding of the digester

In larger biogas units, the dung, urine and other substrate usually enter the plant by pipes, channels, belts or pumps. The available substrate has to enter the digester as soon as it is available to avoid pre-digestion outside the digester. The functioning of the feeding mechanisms has to be checked daily. Separators for unsuitable material have to be checked and emptied. The amounts of substrate fed into the digester may be recorded to monitor the performance of the biogas plant. Smaller plants in developing countries are fed by hand. The substrates often dung and urine, should be thoroughly mixed, plant residues should be chopped, if necessary. Obstructive materials like stones and sand should be removed from the mixing chamber. Simple tools like a rubber squeegee, a dipper, forks to fish out fibrous material, proper buckets and shovels greatly facilitate this work. Filling work is further made easier by smooth concrete stable-floors and a minimized distance between the stable and the plant.

Controlling the overflow

A special problem of small scale fixed dome plants is the clogging up of the overflow point. This can lead to over-pressure (the hydraulic pressure increases with the slurry level in the expansion chamber) and to clogging of the gas outlet if too much slurry flows back into the digester. The overflow point should, therefore, be checked and cleaned daily.

Biogas - Sludge Management

Sludge storage

This to retain the maximum fertilizing quality of digested slurry, i.e. it's nitrogen content, it should be stored only briefly in liquid form in a closed pit or tank and then applied on the fields. Preferably, it should be dug into the soil to prevent losses on the field. Sludge storage is normally affected according to one or the other of the following three techniques

Liquid storage

The effluent outlet of the biogas system leads directly to a collecting tank. Loss of liquid due to evaporation or seepage must be avoided. Just before the sludge is needed, the contents of the tank is thoroughly agitated and then filled into a liquid manure spreader or, if it is liquid and homogenous enough, spread by irrigation sprinklers. The main advantage of liquid storage is that little nitrogen is lost. On the other hand, liquid storage requires a large, waterproof storage facility entailing a high initial capital investment.

The practice of spreading liquid slurry also presents problems in that not only storage tanks are needed, but transport vessels as well. The amount of work involved depends on the distance over which the slurry has to be transported. For example, loading and transporting one ton of slurry over a distance of 500 m in an oxcart (200 kg per trip) takes about five hours. Distributing one ton of slurry on the fields requires another three hours.

Drying

It is only possible to dry digested sludge as long as the rate of evaporation is substantially higher than the rate of precipitation. The main advantage of drying is the resultant reduction in volume and weight. Drying can also make the manual spreading easier. The cost of constructing shallow earthen drying basins is modest. On the other hand, drying results in a near-total loss of inorganic nitrogen (up to 90%) and heavy losses of the total nitrogen content (approx. 50%)

Composting Nitrogen

In this losses can be reduced by mixing the digested sludge with organic material. As an additive to crop residues for composting, biogas sludge provides a good source of nitrogen for speeding up the process. At the same time it enriches the compost in nitrogen, phosphorus and other plant nutrients. Furthermore, the aerobic composting process, by it's temperature, effectively destroys pathogens and parasites that have survived the anaerobic digestion treatment. The ready-made compost is moist, compact and can be spread out by simple tools. With most available transport facilities in developing countries, it is easier to transport than liquid manure.

Factors affecting efficiency of Biogas Plant

1. Climatic temperature (optimum temperature): 35°C
2. Carbon Nitrogen (C/N) ratio of the matter to be digested, optimum value- 25:1 to 30:1
3. pH value of the matter to be digested (7-8)
4. Dilution and consistency of Input

   For optimum anaerobic fermentation, normally 8-10 per cent total solids in feed is required which can be achieved by making slurry of fresh cattle dung in water in ratio 1:1.
5. Loading rate

   Amount raw material fed per day per unit of digester capacity. A daily loading of 16 kg of volatile solids per m^3 of digester capacity produces 0.04-0.074 m^3 of gas per kg of cattle dung fed.
6. Hydraulic Retention time

   For digesters operating in countries of tropical region like india, HRT is usually taken as 40-60 days.
7. Additive supplements

   Some mineral ions and presence of NH_4 from 50 to 200 mg / liter stimulates the growth of anaerobic microbes where as higher concentration of these ions, alkali detergents, antibiotics and organic solvents inhibit the activity of methane producing bacteria.
8. Agitation or mixing of digester contents significantly improves the fermentation.

Security precautions

When operating a biogas plant special attention has to be paid to the following precautions:

- Breathing in biogas in a high concentration and over longer periods of time can cause poisoning and death from suffocation. The hydrogen sulfide contents of biogas are highly poisonous. Unpurified biogas has the typical smell of rotten eggs. Purified biogas is odorless and neutral. Therefore, all areas with biogas operating appliances should be well ventilated. Gas pipes and fittings should be checked regularly for their gas-tightness and be protected from damage. Gas appliances should always be under supervision during operation. Everybody dealing with biogas, in particular children, should be instructed well and made aware of the potential dangers of biogas.

- After emptying biogas plants for repair, they have to be sufficiently ventilated before being entered. Here the danger of fire and explosion is very big (gas/air mixture). The so-called chicken test (a chicken in a basket enters the plant before the person) guarantees sufficient ventilation.
- Biogas in form of a gas-air mixture with a share of 5 to 12 % biogas and a source of ignition of 600°C or more can easily explode. Danger of fire is given if the gas-air mixture contains more than 12 % of biogas. Smoking and open fire must therefore be prohibited in and around the biogas plant.
- The initial filling of a biogas plant poses a particular danger, when biogas mixes with large empty air-spaces. A farmer may want to check with an open flame how full the plant is already and cause an explosion.
- The digester of a biogas plant and the slurry storage facilities should be built in such a way that neither persons nor animals are in danger of falling into them.
- Moved and movable parts should have a protective casing to avoid catching persons or animals.
- Appliances operating on biogas normally have high surface temperatures. The danger of burning is high, in particular for children and strangers. A casing of nonheat- conducting material is advisable.
- The mantle of the gas lamp is radioactive. The mantle has to be changed with utmost caution. Especially the inhalation of crumbling particles must be avoided. Hands should be washed immediately afterwards.
- The piping system can form traps on the farm compound. As much as possible, pipes should be laid some 30 cm underground. Pits for water traps, gas meters, main valves or test-units should be caused by a concrete frame and covered with a heavy concrete lid.

Scope of extension

There seems to have built a general consensus on the ability of slurry to improve the physical and biological quality of soil besides providing both macro and micro-nutrients to crops. These improvements in physical and biological qualities include: improvements in soil structure , improvements in water holding capacity, cation exchange capacity, lesser soil erosion and provision of nutrients to soil micro-flora including nitrogen fixing and phosphorous solublizing organism. In addition bioslurry is free of weed seeds. Anaerobic digestion kills more weed seeds than any other manure processing system (Banik and Nandi, 2004). On the other hand, FYM if left to itself , (open pool manure) looses nutrient most importantly nitrogen and this possess relatively more weed seeds. From the points of view of nutrient recycling, sustained use of agricultural and animal wastes as fuel is not a desirable practice. Biogas technology with its twin benefits of fuel and fertilizer can be of great use in enhancing organic farming, natural farming and overall agricultural productivity. Therefore following extension education message could be given to the farmers for promotion of Biogas technology.

- Use of fuel-wood, agriculture and animal waste for fuel will gradually deplete soil nutrients and thus contributing to the emergence of unsustainable agricultural systems. Although FYM can be directly returned to the soil or composted to complete the nutrient cycle, its utilization in biogas generation has two pronged advantages i.e. energy and better manure.
- Storage handing, treatment and application procedures of all form of organic manures are very important for nutrients conservation and increased crop yield. In no way manures in all these forms should be recklessly exposed to the vagaries of nature as a result of which they lose nutrients.
- Biogas digester does not use up plant nutrients.
- As there may not be congruence between the time of availability of fresh slurry and the time of field application, liquid slurry should be properly stored or composted near the biogas plant.
- No form of biogas slurry can be profitably left spread on the field. It should be ploughed immediately or if the manure is used as top-dresser, it should be covered by soil immediately.
- Biogas plants with toilet attachments can significantly improve health sanitation of community and members.

Conclusion

Bioslurry is used to improve soil fertility, soil structure and crop productivity. It can also be an interesting link between the biogas programs aimed at substituting biomass or fossil fuels for cooking and sustainable agriculture promotion. Bioslurry has so many advantages that it can be referred to as 'bio gold'. Biogas produced from cattle, pig, and buffalo dung (and other excrement, e.g. human), together with the by-product bioslurry, can be a solution to poverty, climate change, poor access to modern energy services, and soil fertility problems. Biogas is a combustible gas produced by anaerobic fermentation of organic materials by the action of methanogenic bacteria.

References

Ahmad, R and Jabeen, N (2009), Demonstration of growth improvement in sunflower (helianthus annuus l.) by the use of organic fertilizers under saline conditions. *Pak. J. Bot.* 41(3), 1373-1384.

Balsari P, Airoldi G, Gioelli F (2005), Improved recycling of livestock slurries on maize by means of a modular tanker and spreader. *Biores Technol* 96: 229-234.

Banik, S. and Nandi, R. (2004), Effect of supplementation of rice straw with biogas residual slurry manure on the yield, protein and mineral contents of oyster mushroom. In: *Industrial Crops and Products* 20 : 311–319.

Department of Science and Technology, Govt. of India, (1981), Biogas Technology and Utilization: A Status Report. New Delhi: DST.

Dikshit, A. K., Pratap, S. (2010), Environmental value of dung in mixed crop-livestock systems. *Ind. J. Animal Sc.* 80 (7), 679–682.

Groot, L. and A. Bogdanski. (2013), Bioslurry = Brown Gold? A review of scientific literature on the co-product of biogas production. Environment and Natural Resources Series. FAO, Rome, Italy. 32p.

Sandeep Kumar, Lal Chand Malav, Mahesh Kumar Malav and Shakeel A Khan (2015), Biogas Slurry: Source of Nutrients for Eco-friendly Agriculture. *International J Ext Res*. 2:42-46.

Warman PR, Termeer WC (2005), Evaluation of sewage sludge, septic waste and sludge compost applications to corn and forage: yields and N, P and K content of crops and soils. *Biores Technol* 96: 955-961.

Section 5: Post Harvest and Value Addition

Improved Agriculture and Allied Practices for Farmers, pp 237-241
New India Publishing Agency, New Delhi, India

21

Grains and Seeds Protection in Storage

Chitra Srivastava

Division of Entomology, ICAR-Indian Agricultural Research Institute New Delhi-110012

Cereals and pulses are the staple food in India, thus it has to be stored by producers in their homes and by traders and government agencies. Variations in the season i.e., warm and high humidity weather is highly conducive to proper development and rapid multiplication of storage insect pests. Almost all insect pests of stored grains have a remarkably high rate of multiplication and within one season, they may destroy 9-10% of the grains and contaminate the rest with undesirable odors and flavors. The type of insects infesting grains vary and include weevils, moths and beetles. Apart from this it is necessary to control or kill egg, larval, pupal and adult stages of the insect. After proper cleaning and drying the produce is stored where damage can be caused due to cross-infestation or infestation already left in the produce. Grains are stored in different types of receptacles and to protect it, different protectants are being used since from the early days. The success in the use of these protectants (organic, inorganic or inert) depends upon the effect on the grain, killing of the insects, practicability, period of storage and lastly the physical and chemical state of treated seeds.

1. Inert dusts

Inert dusts are fine powders chemically inactive but cause considerable insecticidal activity by mechanical abrassing and physical absorption of moisture proofing layers of insect integument. The use of inert dust was initiated sometimes during World War II. Some common inert dusts are ashes, clays, gypsum, lime, magnesite, Nakki, silica, talk, phosphates, diatomaceous earth. It was reported that though most of the dust proved effective and convenient to use but it is difficult to remove it and clean as it goes inside the infested and damaged grains. The dust can only kill larvae and adults moving in between the grains. Once any stage of insects entered inside the grain, it is beyond the reach of the dust.

2. Botanical insecticides

Plants are known to produce a diverse range of secondary metabolites such as terpenoids, alkaloids, polyacetylenes, flavanoids, amino acids, sugars etc. More than 2000 plant species have been reported to possess pest control properties. Pyrethrum is toxic to most insects and act as contact insecticides with quick knock down effect. It is relatively harmless to mammals and can be used safely on edible material. Synergized pyrethrums have been extensively employed earlier to check rice weevil infestation. Formulated as dust, spray. Leaves of Neem (*Azadirachta indica*), its seed powder and oils were used to get protection from storage insects since early days. Azadirachtin, the most important biologically active compound has shown repellent, antifeedant and insecticidal activity. China berry, Bakain (*Melia azedarach*) is a close relative of neem. Pongamia (*Pongamia pinnata*), commonly known as karanj, also possesses various types of bioactivity including antifeedant, oviposition deterrent and toxic effects. The active component has been identified as karanjin, a fluroflavone. Karanj oil as a surface protectant, effectively checked infestation by storage pests like *Callosobruchus maculatus, C. chinensis, Rhyzopertha dominica, and Sitotroga cerealella.* Apart from this, oil treatment is also reported to be very effective to control bruchid infestation in pulses. Now a day's fumigation with phyto-oils is also giving good results to control storage insects.

3. Synthetic insecticides

Malathion was introduced in 1950 by the American Cyanamid Company and was widely used against storage insects. It is powerful, versatile and safe insecticide. It has high toxicity to insects and low toxicity to mammals. Malathion leaves no hazardous chemical residue, off flavors or colors on treated commodities nor it has been found to affect the milling and baking quality. It is effective to all major and minor stored grain insects. It is sold under the name of Cythion and available in various formulations-50% EC, Dust etc. Apart from malathion, some other organophosphates are also very popular. These are Fenitrothion, pirimiphos-methyl, chlorpyriphos- methyl and dichlorvos is widely used to control storage insects. It has contact as well as fumigant action. Its speedy knock down effect, high vapour activity and rapid dissipation of residues from surface make it a suitable insecticide for use. It is toxic to almost all stored grain insects. Presently along with malathion, DDVP is in much use for storage insects.

After this synthetic pyrethroids were emerged with good potential for storage insects. Mainly deltamethrin became much popular. It has proved to be effective even against insects resistant to conventional insecticides. It is marketed as 2.8 EC (Decis) and 2.5 WP (K-obl) by Aventis. Cypermethrin also has given good results for control of storage insects. Recently, cyfluthrin and bifenthrin are also promoted as insecticides to control storage insects.

Management of storage pests

The effective management of storage pests may be ensured by drying the grains properly before storage, storing new grains in the clean go downs or receptacles and plugging all cracks, crevices and holes in the godowns thoroughly. If infestation of grain has already taken place or is imminent on account of a peculiar situation in which the grain is stored, then application of chemicals becomes necessary.

a) Surface treatment

The grains are kept in jute, markin and polypropylene bags. The grains can be protected by spraying insecticides on the surface of the stacks. The go downs are disinfected before storing the grain and new infestation can be prevented by surface treatment. Old gunny bags can be disinfected by dipping them in insecticides and then get it dried. But this is not popular due to residue and resistance problems. Insecticidal spray is recommended on floors, walls, containers and storage facilities.

i) Dose for organophosphates as per malathion—150mg/m^2

ii) Dose for synthetic pyrethroids as per deltamethrin—30mg/m^2

b) Seed treatment

Mixing of insecticides is another method to control storage pests. This method is very popular in other countries. Mixing of insecticides is not permissible in our country due to residue and resistance problem. But insecticides can be mixed in the seeds only. Grain meant for seeds can be protected by mixing malathion 5% dust or 50% EC. Treated seeds must be kept in cool dry places. Also the moisture content of seed should not be above 10%. Otherwise insecticides impair viability and vigor of the treated seeds. In no case insecticide treated seeds be disposed of as food/feed. Deltamethrin 2.4 WP can also be mixed with the seeds.

i) Mixing of organophosphates @ 20mg/kg seeds

ii) Mixing of synthetic pyrethroids @ 1mg/kg seeds

To store the grains/seeds for household purposes storage receptacles are very important. Generally grains are stored in jute bags, plastic bags, polypropylene bags and these bags can be treated with any of the protectants at recommended dose. If the grains are kept in rooms or containers or bags treatment with insecticides is necessary for prophylactic treatment and to check the cross infestation.

For stored grain pests fumigants are the most widely used insect disinfectant. Fumigants are non-persistent and are used to kill spot insect infestation. Due to their penetration ability they kill all stages of insects hidden in the grain or cracks and crevices. The treatment of stored grain and other durable commodities throughout the world continues to rely heavily on fumigation. Mainly two fumigants, methyl bromide and phosphine are being widely used and both are highly toxic to man and strict conditions pertain to their use. Phosphine is usually more preferred owing to its ready availability and ease of application. In fumigation, structure should be properly sealed for complete effect, which is not practical always. Fumigation can control the existing infestation but cannot restrict the cross infestation of insects. Fumigation with periodical prophylactic treatment with insecticides is very useful. Fumigation with methyl bromide is technically more difficult but is achievable in 24 hours

i) Methyl bromide	2kg/100cu.meter space
ii) Aluminium phosphide tablets (3gm)	140/100cu.meter space 3 tablets/tone

c) Hermetic control

For proper development of different insects a particular humidity, certain range of temperature and percentage of oxygen is required. Thus the safety of stored grain from damage caused by insect pests depends on proper management of moisture content of the grain, availability of oxygen and development of temperature gradient. If the moisture content is reduced below 8% , damage due to most of the insects except khapra beetle can be avoided. Only khapra beetle can sustain itself if moisture content is 2%. Initial moisture content before storage can be brought down by sun drying. Each stage of different storage insect require a particular concentration of oxygen for proper survival. If O_2 level is reduced below optimum requirements it is difficult for the insects to survive.

It is an ancient, simple, cheap and effective method of insect management. The natural respiration of insects and grains results in depletion of oxygen level and increase in carbon dioxide. Complete air tightness of storage structure is very important. In structures based on airtight principles serve the purpose of insect control. In hermetic control metabolic activities of insects and microflora act as bio-generators that alter the oxygen and carbon dioxide composition of inter-granular atmosphere so that insects development is arrested. It must be stressed that to kill all insects by hermetic storage takes place a longer time.

d) Modified atmosphere

Disinfestation of stored grain using modified atmosphere involves the alteration of inter-granular storage gases (O_2, N_2, CO_2). Modified atmosphere may be achieved in several ways such as adding CO_2 or N_2 or reducing O_2 and leave no air (vacuum). CO_2 is more effective method than use of N_2 because CO_2 stimulates insect respiration by displacing O_2. Nitrogen kills insect only by displacing oxygen to level of $<1\%$in the air. Modified atmosphere is lethal to insect and can be obtained using nitrogen provided a hypoxic atmosphere of 1% or less is maintained whereas flushing with CO_2, insects are under stress of both due to hyper-carbia and hypoxia. They may provide alterative to phosphine but not to methyl bromide due to comparatively longer exposure. However, quick action can be achieved with the application of CO_2 under reduced pressure or hypoxia. The efficacy of CO_2 increases with temperature and decreases with increase in humidity.

e) Pusa bin

In pusa bin polyethelene sheet save the grain and mud walls of the structure in which polethelene sheet is embedded keeps keep the grain safe from outer temperature and moisture. Material required in this structure is of low cost. It is modern , well developed and low cost storage technology. It is moisture and temperature proof. Without application of insecticides and fumigants. It is rat proof. Seed vigour and seed germination and quality of grain is safe and retained.

Thus for proper storage the grain or seeds should be clean, without insects and dried properly befire feeping in the receptacles. As far as possible the containers should be air tight. Prophylactic treatment should be given as per requirements.

Improved Agriculture and Allied Practices for Farmers, pp 243-246
New India Publishing Agency, New Delhi, India

22

Post Harvest Technique in Fruits and Vegetables for Small and Marginal Farmers

V. R. Sagar

Division of Food Science and Post Harvest Technology, ICAR-Indian Agricultural Research Institute, New Delhi-110012

India produce a wide range of tropical, subtropical and temperate horticultural crops at different places by virtue of its varietals agro- climate conditions. The total annual production of fruits and vegetables is around 135.5 million tones, fruits being 48 million tones and vegetables around 87.5 million tones. India is the second largest producer of fruits and vegetable in the world. They are major sources of important valuable vitamins and minerals. Being highly perishable, fruits and vegetables can be preserved and processed either at farm level or at cottage scale/home scale during market glut season or extra seasonal produce from kitchen garden for later use. Farmers and entrepreneurs can set up their own processing units by adopting simple, low cost innovative processing technologies which will save post harvest losses of these perishable crops and also give good returns to grower and better earning to processor/ entrepreneurs.

Aim of post harvest technology

1. Preservation industry is backbone of the horticulture industry as it takes care of gluts and all possible wastes.
2. To save post harvest losses which are 20-30 percent?
3. To prevent or delay microbial decomposition and self decomposition
4. To make the preserved products more palatable
5. To process into various products having innovative and pragmatic approach for value addition and diversification by making health foods and by utilizing under-exploited raw material.

6. To increase their storage and shelf life.
7. To check contamination.
8. To boost processing industry which is at present 2 percent of total produce,
9. To provide better returns to growers and generate employment.
10. To provide export.

Importance of Processed of Seasonal Fruits and Vegetables.

- Fruits and vegetables play an important role in the health and economy of our country
- They are major source of important vitamins and minerals
- They are protective foods and provide nutrition to our balanced diet
- As they are highly perishable they can preserved for off –season use.

Why to preserve fruits and vegetables

i) To check spoilage caused by various means

Physical changes

Colour

Favour

Texture

Taste

Chemical changes

Carbohydrates

Protein

Fat

Vitamins

Minerals

Microbial changes

Yeast

Moulds

Bacteria

Enzymatic changes

(By inherent enzymes)

Other (rodents, Insects etc.)

ii) To save nutritional losses (vital vitamins and Minerals)

iii) To create employment opportunities and utilize market or kitchen garden surplus in glut season

iv) To process into various products and increase their self life

v) To make them more palatable and acceptable

vi) To store them for off season use.

Modern methods of fruits and vegetable preservation

1. **Physical Method:**	
a) Thermal processing	Pasteurization, Sterilization, processing and packaging
b) Storage at low temperatures	Refrigeration, Freezing
c) Removal of water	Sun- drying, Dehydration, Concentration, Osmotic dehydration
d) Irradiation	Application of UV or ionizing radiation
e) Other means	Carbonation, High pressure etc.
2. **Chemical method**	
• Addition of acid	Pickled vegetables
• Salting or brining	Vegetable/ fruit pickles, lactic acid fermentation and its products.
• Addition of sugar and acid	Fruit preserves, jams, jellies, Marmalades, Candies, etc.
• Addition of chemical	Sodium benzoate, (benzoic acid), Potassium preservatives meta-bisulphite (sulphur dioxide)
3. **Fermentation**	Alcoholic and acetic fermentation –fruit wines, apple cider, vinegar, etc.A judicious combination of one or more than one methods mentioned above for synergistic preservation.
4. **Other method**	Preservation by Carbonations (Carbonated beverages) Preservation by use of high pressure (UHT method for fruit juice preservation)

Advantages of using Appropriate low cost technologies

- Easy to follow and simple to operate
- Economical
- Could cater to indigenous need
- Save losses of fruits and vegetables as well as their valuable nutrients
- Utilize market surplus/garden produce
- Promote horticulture industry
- Boost to the preservation industry
- Benefit the consumers for off-season use
- Provide better returns to growers
- Reasonable price to consumers

Appropriate Methods

Three methods include low cost technologies for preserving fruits and vegetables for checking enzymatic, oxidative and microbial spoilage in the preserved material

- Chemical preservation of fruits and vegetables
  - a) Steeping preservation in a chemical solution
  - b) Preservation by chemical or food additives
    - i) Whole tomato concentrate
    - ii) Fruit pulps/ juices / beverages
- Preservation of vegetable by lactic fermentation
  - a) Individual or mixed vegetable fermentation
  - b) Vegetable juice (black carrot) or kanji
- Pickles/ chutneys/ sauces
- Sun drying/ dehydration
- Preservation by using high concentration of sugar

Improved Agriculture and Allied Practices for Farmers, pp 247-256
New India Publishing Agency, New Delhi, India

23

Importance and Scope of Value Addition of Agricultural Produce

Nishi Sharma and Pratibha Joshi

Centre for Agricultural Technology Assessment and Transfer, ICAR-Indian Agricultural Research Institute, New Delhi-110012

Agriculture shapes the foundation of Indian economy and despite the fact that there has been vast industrialization in most 60 years; horticulture still occupies a place of significance. In India, approximately 20-25% of the horticulture produce goes waste due to improper post harvest operations which reduces the growers share in the present marketing arrangements. The grower is getting hardly 25-35 paisa of a consumer rupee. The Value addition activities where the fruits and vegetables are processed, that fetches higher remunerative prices of the produce. Such opportunities are presently not available in rural areas but could be created through selective mechanization of agriculture and appropriate post harvest management and value addition to the harvested biomass in the production catchments.The agricultural produce involves operations like cleaning, grading, drying, storage, milling, packaging, transport, marketing and utilization. At the end of each operation, value is added to the product. The lowest and the highest monetary values of a food commodity are, respectively, when it is in raw and fresh form and when it is in processed and ready to consume/eat form. Post harvest and food processing technology are commodity and location specific. It prevent avoidable post harvest losses and adding value to the fresh agro-produces. It also creates opportunities for employment and income generation. Integration of production agriculture with on-farm primary processing is needed to have higher and sustainable production, productivity and better quality end products for domestic and export markets. It would help in reducing rural- urban disparity and ensuring household food and nutritional security for all at an affordable cost. The technology is available and a wide range of tools, machines and equipment are used in post-production of agriculture. The globalization of trade may compel many agro processing industries to rely on an imported technology at a high cost and the advantage of such technology may be availed by a few who have an exclusive business interest. Majority of agriculture-based enterprises will continue to depend upon indigenous technology and, therefore, R&D through public support has to be strengthened to become globally competitive and serve small-scale food processing sector of the country.

Objectives of postharvest technology to harvested fruits and vegetables

The three main objectives of applying postharvest technology to harvested fruits and vegetables are:

i) To maintain quality (appearance, texture, flavour and nutritive value)

ii) To protect food safety, and ;

iii) To reduce losses (both physical and in market value) between harvest and consumption.

Effective management during the postharvest period is the key in reaching the desired objectives. The simple, low cost technologies often can be more appropriate for small volume, limited resource commercial operations, farmers involved in direct marketing. Local conditions for small-scale handlers may include labour surpluses, lack of credit for investments in postharvest technology, unreliable electric power supply, lack of transport options, storage facilities and/or packaging materials, as well as a host of other constraints. Fortunately, there is a wide range of simple postharvest technologies from which to choose, and many practices have the potential of meeting the special needs of small-scale food handlers and marketers. Many simple practices have successfully been used to reduce losses and maintain produce quality of horticultural crops in various parts of the world for many years.

There are many interacting steps involved in any postharvest system. While particular practices and the sequence of operations will vary for each crop, there is a general series of steps in postharvest handling systems that are often followed.

- Harvesting and preparation for market
- Curing root, tuber and bulb crops
- Packinghouse operations
- Packing and packaging materials
- Decay and insect control
- Temperature and relative humidity control

- Storage of horticultural crops
- Transportation of horticultural crops
- Handling at destination

IARI Low Cost Storage Technologies for Preservation of Horticultural Produce and Foodgrains

Owing to improper post-harvest operations, especially storage, over 10% of food grains and 30-40% of the fruits and vegetables produced in the country are lost resulting in poor returns to farmers and high cost to consumers. While dry food grains can be stored in bulk without significant damage for relatively longer period of time, storage of fresh horticultural produce, being highly perishable, particularly under the tropical climate in India, is a major challenge needing immediate attention.The spoilage of these commodities can be controlled to a large extent by maintaining proper storage conditions. In this direction, the Indian Agricultural Research Institute has been working to provide indigenous low cost storage technologies for the preservation of different horticultural produce and food grains.

a) Pusa Zero Energy Cool Chamber

The spoilage of fresh fruits and vegetables can be controlled by reducing the storage temperature. Pusa Zero Energy Cool Chamber (Pusa ZECC) can be constructed easily anywhere with locally available materials like bricks, sand, bamboo, *khaskhas* and straw, gunny bags, etc. with a source of water. The Chamber can keep the temperature 10-15°C cooler than the outside temperature and maintain about 90% relative humidity. Small and marginal farmers can easily construct Pusa Zero Energy Cool Chamber (ZECC) near their fields and store a few day's harvest before dispatching it to the market.

Some specific advantage of Pusa ZECC are:

- Can be constructed by an unskilled person
- No mechanical or electrical energy needed
- Allows small farmers to store produce for a few days so that they are not forced to sell at low prices
- Reduces losses and pays for itself in a short time.
- Useful for temporary storage of curd, milk and cooked food.
- Can also be used for mushroom cultivation, sericulture, storage of bio-fertilizers, hardening of tissue-cultured plants, etc.

b) Solar Powered Cooling System for Potato Storage

For the safe storage of potato in the areas where electric supply is erratic or is not available, a Solar Powered Cooling System was developed and evaluated for its performance at IARI.The Solar Powered Cooling System consists of photo-voltaic (PV) solar generator, two batteries, an inverter, a vapour compression refrigeration system and an insulated storage structure. The vapour compression refrigeration system is attached to the insulated storage structure of about 2.5 cubic meter volumetric capacities.This system was evaluated by storing potato safely for about 5 months.The attainable temperature inside the cold store ranged from 8 to 10°C, while the averages of relative humidity were 81% and 86% without storage and with storage of potato, respectively. These are acceptable limits for storage of most of the horticultural produce.The technology has a great scope of adoption in areas with scarce power supply.

System	:	SPV Panel, 14 module @ 35 Wp
Storage batteries	:	2 lead acid dbatteries of 2.26 KWh capacity each
Inverter	:	1250 VA (to convert DC to AC)
System configuration	:	1/6 Hp compressor
		Condenser 0.337m^2
		Evaporator 0-223m^2
Cooling chamber	:	2.5 m^3 volumetric capacity

c) Pusa Concentric Onion Storage Structure

Considerable quantity as well as quality of onion is lost during post harvest period owing to lack of proper onion storage structures in rural area. The farmers not only get less price for their produce but also the environment is polluted because of spoilage of onion.

A concentric onion storage structure (5 tier) of 250 kg (5x50) capacity, made of bamboo and wooden planks, was developed by the Institute. The size of structure can vary depending upon the requirements of the consumers, Pusa Concentric Onion Storage Structure can be used by small and marginal farmers and the consumers in rural as well as urban areas.

Concentric Onion Storage Structure (5 tier) comprises a concentric cylinder made of bamboos, a wooden base, an inlet and an outlet in each tier, four supporting legs and a top cover. The inner and outer walls are made of 25mm diameter of bamboos. The base of each tier is made of 740mm x740mm perforated wooden planks.The whole structure is supported on four 50mm diameter bamboos one at each corner. The height of bottom base above the found level as well as the height of each tier is 450 mm.Each tier of the structure has one inlet at the top for filling and one outlet at the bottom for emptying. The top tier has a folding cover made of bamboo solits. The cost of storage of onion in this structure has been estimated as Rs.2.21/kg.

Some specific advantage Pusa concentric onion storage structure

- Easy to construct and occupies less space in the house or shed
- Peel and pulp of onion is not damaged during loading and unloading
- Onion is not damaged because of compressive load
- A rural artisan can easily construct the structure
- There is reduction in the quantitative and qualitative loss of onion
- Fungal infection and spoilage of onion is minimized
- The environment pollution is minimized

d) Solar-powered evaporatively cooled vegetable vending cart

A solar-powered vending cart was designed and developed for storage of fruits and vegetables. Provision has been made for making the lower section evaporatively cooled so that fruits and vegetables can be kept safely for a longer period of time at the retail vendor's level. Provision has been made for cooling pads and channels for application and distribution of water around the additional 8.0 cubic feet storage space for vegetables and fruits. Also, on top of the platform, provision has been made for water dispersal through cotton wicks and two stationary funnel-bucket assemblies. The evaporative cooling system of the cart uses solar power. A SPV 100 Wp/12 V solar module is used as the power source. It is used to run two forced air fans (DC, 12 V, 0.7 A, 8.4 W each) for the evaporative cooling system and LED bulbs to illuminate the cart during night. A 12 V/7 AH battery is also used to store solar power during the day and to run the system during night or on a cloudy day when solar radiation is not adequate. This system is an excellent alternative for short duration storage of fresh fruits and vegetables at a low cost. It not only reduces the storage temperature but also increases RH which is essential for maintaining the freshness of the commodities. In the solar-powered vending cart, fruits and vegetables can be stored safely for a longer period of time at the retail vendor's level (Samuel *et al.,* 2016).

Processing and value addition

Value added refers to the additional value created at a particular stage of production or through image and marketing. Value added agriculture is a process of increasing the economic value and consumer appeal of an agricultural commodity. It is an alternative production and marketing strategy that requires a better understanding of the rapidly changing food industry and food safety issues, consumer preference and effective management. It may not be inferred that value addition, means only processing a raw material into some form of canned food. There are various ways of adding value to a commodity. Further, in India, the scope of Value addition is tremandous for the reason of availability of raw material as well as the large market size. The two major goals of PHT are loss prevention and value addition to the raw food commodities through preservation and processing. Raw food materials are cleaned, graded and then conditioned either for storage or processing. Processing is done to make raw commodities edible through primary and secondary processing and ready to eat through tertiary processing. At every stage of processing, value is added to the product. Estimated value additions to the raw food materials through primary and secondary/tertiary processing in India are 75% and 25% respectively. It therefore, shows that primary processing has a greater role to play in improving the economic benefits to the farmers. For value addition, agricultural produce accounts for a fairly significant share of the country's acreage and output of many crops. Notable are cereals, millets, fruits and vegetables and other cash commodities grown in different parts of India. While expanded profitability is a fundamental part of an energetic improved post-harvest handling and processing is essential to ensure high-quality products reach the markets. Too often, even when the yields are high, producers lose income due to poor post-harvest practices. Horticulture deals a large group of crops. Therefore, cultivation of crops which belong to us and possess great medicinal, nutritional, health promoting values. India as second largest producer of fruits and vegetables, only 10 per cent of that horticultural produce is processed, but other developed and developing countries where 40-80 per cent produce is value added. Horticultural crops provide varied type of components, which can be effectively and gainfully utilized for value addition like pigment, amino acids, oleoresins, antioxidants, flavors, aroma etc. Post harvest losses in horticultural produce are 5 to 30 per cent which amounts to more than 8000 crore rupees per annum. If we subject our produce to value addition the losses can be checked. Horticultural crops are right material for value addition because they are more profitable, has high degree of process ability and richness in health promoting compounds and higher potential for export. Therefore, horticultural crops are right material for value addition in the present context of agricultural scenario and we should go for new product development to be unique and novel.

Value Addition in Horticultural Crops

ICAR institutes and SAUs have generated several technologies including post harvest management practices for extension of shelf life, long distance marketing, on-farm storage, alleviation of chilling injury. Value added products from horticultural crops viz. Ready to Serve heath drinks, jamun juice concentrate, dehydrated carrot shreds, ripe mango slices, ginger-onion-garlic powder, dehydrated onion slices and powder, frozen red capsicum, ginger slices; and processed products from grains like quick cooking dhal, roasted soy nut, extruded bajra puffs, sorghum flakes etc.

a) Pusa Pearl Puff

- It is ready-to-eat crunchy food product.
- It contains about 13% protein, 5 mg/100g iron and 3 mg/100g zinc.
- Consumption of 100g of this product can meet approximately 25% of the Recommended Dietary Allowanus (RDA) of Fe and Zn in women.
- The product could to be a boon to the huge human population suffering from celiac diseases due to gluten allergy.

b) Osmo-dehydrated Carrot shreds/ Slice

- High in β- carotene
- A value added product in infant foods and other culinary preparations.
- Can be easily reconstituted to carrot dessert
- Product retains natural colour, flavor and texture
- Promising base material for the development of other value added products
- Presence of xanthophyll in yellow carrot as bioactive lutein thus protecting eye from age related macular degeneration and cataract
- Good source of abundant lycopene in red carrot in addition to beta-carotene and alpha carotene as a functional ingredient
- preventing prostate cancer and thus scavenging free radicals from body
- Contain major hydrophilic antioxidants as phenolics, flavonoids and anthocyanin

Flowchart of Osmo-dehydrated Carrot shreds/ Slice

Carrot

↓

Grading, washing and cutting into slices (0.2-0.25 cm)

↓

Blanching in 20% sugar solution at 80-85°C for 4-5 min followed by dipping into 0.2% potassium metabisulphite solution for 10 min

↓

Osmotic diffusion of 50-55% sugar solution at 50-55°C for 90 min

↓

Cabinet drying of carrot slices I stage of drying at 60°C for 3-4 hrs

↓

II stage of drying at 50-55°C for 8-10 hrs

↓

Cooling at room temperature

↓

Storage in air tight plastic containers at 10-15°C for 4-6 months

c) Nutra Aonla Candy

- Novel red amla candy combines the biologically active components from amla and black carrots.
- Amla fruits are known for pharmaceutical applications in ayurvedic system and are high in ascorbic acid, ellagitannins and other phenolics.
- Black carrots are an exceptionally good source of acylatedanthocyanins and possess high antioxidant activity.
- Red amla pieces in syrup enriched with natural colorant has high stability for anthocyanins for 9 months at low temperature.
- Dried candy processed through DIS process, dried and vacuum packed have good shelf for 6 months.
- Incorporation of natural red anthocyanin provides a high sensory appeal to the amla products

d) Tomato Paste

Tomato are rich source of lycopene, an antioxidant with immune stimulatory properties bcarotene, phenolics such as ferulic, chlorogenic and caffeic acid Good source of ascorbic acid, and vitamin E.

Tomato Juice (Strained), Cooking to desired consistency

↓

Judging the end point for puree or paste

(tomato solids by hand refractometer or

Reduction in volume by measuring stick)

↓

Filling hot into bottles or cans (82-88°C)

↓

Pasteurization in boiling water for 20 minutes

↓

Cooling

↓

Storage in ambient temperature

↓

Packing and storage

Pickling and Fermentation of Vegetables

The preservation of food in common salt or in vinegar is known as pickling. It is one of the most ancient methods of preserving fruits and vegetables. Pickles are good appetizer and add to the palatability of meal. Pickles help in stimulation of gastric juice and thus help in digestion. Several kinds of pickles are sold in the Indian market. Mango pickles rank first followed by cauliflower, onion, turnip and lime pickles. Pickles are either manufactured at cottage or home scale as well as commercially manufactured and exported. Pickling is the result of fermentation by lactic acid bacteria which are generally present in large numbers on the surface of fresh vegetables and fruits. These bacteria can grow in acid medium and in the presence of 8-10% salt solution, whereas the growth of majority of undesirable organisms is inhibited. Lactic acid bacteria are most active at 30°C, so this temperature must be maintained as far as possible in the early stage of pickle manufacturing process. When vegetables are placed in brine, it penetrates into the tissues of vegetables and soluble material present in vegetable tissues diffuses into brine solution by osmosis. The soluble material basically includes fermentable sugars and minerals. The fermentable sugar is broken down by lactic acid bacteria which convert them into lactic and other acids. The acid brine thus formed acts upon vegetable tissues to produce characteristics taste and aroma of pickle. In the dry salting method, several alternate layers of vegetables and salt are kept in vessel which is covered with acloth and a wooden board is allowed to stand for about 24 hrs. During this period, sufficient juice comes out from vegetables to form brine. The growth of majority of spoilage organisms is inhibited by brine containing 15% salt. Lactic acid bacteria which are salt tolerant, can thrive well in brine containing 8-10% salt. It is therefore, advisable to place the vegetables in 10% salt solution for vigorous lactic acid fermentation. In dry salting method of fermentation, fermentation process starts and carbon dioxide begins to evolve. The salt content is gradually increased to 15% for steady fermentation process. Under favourable conditions, fermentation is completed in 7 to 10 days.

References

D. V. K. Samuel, P. K. Sharma and J. P. Sinha (2016). Solar-powered evaporatively cooled vegetable vending cart. *Current Science*. 3(12): 2020-22.

http://planningcommission.nic.in/plans/stateplan/sdr_assam/sdr_assch7.pdf

http://www.agriassam.in/agriHorti_profile/Profile_ofAgri-HortwiSector_ofAssam-June2012.pdf

IARI, Low Cost storage technologies for preservation of horticultural produce and food grains. http://iari.res.in/

Nawab Ali, Post-harvest technology for employment generation in Rural sector of India, Central Institute of Agricultural Engineering, Nabibagh, Berasia Road, Bhopal-462 038, MP, India.

The National Cooperative Development Corporation

Improved Agriculture and Allied Practices for Farmers, pp 257-266
New India Publishing Agency, New Delhi, India

24

Packing House Operations of Horticultural Crops

R.R. Sharma

Division of Food Science & Postharvest Technology, ICAR-Indian Agricultural Research Institute, New Delhi-110012

Most of the horticultural produce is of perishable nature and must be protected from postharvest losses, which start from the field itself. Therefore, adoption of proper packing house operation plays a very crucial role in minimizing postharvest losses

Although, we have reached new heights in the production of fruits comparable to China, Brazil and USA and in vegetables to China, yet we are far behind in postharvest management of these perishables, primarily due to lack of appropriate infrastructure. Test studies conducted in Australia have indicated that the distribution cost of perishable horticultural produce is assumed 5 to 9 times, the production cost. Further, packaging accounted for 25-50% of the wholesale price of fruits and vegetables. In India, huge losses of horticultural produce (25-30%) after harvest are encountered mainly due to non-existence of packing house concept.

Packing house operations essentially begins with the use of proper field containers, field transport, temperature protection of harvest produce, sorting, sizing, grading, packing, which are essential preparatory steps to storage, shipment and subsequent marketing. Depending on the kind of produce, additional activities viz. degreening, curing, washing, ripening, waxing, precooling etc. are done. The sequence of activities varies with the crops. A considerable quantity of postharvest losses of horticultural produce could be minimized by adoption of various packing house operations described below:

Use of proper field containers

Selection of proper field container should be done depending on the type of fruit. For example, field bins made up of ¾" plywood, smooth or coated inside and vented on sides and bottom having 120 cm x 120cm x 61cm are commonly used for many fruits. Perforated plastic crates can also be used. The bin surfaces should be clean, smooth and undergo periodic disinfection. Careful field supervision is the most critical factor in protecting fruits from injury. There are many causes of physical injury viz. improper picking procedure, excessive dropping of fruit into baskets, overfilling of these containers etc. Very soft, delicate fruits may be harvested, stored and packed directly into the shipping container by the picker (strawberries, blue berries).

Transport from field

Two major types of damages occur during field transport. These are impact bruising and compression bruising. Compression bruising can be minimized by restricting the depth of loading in the field container. However, the compression bruising could be minimized by adopting following steps

i) Avoiding rough handling by proper supervision of truck / trailer loading and unloading.

ii) Reduction of air pressure of the transport vehicle to reduce motion transmittal to the fruit.

iii) Use of air suspension system in the newly purchased transport vehicle instead of spring suspension. Tests have shown that 50% of reduction in damaging motion with air suspension system.

Temperature protection

- Shading of harvested fruit to minimize warming and sun scalding.
- Avoid harvesting in mid-day.
- If tarpaulin is used to cover the loads of fruit during transport, it should be light colour (white or silver are good) and clean. It should be supported so as to maintain air space over the load.

Degreening

It is the process of decomposing the green pigment in fruits by applying ethylene or similar metabolic inducers to give a fruit it characteristic colour as per the consumer preference. This process is applicable to banana, mangoes, and citrus fruits. The length required to degreen a fruit depends upon the degree of natural colour break and maturity.

Pre-cooling

Precooling refers to the process of removal of field heat. High temperatures are detrimental to the keeping quality of perishable horticultural produce.

Therefore, quick removal of field heat and heat of respiration are essential for slowing down the physiological and metabolic activities of the harvested produce in order to achieve better shelf life, minimize susceptibility to attack of microorganisms, reduce water loss and load on the refrigeration system of the transport vehicle or ship's cargo space. This can be achieved by various techniques viz. aircooling, forced air cooling, hydrocooling, vacuum cooling, package icing etc.

Washing

A clean produce is always demanded by the consumer. Most of the fruits and vegetables are washed after harvesting in order to improve the appearance and removal of primary inoculam load of microorganisms. Numerous conflicting reports on the advantages and disadvantages of washing of fruits and vegetables are available. For example, washing improves the shelf life of banana by delaying the ripening. However, reports also reveal that washing encourages the rotting in many fruits and vegetables as in potato, cucumber etc. Therefore, proper judgments must be made before determining the washing requirement of any horticultural produce by using proper disinfectant/fungicide in the washing solution.

Drying

This is an essential step after washing and must be followed in order to remove the excess of surface water from fruits and vegetables, which would otherwise encourage microbial spoilage. This can be done by passing the fruits and vegetables over sponge rollers.

Waxing

Fruits and vegetables have a natural waxy layer on the outer surface, which gets partly removed during harvesting, rough handling and washing process. An extra discontinuous layer of wax should be applied artificially with sufficient thickness and consistency for the following objectives:

- Replacement of natural wax.
- To reduce the water loss during shipment and storage.
- To give a cosmetic appeal to the consumer.
- To cover up the minute injuries caused during postharvest handling.
- To use as a carrier of fungicide/inhibitor for better shelf life.
- To protect the product against decay organisms.

Vegetable wax, carnauba wax, bee wax, thermoplastic terpene resin, shellack resin are common types used for the preparation of wax emulsion.

Recently, sucrose polyester waxes are gaining popularity due to their water solubility and biodegradability. Waxes may be applied by foaming, spraying, dipping or bruishing. Foaming is a very satisfactory method since it leaves a very thin coating.

Grading

Grading is necessary to get suitable returns with respect to the quality attributes of the commodity. Most countries have their own set of standards for domestic trade. For international trade standards have been defined by the Economic Commission for Europe in 1954 and by the Organization for Economic Corporation and Development (OECD, 1970). Grades are based on soundness, firmness, cleanliness, size, weight, colour, shape, maturity, freedom from disease, mechanical injury, insect damage etc. Grading may be done manually or mechanically.

Sizing

Sizing is primarily done in order to achieve uniformity in the contents.

Disinfestation treatments

One of the most important pests in crop production are fruit flies. Papaya, citrus, cucurbits, mangoes, melons and many other fruits are highly susceptible for fruit fly attacks. Postharvest infestation methods for these fruits have been developed. There are two major approved disinfestations treatments commercially followed. These are: vapour-heat treatment (VHT) and EDB-fumigation treatment.

Packing

Packaging of fresh fruits and vegetables is one of the important steps in the long and complicated journey from grower to consumer. Packaging plays a vital role not only for food preservation and protection during storage, handling, transportation and distribution but has assumed a multifunctional role by serving as a symbol of value addition, and assurance of quality and quantity/ number and also a conveyor of convenience and thus an instrument or tool for marketing. In recent years, the importance of packaging has been increased to a great extent. This is mainly due to increase of consumer awareness and willingness to pay for hygienic and value added package. For packing, different types of packages are used as under:

1. Sacks and nets

   The material used for sacks and nets may be woven natural fiber (jute, kenaf, sisal, cotton), woven synthetic (Polypropylene, polyethylene), knitted natural fabrics (cotton), knitted synthetic (polyethylene) or non-woven synthetic (propylene).

2. Wood pallets

   Pallets were first used during world War –II as an efficient way to move goods. Depending on the size of produce package, a single pallet may carry from 20 to over 100 individual packages. Because these packages are often loosely stacked to allow air circulation, or are bulging and difficult to stack evenly, they must be secured (unitized) to prevent shifting during handling and transit.

3. Pallet bins

   Substantial wooden pallet bins of milled lumber or plywood are primarily used to move produce from field or orchard to the packaging house. Depending on the application, capacities may range from 12 to more than 50 bushels. Although the height may vary, the length and width is generally the same as a standard pallet (48 inches x 40 inches). Moreefficient double-wide pallet bins (48 x 80 inches) are becoming more common in some produce operations.

4. Wire-bound crates

   Although alternatives are available, wooden wire-bound crates are used extensively for snap beans, sweet corns and several other commodities that require hydro-cooling. Wire-bound crates are sturdy, rigid and have very high stacking strength that is essentially unaffected by water.

5. Wooden crates and lugs

   Wooden crates, once extensively used for apples, stone fruits and potatoes have been almost totally replaced by other types of containers. The relative expense of the container, a greater concern for tare weight, and advances in material handling have reduced their use to a few specialty items, such as expensive tropical fruit. The 15, 20 and 25 pound wooden lugs still used for bunch grapes and some specialty crops are being gradually replaced with less costly alternatives.

6. Wooden baskets and hampers

   Wire-reinforced wood veneer baskets and hampers of different sizes were once used for a wide variety of crops from strawberries to sweet potatoes.

They are durable and may be nested for efficient transport when empty. However, cost, disposal problem, and inefficient palletilization have severely limited their use to mostly local growers markets where they reuse baskets many items.

7. Corrugated fibre board

   Corrugated fiberboard (often mistakenly called cardboard or pasteboard) manufactured in many different styles and weights. Because of its relatively low cost and versatility, it is the dominant produce container material and will probably remain so in the near future. The strength and serviceab8lity of corrugated fiber board have been improving in recent years. Most corrugated fiber board is made for three or more layers of paperboard manufactured by the kraft process. In the recent years, large double-wall or even triple wall corrugated fiber board containers have increasingly been used as one way pallet bins to ship bulk produce to processes and retailers. Mangoes, apples are primarly packed in CFB in India. Cabbage, melons, potatoes, pumplkins and citrus fruits have all been shipped successfully in these containers. The containers cost per pound of produce is as little as one fourth of traditional size containers. Some bulk containers may be collapsed and re-used.

8. Pulp containers

   Containers made from recycled paper pulp and a starch binder are mainly used for small consumers packages of fresh produce. Pulp containers are available in a large variety of shapes and sizes and are relatively inexpensive in standard size. Pulp containers can absorb surface moisture from the product, which is a benefit for small fruit and berries that are easily harmed by water. Pulp containers are also bio-degradable, made from recycled materials, and recyclable.

9. Paper and mesh bags

   Consumers packs of potatoes and onions are about the only produce items now packed in paper bags. The more sturdy mesh bags has much wider use. In addition to potatoes and onions, cabbage, turnip, citrus fruits, and some specialty items are packaged in mesh bags. Sweet corn may still be packed in mesh bags in some market. In addition to its low cost, mesh has the advantages of uninhibited airflow. Good ventilation is particularly beneficial to onions.

10. Plastic bags

    Plastic bags (polyethylene film) are the predominant material for fruit and vegetable packaging. Besides the very low material costs, automated bagging machines further reduces packing costs. Film bags are clear, allowing for inspection of the content, and readily accept high quality graphics. Plastic films are available in a wide range of thickness and grades and may be engineered to control the environmental gases inside bags. The film material 'breathes" at a rate necessary to maintain the correct mix of oxygen, carbon dioxide, and water vapour inside bags. Since each produced items has its own unique requirement for environmental gases, modified atmosphere packaging material must be specially engineered for each items.

11. Shrink wrap

    One of the newest trends in produce packaging is the shrink wrapping of individual produce items. Shrink wrapping has been used successfully to package apples, pomegranates, potatoes, sweet potaoes, onions, sweet corn, cucumbers and a variety of tropical fruits. Shrink wrapping with an engineered plastic wrap can reduce shrinkage, protect the produce from diseases, reduce mechanical damage and provide a good surface for stick-on labels.

12. Rigid plastic packages

    Packages with a top and bottom that are heat formed from one or two pieces of plastic are known as clamshells. Clamshells are gaining popularity because they are inexpensive, versatile, provide excellent protection to the produce, and present a very pleasing consumer package. Clamshells are most often used with consumers packs of high value produce items like small fruits, berries, mushrooms, etc., or items that are easily damaged by crushing.

Recent developments in food packaging techniques

- Vacuum packaging
- Active and intelligent packaging
- Modified atmosphere packaging
- Aseptic packaging

Vacuum packaging

Modification of packaging atmosphere from normal atmosphere (78% nitrogen, 21% Oxygen) is widely used as effective modern packaging technologies which can significantly extend shelf life and improve quality level of most food products. The modification of internal package atmosphere may take place at the level of total pressure and/or partial pressures of component gas components. When the total pressure inside the package becomes lower than that of outside environment as a consequence of a more or less evacuation of the internal atmosphere, it is commonly referred to as 'vacuum packaging'.

Active packaging

Fresh foods just after harvest are still active biological systems. The atmosphere inside a package constantly changes as gases and moisture are produced during metabolic processes. The type of packaging used will also influence the atmosphere around the food because some plastics have poor barrier properties to gases and moisture.

The metabolism of fresh food continues to use up oxygen in the headspace of a package and increases the carbon dioxide concentration. At the same time water is produced and the humidity in the headspace of the package builds up. This encourages the growth of spoilage microorganisms and damages the fruit and vegetable tissue.

Many food plants produce ethylene as a part of their normal metabolic cycle. This simple organic compound triggers ripening and aging. This explains why fruit such as bananas and avocados ripen quickly when kept in the presence of ripe or damaged fruits in a container and broccoli turns yellow even when kept in the refrigerator.

Intelligent packaging

It is defined as the packaging that monitors the conditions of packaged foods to give information about the quality of the packaged food during the transport and storage. It is also defined as a packaging system that is capable of carrying out intelligent functions (such as detecting, sensing, recording, tracing, communicating, and applying scientific logic) to facilitate decision making in order to extend shelf life, enhance safety, improve quality, provide information and warn about possible problems.

Intelligent packaging can play an important role in facilitating the flow of both materials and information in the food supply chain cycle. An intelligent packaging system consists of 4 components viz. smart package devices, data layers, data processing and information highway. The smart package devices are largely responsible for giving birth to the concept of intelligent packaging since they impart the package with a new ability to acquire, store and transfer data. The data layers, data processlllg and information highway are collectively referred as decision support system.

Controlled and modified atmosphere storage

Controlled atmosphere usually indicates monitoring and control of gaseous composition.This is the case with bulk stores for fruit and sometimes with transport containers. It is not practical with small packages in the distribution system.

The term modified atmosphere is used when the composition of the storage atmosphere is not closely controlled. The initial atmosphere is intentionally adjusted to give a gas mix as close as possible to that which will optimize the shelf life. Subsequent movement of gases and moisture into and out of the package is controlled only by the ability of the packaging film to act as a barrier.

In modified atmosphere packaging, the atmosphere inside the packing is modified as per requirements of the commodity.

Aseptic packaging

Heat is the most important means for food preservation. A typical thermal process consists of heating, holding and cooling of a food product. Thermal processes may be classified based on whether the inactivation of microorganisms occurs before or after the food is filled inside the container. The former includes hot filling and aseptic packaging, in which a heat processed food is filled and packed in a clean environment. In aseptic packaging, the food and the package are sterilized (by heating and subsequent cooling) before the filling operation and the recontamination of the heat processed foods is prohibited completely by maintaining the environment and containers aseptic.

Section 6: Agricultural and Allied Activities

Improved Agriculture and Allied Practices for Farmers, pp 267-273
New India Publishing Agency, New Delhi, India

25

Adopt Apiculture for Productivity and Prosperity

Debjani Dey

Division of Entomology, ICAR-Indian Agricultural Research Institute New Delhi - 110012

Honeybees are marvelous insects known to mankind since the prehistoric times. The art of their scientific management is called as apiculture or beekeeping and the place where hives are maintained is called apiary. Beekeeping is a decentralized, forest and rural agriculture based industry, which does not require any raw material. The raw material needed is in the form of nectar and pollen, which is freely available from flowers in nature. Bee hives neither demand additional land space nor do they compete with agriculture or animal husbandry for any input. The beekeeper only needs to spare a few hours in a week to look after his bee colonies. Beekeeping is therefore ideally suited to him as a part-time occupation. Beekeeping constitutes a source of sustainable income for the rural and tribal farmers. It also provides them valuable nutrition in the form of honey. However, in addition to honey, production of bees wax, propolis, pollen, royal jelly, pedigree queen bees, package bees and renting out honey bee colonies for crop pollination are some of the potential areas of apicultural diversification in India (TOI, 2012).

India is a vast country with varied climates and ecological conditions ranging from tropical to subtropical in its southern, central and eastern regions, from sub-temperate to temperate along its north and north west and semi arid to desert conditions towards the west. The major geographical regions facilitating beekeeping development are classified into: 1. Southern peninsular region; 2. North east region; 3. Indo Gangetic plains; and 4. Northern hill region (Thakar, 1976). The geographical position of India and the related agro climatic condition favor the growth of a wide variety of flora both natural and cultivated. The extensive area of forest and millions of acres of cultivated land sustains a large proportion of insects and honeybees. Realizing the viability of beekeeping, the Khadi and Village Industries Commission took up beekeeping as a village industry and extensively organized beekeeping programmes throughout India and now there are more than a million bee colonies reared by lakhs of beekeepers (Thomas *et al.,* 2002). Different societies in Asia, Europe, and North America have evolved their own beekeeping methods, investing in such diverse areas as bee genetics, hive design, management operations, managed pollination, and honey processing and marketing. Scientists and development workers have studied it for more than 150 years; continuous selection and breeding have improved the bee's genetics; and Langstroth's hive design which is based on optimising bee space (the space required for the movement of bees) in the hive has helped beekeepers and bee enterprises to produce more honey in a sustainable manner.

It is perhaps the only industry, which besides production of honey, royal jelly, propolis, beeswax, bee venom and several medicinal compounds, improves crop productivity through pollination. It maintains the stability of ecosystem and environmental quality, which in turn helps in conservation of biodiversity.

Beekeeping has tremendous scope for development of ancillary industries and its untapped potential remains to be explored for increasing opportunities for gainful employment and income generation in rural areas. Each insect species has its own role in the ecosystem and its loss could affect the complexities and abundance of other organisms. Although production of honey has been the major aim of this industry but beekeeping today also includes production of beeswax, bee collected pollen, bee venom, royal jelly, propolis, as also of package bees, queen bees and nucleus colonies etc. All these are possible only with a proper management of bees, utilization of the local plant resources and adapting to the local climatic conditions. Honey bees, apart from yielding honey and wax, and acting as pollinators, also yield valuable products like, royal jelly, bee pollen, propolis and bee venom having nutritional / medicinal values (Viraktamath *et al.,* 2013). Intensification and diversification are the present pressing needs for the sustenance of Indian agriculture and are likely to be vigorously followed in future in order to support the burgeoning population. Bees constitute an important component of agro-ecosystem as they provide free ecosystem services in the form of pollination which not only enhances the productivity of agricultural crops but also helps in conservation of biological diversity through propagation of wild flora besides, providing honey and other hive products. Apiculture (Beekeeping) as non land based income and employment generating activity is fast becoming a prime component of present day strategies for integrated rural development and off farm employment. Intensive agriculture, increasing monoculture, massive deforestation and use of pesticides are the major threats to global bio-diversity. The sustainable increase in crop production lies not in increasing the area under cultivation but in increasing the production levels of cultivated areas. Honeybees can play a significant role in this direction. They act as cost free bio-inputs in farming and forest systems and increase cross pollination in cultivated as well as wild plant species. They maintain gene flow in ecosystems by cross-pollination. This leads to increased hybrid vigour, fruit and seed set and hence higher productivity, at the same time conservation of wild flora and sustenance of wild life.

Commercial outlook is now being developed as a result of the introduction of the European honeybee, *Apis mellifera*. Even today the average Indian beekeeper does not have the required knowledge and technology base for efficient running of modern beekeeping. In order to fully utilize the potential offered by this industry in our country large-scale training programmes are necessary to create a new cadre of beekeeper-farmers. In order to preserve the valuable resources, conscious efforts are necessary to stop deforestation in potential beekeeping areas, to conserve biospheres suited to honey bees, to establish bee plant nurseries and to regulate insecticide application in farms and orchards.

The success of bee keeping depends upon the understanding of the biology, behavior of honeybees, proper management techniques and latest equipment for handling them. It is advised that a beginner should start with a few hives, which can be gradually increased to make the venture profitable. To start beekeeping with 5 hives an initial investment of approximately Rs 15,000 is required which includes non-recurring expenses of around Rs 12,000. The income generated during the first year itself by selling honey and other bee products recover the amount spent in establishing the apiary.

Bees are estimated to pollinate 16% of the total of 0.25 million flowering plant species known so far. One-third of human diet is said to be derived from products of bee-pollination. About 90% of the word's plant food production is mainly based on 82 products derived mainly from only 63 plant species. The importance of bees can be realized from the fact that for 39 of these plant species bees are the major pollinators (Thakur, 2012). Pollination is a valuable ecosystem service and also of high economic value through enhanced agricultural production especially of fruits, seeds, vegetables, fibre crops and nuts (Costanza *et al.,* 1997; Gordon and Davis, 2003). There are at least four species of true honeybees and three species of the stingless bees native to India. Several sub-species and races of these are also known to exist. The exotic honey bee, *Apis mellifera* introduced during the sixties of the last century has established well in our country. Together these represent a wide variety of bee fauna that can be utilized for the development of honey industry in the country. They are *Apis florea*, the little bee; *Apis cerana indica*, the common Indian bee; *Apis dorsata* the giant bee; and *Tetragonulla irridipennis*, the stingless bee. Besides, these there are many solitary bee-species such as *Bombus, Ceratina, Halictus*, etc. Out of all these, only *Apis cerana indica* and *Apis mellifera* bees can be domesticated.

Beekeeping can be started by anyone even with a single colony, which can be increased to hundreds later. It is ideally suited for the rural areas as it utilizes only the available natural resources which otherwise go waste. A beekeeper needs little investment for beekeeping as no sophisticated machinery is needed.

Some facts about bee pollination are:

- More than 50% of the existing species of plants propagated by seeds are dependent upon insects for adequate pollination.
- Only 15% of the 100 or so crops that feed the world are pollinated by domestic honey bees while 80% are pollinated by wild bees and other wild life.
- Value of additional yield obtained due to bee pollination alone is 15-20 times more than the value of all the hive products put together.
- In India 50 million hectares of land is under bee dependent. It has been estimated that bees are gainfully tapping only about 1/4th of the available floral resources of the country.

The honeybees can be obtained from the following sources:

a) Commercial bee-keepers or Government bee farms
b) Individual beekeepers
c) Capturing of stray swarms.

Requirements for Setting up an Apiary

Areas with abundant flowering plants eg., forest sites, agricultural farms and fruit orchards are ideal for beekeeping.

1. Installation of the apiary

It depends on a series of factors which are:

Vegetation: An apiary is established in a zone where nectar yielding flowers are abundant. Most of our evergreen or semi-evergreen forests, plantations, agricultural farms having crops like eucalyptus, shisham, pongamia etc., orchards like litchi, citrus, mango, peach, plum, apricots, apple and gardens can be selected for as suitable sites for beekeeping. When plenty of flowers are present, 4 colonies / hectare are recommended.

Water availability: Availability of water is necessary at the periphery of the apiary. The minimum water requirement is around 45 ml/beehive/day during winter and 1000 ml/beehive/day during summer.

Orientation of the colonies: The best orientation is South, SE and SW functionally with the direction of winds. Excessive wind causes hindrance in the exit and entrance of bees to the colony.

Set up: The beehives are placed on horizontal stands according to the ground, with a certain inclination to the tipper. It helps exit of water and strange particles out from the colony. The colony must be isolated from the ground to avoid humidity. Ant pans should be placed below the legs of the stand to prevent ants from reaching the beehives.Distance between two apiaries should be three to four kilometers.

2. Bee-equipments

There are several types of indigenous and traditional hives including logs, clay pots, wall niches, baskets and boxes of different sizes and shapes. In modern beekeeping, the combs are built on wooden frames that are moveable. This facilitates inspection and management of bee colonies. Three types of moveable frame hive are in common use: the Newton type along with its standardized ISI version, the Jeolikote Village type, and the Langstroth type. Besides the hives, the beekeepers need equipment and implements like the hive stand, nucleus box , smoker, honey extractor, bee veil, apron and hand gloves for protection from bee-sting etc. The honey industry also needs equipment and machinery for handling and processing of honey, beeswax, for manufacture of comb foundation sheets, and for other operations. Therefore, bee keeping also has tremendous scope for setting up of several ancillary industries.

3. Basic steps for beekeeping

- Beekeeping should be started with a single colony / few colonies, which can be increased later as experience is gained
- Bee hives should be kept in a place which is not affected by strong winds, sun and away from crowed places.
- A good source of water should be available nearby especially during summer.
- Hives should be at least 6-10 feet apart and on stands.
- Ant pans should be used to prevent ants and other insects from reaching the hive.

- Hive entrance should preferably face east.
- Ground around hives should be kept free of grass, weeds, black ants, white ants etc.
- Colonies should be protected from bee predators like wasps, lizards, bee eating birds etc.
- Hives should not be opened too often and several operations should be accommodated at one examination which should be carried out on a bright sunny day
- Bees should not be disturbed on cold, rainy or windy days or at night
- A beginner should wear protective clothing while examining a bee colony and should handle the frames gently without jerking

Conclusion

Beekeeping contributes greatly to the agricultural economy of the country. Therefore steps should be taken towards the conservation of the bees by controlling deforestation, taking up measures like afforestation and social forestry, which can in turn create proper habitats for honey bee populations. Insecticide application, which is a regular practice in farm and orchards crops, should be regulated as a measure to conserve bees and protect beekeeping. In certain cases the losses due to killing of pollinators, is several times more than the losses due to pest infestation. Such aspects should be brought to the notice of the farmers and they should be educated in integrated pest control measures that ultimately help him realize better and improved crops. Honey industry in the country can well become a major foreign exchange earner if international standards are met. Beekeeping has thus contributed to rural development through the centuries by supporting agricultural production, providing honey, wax, and other products for home use, and by providing income for both farmers and the landless. Beekeeping has holistic benefits that relate to health (being used as a medicinal product and as food), the economy (directly through sales of honey and other bee products, and indirectly through increased productivity of pollinated crops, as well as bee enterprise activities), employment (in honey production and pollination services), and the environment (by ensuring pollination of wild species). To derive maximum benefits from beekeeping it should be diversified for production of various bee products and multiplication of colonies for sale. This will increase per hive productivity leading to additional income generation and ultimately upliftment of rural masses by securing sustainable livelihoods.

References

Costanza, R., d'Arge, R., de Groot, R., Farber, S., Grasso, M., Hannon, B., et al (1997) The value of the world's ecosystem services and natural capital. *Nature* 387:253–260

Gordon, J. and Davis, L. 2003. Valuing honeybee pollination: a report for the Rural Industries Research and Development Corporation. RIRDC Publication No 03/077 RIRDC Project No CIE-15A:1-42

Thakar C.V., Practical aspects of bee management in India with Apis cerana indica: Apiculture in Tropical Climates, 1976, pp. 51-59.

Thakur, M. 2012. Bees as pollinators – Biodiversity and Conservation. International Research *Journal of Agricultural Science and Soil Science* 2(1): 1-7.

Thomas D., Pal N., Subba Rao K. (2002). Bee management and productivity of indian honeybees. Apiacta. Http://www.fiitea.org/foundation/files/2002/d.%20thomas.pdf

TOI, 2012. Diversified beekeeping is more remunerative: Experts. Aug 18, 2012.

Viraktamath, S., Fakrudin, B., Vastrad, A.S., and Mohankumar, S. 2013. Monograph on morphometry and phylogeography of honey bees and stingless bees in India. Network project on honeybees and stingless bees, Department of Agricultural Entomology, University of Agricultural Sciences, Dharwad, Karnataka.

Improved Agriculture and Allied Practices for Farmers, pp 275-282
New India Publishing Agency, New Delhi, India

26

Commercialized Mushroom Cultivation-An Agricultural Industry for Rural Masses

R.K. Sharma

Division of Plant Pathology, Indian Agricultural Research Institute New Delhi - 110 012

Mushrooms are fleshy fungi constituting a purely vegetarian diet, which is very tasty and nutritious. Some mushrooms are edible whereas others are poisonous. So far, more than 1600 types of mushrooms are known of which, nearly 100 have been accepted as food worldwide. Diversification in any farming system imparts sustainability. Mushrooms are one such component that not only impart diversification but also help in addressing the problems of quality food, health and environment related issues. One of the major areas that can contribute towards goal of conservation of natural resources as well as increased productivity is recycling of agro-wastes including agro-industrial waste. Utilising these wastes for growing mushrooms can enhance income and impart higher level of sustainability (Manjit Singh *et al.*, 2011).

Mushroom cultivation activities can play an important role in supporting the local economy by contributing to subsistence food security, nutrition, generating additional employment and income through local, regional and national trade; and also offering opportunities for processing enterprises. Among the 33 globally cultivated types of mushrooms 3 mushrooms i.e. white button, oyster and paddy straw mushrooms are popularly grown in India. However, sporadic cultivation of black ear and milky mushrooms has also been started in recent years. Romans regarded them as food of God, while Chinese termed these as elixir of life. Cultivation of mushroom is a vibrant example of small-scale farming in many developing countries that has alleviated rural poverty and improved diversification of agricultural production (Godfrey *et al.* 2010). Most of the mushrooms species have unique ability to degrade cellulose, hemi-cellulose and lignin of different agro-wastes and organic waste materials and are effective means of recycling these waste materials into edible biomass of high nutritive and medicinal value. Crop residue provide little economic return to the farmers, however if used in mushroom cultivation, it can lead to integrated rural development by increasing income and self employment. Primary investment is one of the major factors determining the capacity to produce mushroom (Mohammad *et al.* 2017). Mushroom constitutes low calorie but nutritious diet as it is rich in protein, vitamins and essential minerals. In general, mushroom fruit bodies on dry weight basis contain 55% carbohydrate, 32% protein, 2% fat and rest minerals and vitamins. These are excellent source of thiamine (vitamin-B_1), Riboflavin (B_2), Niacin, Pantothenic acid (Vitamin B. Complex), biotin, folic acid and Vitamin C, D, A, and K to some extent. Moreover, these vitamins and minerals are retained even after cooking.

Minerals present in mushroom are phosphorus, potassium, copper and iron. Besides these, sodium, calcium and magnesium are also present in traces amount. Mushroom protein contains all nine essential amino acids required for human growth. The digestibility of mushroom protein is 71-90% (FAO, 1971).In India, mushroom production is little over 1,00,000 tones and mainly comes (nearly 90-92%) from white button mushroom while rest comes from oyster and paddy straw.

Cultivation season of different mushroom in North Indian Plains

- October to March - White button mushroom (*Agaricus bisporus*)
- May to July - Paddy straw mushroom (*Volvareilla* spp.)
- Mid August to mid April - Oyster mushroom (*Pleurotus* spp.)
- February to April - Milky mushroom (*Calocybe indica)*

Table 1: Temperature requirements for the growth of different species of mushrooms.

Mushroom Species	Temperature °C	
	Mycelial growth	Fruiting
Agaricus bisporus	22-25	14-18
A. bitoriquis	26-28	24-30
Auricularia spp.	20-34	12-30
Lentinula edodes	20-27	10-20
Pleurotus eryngii	18-22	14-18
P. flabellatus	25-32	22-26
P.florida	25-32	18-22
P. sagor-caju	25-32	22-26
Volvariella volvacea	28-35	28-35
Calocybe indica	28-35	28-35

Cultivation of Oyster mushroom

Oyster mushroom is commonly called as *dhingri* in India. Its cultivation can be done on number of agricultural and other organic waste materials. The important substrates includes straw of different cereals, sugarcane bagasse, cotton waste, jute, coir pith (coir waste), groundnut pod shell, small wood pieces, saw dust, maize cob and banana pseudo stem, etc. The choice of organic waste depends upon the wide spread availability of these materials.

Substrates preparation, Filling and spawning

It is commercially cultivated generally on wheat or rice straw. The long pieces of straw (4-6 cm) are dipped in water for 16-18 hrs and 36 hours for rice and wheat straw, respectively. Straw should be chemically sterilized by mixing 7g bavistin with 125ml of formalin in 100 liter of water. Alternatively, straw can be treated in hot water by dipping it in hot water (80^{0}C) for 2 hours. After overnight soaking, spread the straw on pucca floor for removal of the excess water. Mix the spawn thoroughly @ 2% (w/w basis) of the wet substrate or 5% of dry straw. Mixing should be done under shade. Straw mixed with spawn is then filled in polythene bags (45 x 60cm size) up to 2/3rd of its capacity. These bags are then tied and holes of 2-3 mm size are made in them at a space of 5 cm. Keep these bags at 22-26^{0}C in the mushroom house for spawn run. Maintain mushroom house temperature between 22-26^{0}C and relative humidity between 80-85 %. Within 15-18 days, white cottony growth of mushroom mycelium spread in these bags. These bags are then opened and kept on racks with 25-30 cm spacing. Water is sprinkled over them in morning and evening. In next 14-15 days, it starts pinning which is then converted into fruit bodies in next 4-5 days. Harvesting is done before the margin of the fruit bodies start curling inwards. Average yield of 60-70% of dry straw can be taken easily.

Cultivation of paddy straw mushroom (*Volvariella volvacea*)

This mushroom, first cultivated in China is also known as Chinese mushroom. It is dark-gray in colour and best cultivated in hot humid climate with temperature between 28-39^0 C and relative humidity between 80- 90%. It can be cultivated in and around Delhi between mid April to mid-August.

Open air cultivation

In this method, foundation beds of 100 x 60 x15-20 cm are made with the help of bricks or mud under the shade, to save them from rains or direct sunlight. Paddy straw bundles of 7-8 cm diameter and 70-80 cm. length are made by tying them at one end. These bundles are soaked in water for 16-18 hours in a water tank. Chemical sterilization of the straw with bavistin 7g and formalin 125 ml in 100 liter of water is mandatory. While soaking, cover the water tank with the polythene sheet. Drain off excess water on a pucca floor. A bamboo frame exactly of the size of the bed or foundation is kept over bed. Now place four bundles of paddy straw (water soaked) side by side over bamboo frame. Keeping tide end of the bundles in same direction, place another set of 4 bundles over it with tide ends in opposite direction. This complete the first layer of bundles. Scatter the grain spawn about 8-12 cm from the edges of the bundles and dust it with powdered arhar dal or gram flour. Wheat bran or rice bran can also be used. Make second, 3rd and 4th layer of bundles in the same way and scatter the spawn. Finally bed is covered with the transparent polythene / alkathene sheet.

Care of bed, fruiting and harvesting

Within 7-8 days,mushroom mycelium permeates the straw if the temperature between 32-36^{0} C is maintained.

At this stage, remove the plastic sheet. Mushroom fruiting occurs in nearly 18-20 days after spawning at favourable moisture and temperature conditions. Fruiting is continued for another 10-12days. In paddy straw alone, yield of 12-14 kg / 100 kg of wet substrate can be obtained. Harvesting of the mushroom should be done when volva is just broken exposing the inside of mushroom. Drying of mushroom can be done under shade or in sunlight.Paddy straw mushrooms are very delicate in nature and can be stored under refrigerated condition for 2-3 days only.

Cultivation of white button mushroom (Agaricus bisporus)

White button mushroom is most commonly grown mushroom in north India during winter from October to mid March.

Raw materials & Ingredients of compost: Different formulations have been given by different organizations /workers/ institutions for making the compost. Most common compost formulation is given below.

Wheat straw/ rice straw	300 kg/ 400 kg
CAN(calcium amm. Nitrate)	9 kg
Urea	3 kg
Murate of potash	3 kg
Super phosphate	3 kg
Wheat bran/rice bran	15-20 kg
Gypsum	30 kg
Lindane dust	250 g
Molasses	5 kg

Compost preparation : can be prepared by using any one of the methods.

Long method: Spread the wheat straw in a thin layer of 8-10 inches thickness over floor of the composting yard. Sprinkle water over the straw to wet the straw thoroughly. Mix Urea, CAN, wheat bran and other ingredients except insecticides and gypsum in the pre wetted straw. Heap the mixed straw into a pile with the help of stack mould of the size of 1.25 m width x 1-1.25 m height x adjustable length depending upon the quantity of straw. The entire pile is opened and spread over composting yard on 3rd or 4th day for at least 45-60 minutes. This is called turning and is repeated every 3rd or 4th day. During 3rd or 4th turning gypsum is added. Add insecticide during 5th or 6th turning. Each turning should be uniform with thorough mixing of the straw. Open the pile after 6th turning and check the smell of ammonia if it persists, remake the pile and leave it for another 2-3 days. In this way, compost will be ready in 21-24 days.

Pasteurization method- This is done in 2 phases. First phase is completed in the composting yard while the second phase is completed inside pasteurization tunnel or bulk chamber with the help of steam for conditioning of the compost.

Phase-I

It involves wetting of the straw and followed by mixing of ingredients in the straw as in the long method. But in this case turning is given after every 48 hrs (2 days). During 3rd turning or on 6th day total amount of gypsum is added in the compost. Give 4th turning on 8th day, and fill the compost in pasteurization tunnel on 10th day.

Phase-2

In pasteurization tunnel temperature of 48-50^{0}C is maintained for 2-3 days. With the flow of steam, temperature of the tunnel is raised to 58-60^{0}C and maintained for 6 hr. Fresh air is then allowed to come in through ventilation. Once the temperature of tunnel comes down to 50-52^{0}C, it is maintained for 3 days. After it, the tunnel is cool down to 25-28^{0}C. By this method compost is prepared in 19-20 days. Fully prepared compost is dark brown in colour, with no trace of ammonia and no unpleasant odour and smells like fresh hay. Water content in the range of 68-70% in the compost is best for mushroom growth.

Filling of the compost: Fully prepared compost is spread over composting yard for overnight. Next morning when its temperature comes down to 25-27^{0}C fill it in trays, racks or poly-bags after thorough spawning. Mixing the mushroom seed or spawn in the compost is called as spawning.. After filling it is compressed lightly and leveled. Four racks can be made one over other. Lower racks rack must be 20-25 cm above the ground level and is followed with 45-50 cm surface to surface distance between racks. Mushroom seed is called spawn and nearly 160-170g of spawn is used per square meter area of the compost of 15-16 cm thickness Under ideal conditions (22-26^0 C temperature), 1-1.5% spawn of the dry straw(w/w) or 0.50to 0.75% of the wet compost is enough. In racks surface spawning or double layer spawning can also be done. After spawning cover the compost with old newspaper.

Spawn running

Maintain the temperature of the mushroom house between 22-26^{0}C and R.H. at 80-85%. In these conditions, the compost surface is covered with white cottony growth of the mycelium within 14-15 days of spawning. This condition is called spawn run.

Casing

After the completion of spawn run casing is done by removing the news paper sheet from the racks or trays after covering it with 4-5 cm thick layer of casing mixture and sprinkle some water. Most common casing material used is mixture of .FYM + 2-3 years old spent compost (1:1 ratio) after chemical or steam sterilization. After casing maintain the mushroom house at 24-25^{0}C for another 4-5 days. Within 4-5 days white mycelium will spreads in the casing soil after which lower down the temperature to 18^{0}C and maintain it between 14-18^{0}C during rest of the fruiting period. Watering should be done as and when required to maintain relative humidity at 80-85% throughout the cropping period. Under favourable conditions, fruiting will start after 15-20 days of casing. Within 4-5 days pinhead will attain the shape of white buttons.

Harvesting

It is done when the cap size of mushroom is 3 - 4.5 cm diameter with the help of knife or hand. Soil from the stem is removed with the help of knife. The yield in bag cultivation method is estimated to be 10-15 kg/100 kg of compost. On the other hand with pasteurised compost 16-20 kg of fresh mushroom can be harvested per 100 kg of compost in 6-8 weeks.

Conclusion

Quality food, health and environment are the major concernsfacing our country. Mushroom cultivation helps to address the issueof nutritional security and also provides solution for proper recyclingof agro-wastes. In addition to good quality protein, no cholesterol,high fibre, low sodium, good quantity of vitamins and minerals, themushrooms also have bioactive compounds. Three species of mushroom namely white button, oyster and paddy straw mushrooms are mainly cultivated in India. Different species require different substrates and cultural conditions for optimum growth and cultivation. Details from substrate preparation to spawning and growth conditions have been presented for three commonly cultivated species.

References

Godfrey EZ, Siti MK, Judith ZP (2010) Effects of temperature and hydrogen peroxide on mycelial growth of eight Pleurotus strains. *Scientia Horticulture*, 125:95-102.

Manjit Singh Bhuvnesh Vijay Shwet Kamal G.C. Wakchaure (2011). Mushrooms Cultivation, Marketing and Consumption. Directorate of Mushroom Research (Indian Council of Agricultural Research) Chambaghat, Solan –173213 (HP). Pp 274.

Mohammad N. E., Ahmed R., Mohammad S.A., Mohammad S.R. Ahmed K.U. (2017), Mushroom Cultivation as a Small-Scale Family Enterprise for the Alternative Income Generation in Rural Bangladesh. *International Journal of Agriculture, Forestry and Fisheries* 5(1): 1-8.

Improved Agriculture and Allied Practices for Farmers, pp 283-298
New India Publishing Agency, New Delhi, India

27

Agricultural Mechanization Technologies for Enhanced Food Production

Indra Mani Mishra

Division of Agricultural Engineering, ICAR-India Agricultural Research Institute New Delhi-110012

Mechanization is viewed as package of technology to increase land productivity by improved timeliness of operations, reduced crop losses and improved quality of agro-produce to increase efficiency of inputs used through their efficient measurement and placement, to increase labour productivity by using labour saving and drudgery reducing devices and reduce cost of cultivation. echnology and machinery enhanced the ability, quality, accuracy and efficiency of the human being. By using technology in any field the rate production and quality automatically increases. The country witnessed unprecedented growth in agriculture which has helped India to graduate from hunger to self sufficiency in food grains by increasing the food grain production from 51 million tonnes to 208 million tones. The estimated contributions of improved agricultural tools and equipment in augmenting total food production of India is encouraging; savings in seeds: 15-20%; savings in fertilizers: 15-20%; savings in time: 20-30%; reduction in labour: 20-30%; increase in cropping intensity: 5-20%, and increase in productivity 10-15 per cent. The major issues before Indian agriculture with respect to mechanization are; tractorization vs mechanization, custom hiring and contract farming, horticultural mechanization, effective extension activity, sugarcane mechanization, hill agriculture and industry institution interaction. The main technologies which can be taken to farmers' field for enhancing productivity and production Aqua-ferti seed drill, two row powered seeder, okra planter, tractor mounted inclined plate planter, tractor mounted rotary tiller (Rotavator), tractor mounted zero-till drill, tractor mounted raised bed planter, tractor mounted sett cutter planter for sugarcane, pneumatic planter, vegetable transplanter, low land manual rice seeder, self-propelled rice transplanted, self-propelled power weeder, tractor operated aero blast sprayer, tractor mounted vertical conveyor reaper, maize dehusker-cum-sheller, tractor operated straw reaper (combine) and tractor operated fodder harvester. These technologies need to be introduced after necessary modifications and adjustments, depending upon the agroclimatic and terrain and soil conditions, through proper mode e.g. as individual /group ownership, on custom hiring basis or mechanization package for contract farming.

Impact of Farm Mechanization on the Farmers

The effects of the farm mechanization on the farmers are in the form of new seed, fertilizer technology, new cultural techniques of farming, modern farming implements and changes in the timing of operations. Typically, however, improvements in technology also increase the productivity of capital and alter the technological rates of substitution of capital for manpower, reducing the amount of capital that is necessary to replace a unit of manpower at particular levels of output. Other innovations make it possible to reduce the amount of manpower in relation to land needed to produce specified levels of output. Mechanization affects the coat structure of agricultural production by:

- Saving labour (manual and bullock)
- Easing jobs
- Increasing yield
- Saving land
- Facilitating the opening up of new land
- Conserving natural resources

Sequence of mechanization of different operations

Farm mechanization helps in effective utilization of inputs to increase the productivity of land and labour. Besides it helps in reducing the drudgery in farm operations.The early agricultural mechanization in India was greatly influenced by the technological development in England. Irrigation pumps, tillage equipment, chaff cutters, tractors and threshers were gradually introduced for farm mechanization. Farm operations requiring high power inputs and low control are mechanized first (tillage, transport, water pumping, milling, threshing, etc.). Farm operations requiring medium levels of power and control are mechanized next (seeding, spraying, intercultural operations, etc.). India being a vast country having wide variation in cropping pattern, climate conditions and moisture availability there have been enhancement in farm power despite limited irrigation facilities in states like Andhra Pradesh, Assam and Uttarakhand.

Table 1: Sequence of Mechanization of Different Operations

Sequence	Operation		
	I	II	III
Type of operation	High power low skill	Medium Power Medium Skill	Low Power High Skill
Stationary	Grinding, milling, crushing, water pumping, threshing	Grinding by size, cleaning	Grinding by quality
Mobile	Land preparation	Seeding of grain,	Transplanting, and
	Transport	Harvesting of grain	Harvesting of cotton, Sugarcane,
		Fruits and Vegetables	

Mechanization and Productivity

The estimated levels of mechanization of various farm operations in India are: 40% tillage, 30% seeding/planting, 37% irrigation, and 48% threshing of wheat, threshing of rest of the crops 5%, and plant protection 35 per cent (CIAE, 2008).

The state of agricultural mechanization in India may be characterized with large variations in terms of power availability varying from 0.60 kW/ha in Orissa to 3.5 kW/ha in Punjab during 2001.The average farm power available was about 1.46kW/ha comprising of about 84% from mechanical and electrical sources and 16% is from animal power and human labour (IASRI, 2006). There is a strong linear relationship between power available and agricultural output on per ha basis, Figure 1.

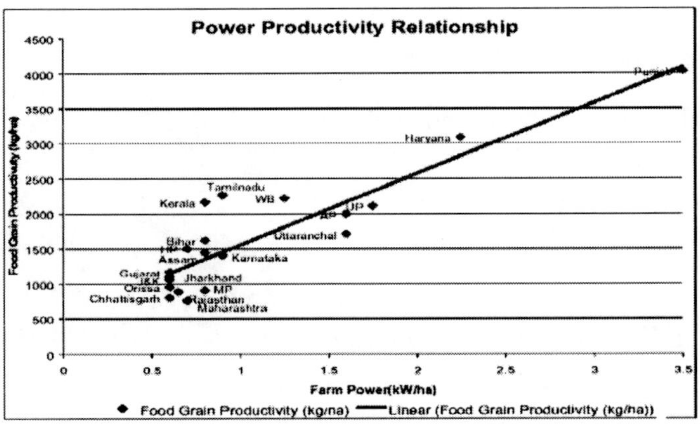

Fig. 1 Relationship between Farm Power and Productivity over different states

Challenges

Horticultural mechanization

Except for potato, mechanization of horticultural crops particularly vegetables is yet to pick up. Planting/transplanting, plant protection and harvesting are key operations, which are labour intensive. Some designs have been developed by R&D institutions; however, introduction of suitable machinery either through new development or by modifying imported designs is required.

Sugarcane mechanization

A number of equipment have been designed by R&D institutions however, use of specific sugarcane machinery e.g. sugarcane planter, either semi-automatic or automatic, and sugarcane harvester is negligible at farmers' field. Generally, farmers own a tractor drawn four bottom ridger, for planting of sugarcane, and tiller for inter-culture operations. Sugarcane harvesting is largely manual except in some sugar mills where cut and load type chopper harvester with engine power 240 hp is in use. Sugarcane is transported to the mill by three modes, namely, trucks, tractor-trailers and animal carts. Pneumatic wheel animal carts driven by buffaloes are common in western U.P. Sugarcane being a labour intensive crop; a comprehensive mechanization package is required both for improved cultivation and better sugar recovery.

Useful Engineering Technologies

1. Two row powered seeder

   It is a two row machine operated by3 hp power tiller. It is suitable for hill region for sowing of all crop such as wheat, maize etc. The field capacity of the machine is 0.2 ha/h.

2. Okra Planter

   It is a three row 35 hp or above tractor drawn equipment. It is suitable for planting okra seed. In this equipment separate seed box and inclined measuring unit is used for every row. The field Capacity of the machine is 0.2ha/h. The advantages of machine are :

   - Saving of time and water as planting is done on ridges. This also increases the yield of the crop.

- Normally 2-3 seeds are planted at one place, thus it reduces seed rate.
- Row to row and plant to plant spacing is maintained.
- This implement covers one hectare of area in 5 hours, thus it saves 77 man-days and 35% of operational cost as compared to traditional method.

3. Tractor mounted inclined plate planter

   It is a six row 35 hp or above tractor drawn equipment. It is suitable for planting groundnut, gram, soybean, mustered etc. Row to row distance can be controlled and planting of different seeds in different rows is possible. In this equipment separate seed box and inclined measuring unit is used for every row. It was designed and developed by CIAE, Bhopal.

4. Tractor Mounted Rotary Tiller (Rotavator)

   It is also suitable for incorporating straw and green manure. Pulverizing of soil is more uniform and better because impact of revolving blades of rotavator sheared the soil and made the soil fine. The field capacity is 0.25ha/h, saving in fuel is 25-40%, saving in time for preparation of soil for wheat sowing is 40-60 % and saves 20-30% water in case of paddy.

5. Tractor mounted zero-till drill

   It is suitable for sowing of wheat directly in rice-harvested fields without preparing the seedbed. The field capacity is 0.3 ha/h.

6. Tractor mounted raised bed planter

   It is used for sowing maize, groundnut, cotton etc on the beds. The field capacity is 0.26 ha/h. It saves about 20-30% water and 20% seeds.

7. Tractor mounted sett cutter planter for sugarcane

   It is a three row tractor drawn sugar cane planter which can cut whole sugarcane into pieces of uniform size, place it in the furrow prepared by ridger, setts are covered by soil and pressed in one operation. The field capacity is 0.2 ha/h

8. Pneumatic Planter

   It is a six row (2 or 4 optional) 35 hp or above tractor drawn equipment to plant single seed at predetermined seed/row spacing. It is suitable to plant seeds of mustard, sorghum, soybean, cotton, pigeon pea, maize, groundnut, okra etc. It consists of main frame, aspirator blower, disc with cell type metering plate, individual hopper, furrow openers, pto driven shaft, ground drive wheel etc.

9. Vegetable Transplanter

   It is suitable for transplanting of vegetable seeding like tomato, chilly, cauliflower, cabbage, brinjal etc. The field capacity is 0.5-1.0 ha/h and fuel consumption is 3.5-4.5 lit/ha.

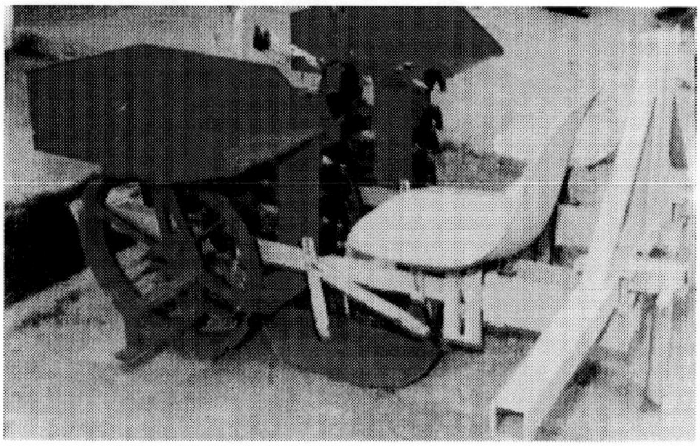

10. Low Land Manual Rice Seeder

    It is suitable for sowing pre-germinated paddy in puddled fields. The field capacity is 0.12 ha/h .

11. Self-propelled Rice Transplanter

    The row of rice as recommended by the agronomist is about 20 cm and, therefore, working width of the machine must be a multiple of 20 thus fixing the size to 1.0, 1.2 and 1.6 m covering 5, 6 and 8 rows of rice in the field. Wider machine are more time consuming in turning, loading trays with seedlings and thus result in lower field efficiency.

12. Self-propelled power weeder

    It is operated by power tiller. Its field capacity ranges from 0.1 to 0.13 ha/h. It saves 90% time in the operation and 30% in the cost of operation as compared to manual weeding using *khurpi*.

13. Tractor operated aero blast sprayer

It is useful for spraying on horticultural trees and tall crop such as cotton, sunflower, pigeon pea etc. Machine can cover about 1.7 ha/h at a speed of 1.5 km/h. The rate of application can be varied from 100 to 400 l/ha depending upon the value setting.

14. Tractor Mounted Vertical Conveyor Reaper

It is used for harvesting cereal crops like wheat and paddy. The field capacity of the machine is 0.4 ha/h.

15. Maize Dehusker-cum-sheller

It is operated by tractor or engine. The percent recovery of grain is 97 % whereas in manual threshing it is only 78.4%. There is a saving of about 60-70% in cost and about 90% in labour as compared to the traditional methods.

16. Tractor operated straw reaper (combine)

It is used to recover wheat straw after combine operation. The capacity of machine is 0.4 ha/h at a speed of 2.5km/h. Straw recovery is about 55-60%. There is an additional grain recovery of 50-100 kg/ha. The cost of grain recovered is almost equal to the amount paid for hiring the machine.

17. Tractor operated fodder harvester

The machine is capable of harvesting, chopping and loading of fodder crops such as maize, *Bajra*, oats in a trailer, attached behind the tractor. The capacity of the machine is 0.20 ha/h at speed of 2.5-4.0 km/h.

18. Pusa Wheel Hoe

Pusa Wheel Hoe is a very simple and useful implement for weeding in between lines of standing crops by which very efficient and effective weeding is possible.This equipment is foldable which enables ease while transporting. Advantages of wheal hoe includes

- Useful for weeding and inter-cultural operations in row crops.
- Angle of operation could be adjusted according to the need.

- This can be used by giving forward and backward motion in standing posture, which results in less fatigue.
- Cost effective for weeding and inter-cultural operations.

19. Pusa Aqua Ferti-Seed Drill

Lack of proper soil moisture at root depth zone at the sowing time of rabi crops such as wheat, gram, mustard etc. makes sowing difficult, which affects the germination and growth of crops. This machine has been invented for dryland areas, which is more than 60% in the country. The machine makes possible the application of aqueous fertilizer, i.e.,diluted solution of urea and DAP etc., along side the seed, which helps in better germination and initial development of the crop/seedling. It is operated by 45 hp or tractor. It is suitable for sowing of wheat in dryland area.

Advantages includes

- Timely sowing of rabi crops like wheat, gram, mustard etc. in rainfed areas.
- Useful for dry land areas. An increase of 53% in germination and 35% in yield have been observed in wheat crop in comparison to traditional sowing.
- Uniform distribution of aqueous fertilizer helps in increased germination.
- Efficient utilization of water and fertilizer with seed and consumes about 10 m (10,000 litres) of water/ ha.

20. Pusa Pre-germinated Paddy Seeder

The direct seeding of wet paddy on puddled soil has gained popularity in the recent time at National and International level. The direct paddy seeding gives faster and easier crop establishment, reduced labour use, lesser drudgery, earlier crop maturity, more efficient water use due to

reduced crop duration and increased benefit-cost ratio. Pusa Pre-germinated Paddy Seeder is available in two forms - three and six rows.

Advantages includes:

- Useful equipment for areas, where labour is a problem.
- Delayed sowing of paddy could be possible.
- Paddy sown with this machine matures 10-15 days earlier in comparison to transplanted paddy.

Seed may be dipped in salt solution (1-2 tea spoon salt/ litre of water) so that damaged seeds float on the surface of water. Floated seeds may be thrown out and remaining seeds may be washed with water and soaked in water for 24 hours. After removing the water, the soaked seeds is spread on gunny bags and covered with wet gunny bags around 24-48 hours for germination of seeds. For proper sowing, the maximum length of plumules should be 0.5 cm. After that, it is sown in lines with this machine.

21. Pusa Power Operated Seed Drill

Pusa Seed Drill is a walk behind equipment and useful for small and marginal farmers, particularly those of hilly areas, for sowing of different crops. This is relatively light weight equipment and can easily be shifted from one place to other. For cultivation in small plot sizes, particularly in hills, the tractor drawn seed drill is not suitable and under such situation, small sized machines are useful. Useful for farmers with small land holdings and for hilly regions. Seeds and fertilizer can be applied simultaneously. Row to row spacing and uniform seed rate is maintained. Saves seeds, time and increases yield by sowing through this machine. Height of furrow opener can be adjusted. Different row crops can be sown with this machine.

22. Pusa Solar Powered Knapsack Sprayer

Knapsack Sprayers are widely used by small and marginal farmers to spray plant protection chemicals by operating lever @ 16-20 stroke/minute which induces fatigue to workers along with greater variation in spray pressure causing inconsistency of application and in turn adversely affecting pest control and in long run it may lead to ground water pollution. In light of the above problem, solar powered spraying system has been developed which can be even fitted with the 2 existing sprayer. The diaphragm type pump is able to create 4 kg/cm pressure without pulsation, which is sufficient for fine & uniform spray. Advantages includes:

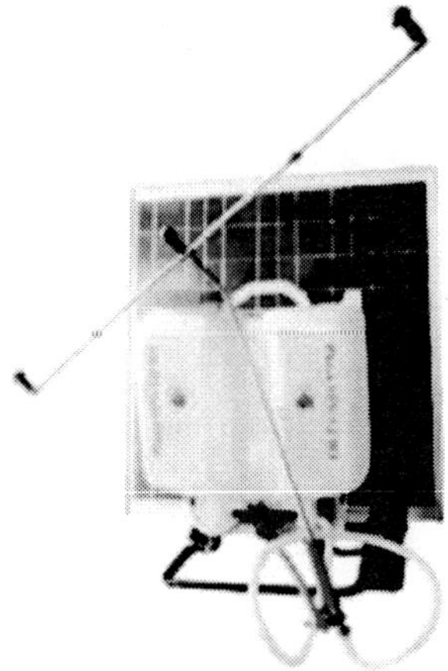

- Equipped with constant pressure system which gives quality spray i.e. uniformity, fine spray, effective & efficient control of pest is achievable.
- The acid-lead or Li-O battery can be charged by SPV panel of 60W. The fully charged battery is able to operate the sprayer for 6 hour continuously, which is adequate for a day operation.
- It enhances field capacity considerably and makes it suitable for row crop applications.
- A 100 mesh filter is provided just before nozzle, which enhances spray quality by filtering the spray material and prevents nozzle from clogging.
- Over and above, it is able to reduce farmer's drudgery as the lever operation has been replaced by solar powered

23. Pusa Basmati Paddy Thresher

An axial flow Pusa Basmati Paddy Thresher has been developed for threshing of basmati paddy without visible or invisible mechanical injuries with high labour efficiency, timeliness in completion of operation and also economizing the operation. The optimum rotation speed for cylinder is about 25 m/s. The high density polyethylene laminated tempered mild steel on spike tooth is found effective in reducing invisible injury to kernels. Advantages includes:

- Significant reduction in cost for threshing (up to 83 %) than conventional method of threshing.

- It reduces labour requirement by 88% than conventional method with 4% point enhanced recovery.
- This machine can be towed to field hence, it further reduces labour and power requirement for transporting material to threshing site. It also reduces the losses which occur in handling of bulk material.
- Due to higher capacity, timely completion of the operation also enables farmers to complete other required farm operation for consecutive crop in time and harness its benefits.

24. Pusa Okra Seed Extractor

The Pusa Okra Seed Extractor facilitates the extraction of okra seed for seed purposes with minimum damage and high extraction efficiency. In this machine feeder unit, threshing unit and cleaning unit are put in frame. Various units of the machine are operated through belt and pulley drive system.

25. Pusa Vegetable Seed Extractor

Traditional method of vegetable seed extraction is very difficult and time consuming with possible health hazards to labourers. Pusa Vegetable Seed Extractor is an ergonomically efficient machine, which increases the efficiency of seed extraction process. The machine is available in both manually operated and power operated models. The manual machine is of low cost and it includes a rough-surfaced rotating cylinder, a metallic sieve and a hopper. The motor operated model of the machine is also available. Advantages includes:

- Suitable for efficient extraction of seeds of tomato, brinjal ash gourd and bottle gourd.
- Use of this machine prevents skin problems/injuries to hands and legs, which are prevalent in traditional methods.
- Vegetables are directly fed into the machine unlike in traditional method

where vegetables are stored for 24 hours for seed extraction.

- The powered version of this machine is suitable for commercial vegetable seed growers.

26. Pusa Chaff Cutter Safety Interventions

Every rural household irrespective of their socio-economic status having cattle keeps the chaff cutter, which is usually operated twice daily depending upon the number of cattle. Any member of family performs the operation and the chaff cuter is also accessible to children for playing. This is associated with many injuries in the rural households. The severities of injuries are also high as it has sharp blades and feed rollers. Inadvertently, body part coming in contact with the blades may results in amputation. Similarly hands getting trapped between the serrated power rollers also result in crushing injury.

In order to prevent injuries the following interventions have been developed:

a) Blade Guard: It consists of a metal guard made up of mild steel sheet and can be attached to the two existing blade bolts. It can be opened for sharpening of the blades.

b) Flywheel Lock: It is spring loaded mechanical lock that prevents rotation of the flywheel in standby mode.

c) Warning Roller: It consists of a serrated wooden roller that gives a warning when the hand is in danger zone while feeding the chaff cutter.

Advantage

- This prevents injuries during working and non-working conditions.
- It is easy to fabricate and install on existing and new chaff cutter.

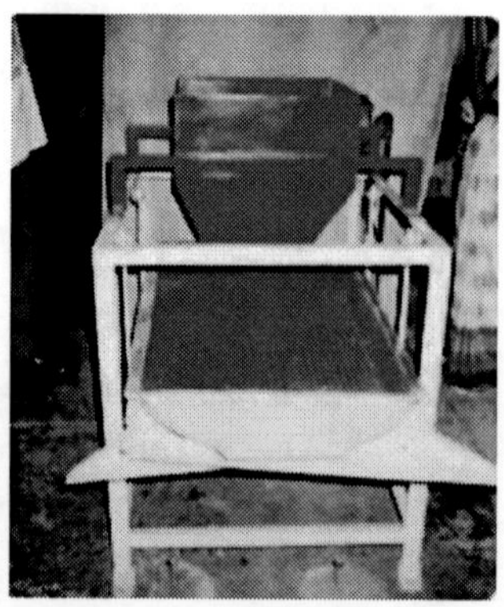

27. Pusa Rice Grading Machine

The Pusa Rice Grading Machine adds quality to the produce. Using this grading machine, a farmer can get different sizes of rice grains. Small farmers can do grading of rice at village level and earn more money.

Advantage

- All varieties of rice can be graded by the machine.
- Grading improves the quality of the product and helps in value addition.
- Grading fetches more value in the market. Requires less labour and cost.

28. Mobile Animal Feed Block Formation Machine

Animal feed blocks can be made by using the Mobile Animal Feed Block Machine in the field itself. The mobility facilitates the formation of feed blocks at a location of convenience. The blocks can be stored for a period of one year. Transportation becomes easy owing to reduction in volume.

Advantages includes:

- Animal feed or straw is reduced to 1/7 of its original volume, which results in less storage and transportation cost.
- Being portable, this machine can be used to make blocks wherever animal feed is available. Nutritious ingredients present in the feed make it a complete animal feed.
- The feed blocks can be stored for about one year so that feed is available to the animal throughout the year, especially in areas where natural calamity occurs.

29. Urea, Mollasses and Minerals Block (UMMB) Forming Machine

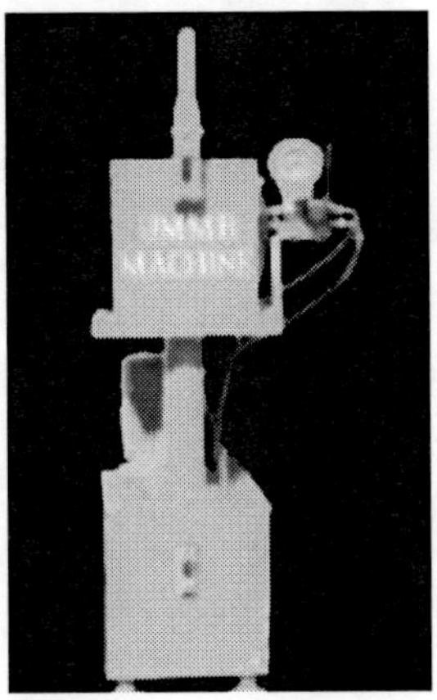

UMMB machine is used for mixing minerals (in powder form) in proper quantity along with animal feed blocks. The following ingredients are used in block formation:

Mollasses : 35 %

Urea : 10 %

Broken rice : 35 %

Groundnut : 10 %

Cement : 6 %

Minerals : 3 %

Salt : 1 %

Advantages of UMMB Machine are

- The blocks fulfill the energy and protein requirement of animals.
- Feeding milch animals with 200-300 g of this feed increases their milk production.

30. Pusa Animal Feed Block Formation Machine

The Animal Feed Block Formation Machine has been developed for making appropriate sizes of animal feed blocks by mixing crop residues with essential nutrients. These nutrient rich feed blocks are very useful in those areas where fodder is not available. The blocks have a shelf life of one year.

Frame contains a feeding chamber, a compression chamber and a feed block die. There are two hydraulic cylinders of different sizes in the machine. The larger cylinder is for compression of feed and the smaller cylinder for opening and closing the gate of compression chamber. Power pack contains a high and low pressure pump, a pressure switch, a solenoid valve, an oil tank, etc. The hydraulic cylinders are operated by high and low pressure pumps. Moving direction of the cylinder is monitored by solenoid valve. Electric control panel regularizes the movement time of hydraulic cylinder. Advantages includes;

- The machine can make different sizes of feed blocks for animal feed as well as blocks of paddy straw.
- Easy to operate by a semi-skilled operator. Using this machine, crop straw/crop residues can be converted into blocks which increase its shelf life.

- It is very economical because transport to far-off places becomes easy

31. Pusa Solar Powered Dryer

A blower is used to suck hot air from the flat plate collector to transfer it to the drying chamber of batch type forced convection dryer. To harness the solar power, a system has also been developed to run blower using solar energy. The DC power of the solar energy is converted into AC power through inverter. A battery is used to provide the backup in the absence of solar power.

Advantages of Pusa Solar Powered Dryer includes

- It is suitable for drying of fruits and vegetables.
- By operating blower with solar energy, no conventional power is required thus saving cost of electrical power.
- It may be used in areas where there is no grid electricity.
- The above solar power pack may also be used for operation of refrigerator (90W), three tube lights (84W) and a Fan (50W) also.

32. Pusa Power Operated Winnower

Cleaning of the produce is one of the important steps in grain processing and value addition. The Power Operated Winnower has been developed for this purpose.

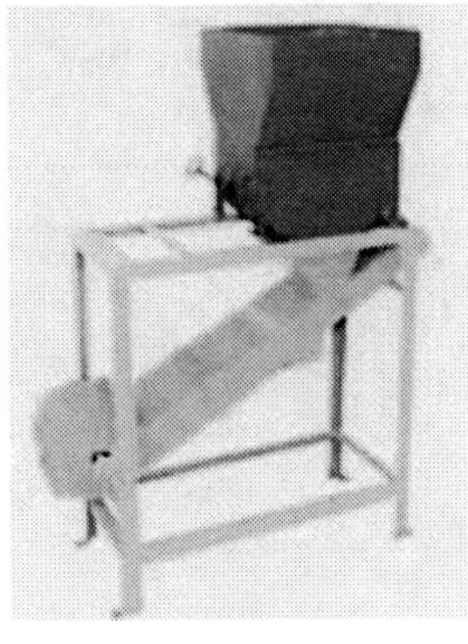

Advantages includes:

- High quality cleaning enhances the quality of produce.
- Exhaust air blows off the dust and unwanted particles and the final product is collected clean.

- All types of grains can be cleaned by using this equipment.
- Two or more mixed grains can be separated by using this machine. Saves money and time.

Conclusion

The improvements in agro technology including that of farm mechanization in India brought about revolutionary enhancement in agricultural production. From a net importing country, India is now producing more than 264 million tons (mt) of food grains. For meeting the food security needs of the country, agricultural productivity and its growth needs to be sustained and further improved. Farm Mechanization provides the technology to facilitate agricultural growth through efficient use of inputs. It can have a great role to play in the development of agriculture. Mechanization is important from tillage and seed-bed preparation, sowing/ planting, inter-culture, fertilizer application, irrigation, harvesting, to post harvest operations. With newer technologies like zero tillage, raised bed planting, residue management, precision agriculture, micro irrigation, mulching, etc., for farming the dependence on farm mechanization has increased. Adoption of mechanization ensures timeliness of agricultural operations reduces cost of production as well as reduces drudgery in carrying out various farm operations. Equipment and implements for various operations like tillage, sowing, irrigation, plant protection and threshing, etc. are generally being used by farmers.

References

Singh, G. 2006. Estimation of a mechanisation index and its impact on production and economic factors ‾a case study in India. *Biosystems Engineering*, 93(1): 99-106.

Singh Gyanendra and Hukum Chandra. 2002. Production and economic factors growth in Indian Agriculture. Central Institute of Agricultural Engineering. Tech. Bull. No. CIAE/200/91.

Singh, S. P., R. S. Singh, and S. Singh. 2011. Sale trend of tractors and farm power availability in India. *Agricultural Engineering Today*, 35(2): 25-35.

Singh, Gyanendra and D. De. 1999. Quantification of a mechanization indicator for Indian agriculture. *Applied Engineering in Agriculture, ASAE*. 5(3): 197-204.

Singh, Gyanendra and Ojha, T.P. 1993. Agricultural Engineering Research Promotion by ICAR during VIII Plan. *Agricultural Engineering To-day* (34): 42-98.

Section 7: Animal Husbandry and Fisheries

Improved Agriculture and Allied Practices for Farmers, pp 299-305
New India Publishing Agency, New Delhi, India

28

Dairy Farming - A Profitable Enterprise for Rural Employment Generation

R.S. Chhillar

ICAR-Indian Agricultural Research Institute, New Delhi-110012

Amongst agricultural commodities milk is the single largest contributor to gross value of output followed by paddy, wheat, oilseeds, pulse and sugar. It is not only that, milk is also least contributor to inflation.Thus it can be said that milk is the lifeline of Indian economy. Unlike developed countries, unique feature of Indian milk production system is that about 70% of it is produced in about 70 million household dairy units in rural areas.Dairy farming has always been a sub-system of agriculture in such household units which entirely depends on the recyclable crop residues. Moreover the milk production system has been mainly confined to landless labourers, small and marginal farmers.Their poor socio-economic condition, religious sentiments for cattle and utilitarian attitude towards dairy farming coupled with poor milk production potential of dairy animals has remained a major impediment in the overall development of the system. Livestock is an important source of supplementary income. Mixed farming has been serving as an insurance against natural calamities, while supporting food security and nutrient recycling. India has over 500 million livestock, which include cattle, buffaloes, sheep and goats. Among them, cattle and buffaloes are popular for milk production. As milk is an important part of the Indian diet and bullock power is essential as draft animal for farming and rural transportation.

India is emerging as a global economic power and the economy is growing at an average of 8- 9 per cent for last couple of years.The Dairy sector is contributing significantly to the national economy and with 112 Million Metric Tonns (MMT) milk productions during 2009-10; it ranks first in the World. It provides gainful employment and supplementary income to large number of people. Dairy sector contribute significantly on poverty reduction in rural areas as income from milk in the total income of under-privileged family is as high as 75 to 80% during drought and the employment generated is relatively high. Dairying has become an important secondary source of income for millions of rural families and has assumed an important role in providing employment and income generating opportunities. A unique feature of Indian dairying is that most of the milk is produced by small, marginal farmers and landless labourers. Dairying has provided employment opportunities to millions of families in the rural areas. Buffalo has been the backbone of rural economy, as this animal forms and integrates part of the farming system in India. It provides milk, meat, skin and also used as a draft animal. The population of buffalo is estimated to be approximately 166.4 millions of which 161.4 million of them are in Asia (97.2%). India is endowed with largest livestock population in the world. India has over 94.13 million of buffalo and their population is approximately 56.6 percent of the total world buffalo population (Aulaka and Singh, 2012 and Sethi, 2003). However, the commercial milk farming is coming up in a big way particularly around metropolitan cities and urban centers where the demand for milk is high. Besides household dairy units of 2-3 milch animals small to medium size dairy farming units are coming up in peri-urban villages also. When household dairy units with utilitarian attitude of farmers have since long been supplementing the income of farmers, the commercial milk farming is bound to be a profitable enterprise if planned, established and managed on scientific lines. The essentials in developing a profitable dairy enterprise are as under:

a) Location and buildings of dairy farm.

b) Selection and breeding of milch animal.

c) Scientific feeding practices for dairy animals.

d) Appropriate health care of animals.

e) Improved management practices.

A number of studies on cost and returns from milk production have been carried out in India (Bhangaroo, 2007; Singh *et al.,* 2009; Meena *et al.,* 2010; Tanwar *et al.,* 2012; Bardhan and Sharma, 2012) Profitability of a dairy enterprise can significantly be increased if a dairy farmer knows the art of combining ideas, facilities, processes, materials and labour for optimum production from dairy animals and proper marketing of the produce. A successful dairy farmer should plan, organize, coordinate, direct and control all his resources by deciding what to do, how to do it and when. For correct decision-making he should be well informed about all kinds of information affecting his dairy business and at least following minimum farm records.

1. Complete inventories at the beginning and end of each year including a summary of all assets, debts and net worth.
2. Production records (including milk and its by products)

3. Current expenditures and receipts (including quantities sold)
4. An annual production and financial summary.
5. An analysis of the annual records to determine strong and weak points of his business and to serve as a guide for wiser adjustments in annual budgeting.

Moreover a dairy farmer should regularly and efficiently monitor the day to day activities of the dairy farm to ensure the following:

i) Proper health care and feeding of newly born and growing calves to raise a healthy replacement stock with higher production potential.

ii) Proper health care and feeding of dairy animals during 40 days pre calving and 60 days post calving period, which is very critical for higher production and reproduction of dairy animals.

iii) Regular observation on the health condition and milk production of each Individual dairy animal.

iv) Regular update of pedigree, health, production and reproduction records of all dairy animals.

v) Timely culling of uneconomic producers from the herd.

If the aforesaid essentials of profitable dairy farming are taken care of medium or large dairy enterprises can give an annual net income @ Rs. 15000/- (Rupees fifteen thousand) per adult milch animal. To illustrate the economics of dairy farm the involved costs and returns of a 20 milch animal unit can be evaluated as under.

Basic Assumption

1. The entrepreneur is having suitable piece of land for establishing the unit with an installed tube-well of quality drinking water and is in a position to spend 20% of the fixed capital expenditure.
2. The unit will have milch animals in the ratio of 60% buffaloes and 40% crossbred cows and the animals in their first or second early lactations will be purchased at an average price of Rs. 18000/- including cost of transportation.
3. Annually the buffaloes and crossbreed cows will be in milk for 275 days and 305 days respectively with a corresponding average daily milk yield of 8 lit and 10 lit.

4. Average sale price per lit of milk is Rs. 16/- for buffalo milk and Rs. 14/- for cow milk.
5. The cost of rearing calves up to the age of disposal/replacement stock is covered by their disposal replacement stock price.
6. The average cost per quintal of dry fodder, green fodder and concentrate ration is Rs. 100/-, Rs. 100/- and Rs. 700/- respectively.
7. Bank loan is available with a simple interest rate of 11% per annum and the loan is to be repaid in 5 years in monthly installments.

From the data presented in the tables it is evident that major portion of the investment in the project is on procurement of milch animals which accounts for 60% of the total investment. Cost of construction accounts for 33.3% of the total cost. Among the annual variable costs the feed cost constitute the major portion (73%) followed by different depreciations (12.5%) labour cost (9%) The total net profit of Rs. 2,41,300 per annum is sufficient to leave a regular monthly income of about Rs. 7700 to Rs. 11200 during the loan repayment period of five years and more than Rs. 20,000 per month subsequently.

Capital Expenditure

(A)	Fixed Capital Investments	20 animal Unit
1.	Cost of animals	3,60000
2.	Cost of construction of buildings for milch animals (2 m^2 covered floor space and 4 m^2 loafing area/ animal) @ Rs. 5000 per animal	1,00000
3.	Cost of construction of office, laboratory, recording room, feed & fodder stores and calf pans, manure pit etc @ Rs. 5000 per animal	1,00000
4.	Cost of appliances and equipments @ Rs. 1000 per animal	20,000
5.	Working Capital @Rs. 1000 per animal	20,000
	Total	6,00000
(B)	**Annual Fixed Cost**	
1.	Depreciation of the cost of animal @ 15%	54000
2.	Depreciation of the cost of building @ 5%	10000
3.	Depreciation of the cost of equipments @ 15%	3000
	Total	67,000

Feeding Schedule for milch animals at the farm

Feeding during location period

	Type of feed	Quantity/animal/ day (Kg)	Cost of feed/ animal (Rs.)
(A)			
1.	Dry fodder	7 kg	7
2.	Green fodder	25 kg	25
3.	Conc. Ration	4 kg	28
	Total		60
(B)	Feeding during dry period		
	Dry fodder	10 kg	10
	Green fodder	10 kg	10
	Conc. Ration	1.5 kg	10
	Total		30

Annual Variable Cost

S.No.	Particulars	Amount (Rs.) 20 animal unit
1.	Cost of feeding dairy animals during location period (275x12x60)+(305x8x60)	344400
2.	Cost of feeding dairy animals during dry period (90x12x30) + (60x8x30)	46800
3.	Cost of labour @ Rs. 2000/month (One labour for every 10 animals and their followers)	48000
4.	Insurance of milch animals @ 4% per annum	14400
5.	Amount spent on Veterinary & breeding aid @ Rs. 2.0/- per milch animal/day	14600
6.	Annual fixed cost	67000
	Total Annual Expenditure	535200/-

Annual Income of Enterprise

S.No.	Particulars	Amount (Rs.) 20 animal unit
1.	**Sale of milk**	
	i) Buffalo milk (275 days x 8 lit x Rs. 16/- x 12)	422400
	ii) Cow milk (305 days x 10 lit x Rs. 14/- x 8)	341600
2.	Sale of compost manure (25 quintals of manure/ animal/year)@ Rs. 25 per quintal	12500

3.	Gross return	776500
4.	Total operational cost	535200
5.	Total Profit	241300
6.	Profit per animal per year	12065

Economics of the Enterprise From I[st] to 5th Year

Year	Gross Profit	Repayment of Principal	Repayment of interest	Loan outsta-nding	Net surplus for family expenditure
Ist	241300	96000	52800	384000	92500
2nd	241300	96000	42240	288000	103060
3rd	241300	96000	31680	192000	113620
4th	241300	96000	21120	96000	124180
5th	241300	96000	10560	NIL	134740
Subsequent Years	241300	-	-	-	241300 or (Rs. 20108 Per month)

Conclusion

The Indian dairy industry has grown consistently ever since the White revolution of the 1970s, making India, the world's largest producer of milk with 17% global share. With an annual production of 146 million tonnes of milk India generates approximately USD 70 Billion of revenue. The Indian dairy market is expected to double within the next decade, primarily driven by over 15-20% growth in value added dairy segment. To leverage this high growth potential and to meet the rising demand, a sustainable and strong dairy farming base will be critical. For achieving this, it becomes critical to address key challenges faced by the industry such as, low animal yields, ineffective breeding, improper feed and fodder management, deficient veterinary care, poor farm management and low financial inclusion among others. With dairy farming in India dominated by smallholder farmers, with an average herd size of less than 2, it becomes all the more challenging to address these issues, in the specific context of making small holder dairy farming globally competitive. The cost of milk production from dairying is an important aspect for producers, consumers and policy makers to provide effective linkage between the milk producers and consumers, so that the producer get remunerative price for their milk and consumers get milk and milk products at reasonable rate.

References

Aulaakh, G. S. and Singh, Rajbir, (2012). Adoption of recommended management practices by the buffalo owners. *Indian J. Dairy Sci.* 65(5): 431-434.

Bardhan, D. and Sharma, M. L. (2012). Economics of milk production in Kumaun region of Uttarakhand. *Indian J Dairy Sci.* 65(5): 416-422.

Bhangaros, K.S. (2007). An economic analysis of dairy farming in rural Punjab: A case study. *Journal of Agriculture Development and Policy.* 17(1&2): 26-28.

Meena, G. L.; Jain, D. K. and Chandel, B. S. (2010). Economic analysis of milk production in Alwar disrtrict of Rajasthan. *Journal of Dairying Foods and Home Science.* 29(1): 1-7.

Singh, S. P.; Gill, M. S.; Gangwar, B. and Singh, M. P. (2009). Livestock in irrigated farming systems of Uttar Pradesh. *Indian J. Animal Sci.* 79(9): 925-931.

Tanwar, P. S.; Kumar, Yogendra and Sankhala, Gopal. (2012). Economics of milk production among member and nonmember families of dairy cooperatives in Jaipur (Rajasthan). *Indian J. Dairy Sci.* 65(5): 405-409.

Improved Agriculture and Allied Practices for Farmers, pp 307-314
New India Publishing Agency, New Delhi, India

29

Fishery Based Entrepreneurial Opportunities in India

Usha Moza

Fisheries Division, ICAR, Krishi Anusandhan Bhavan-II, New-Delhi. 110 012

India is endowed with 8118 kilometers of coastline and access to 5.3 lakh square kilometers of continental shelf to facilitate marine fishing.195210 kilometers long river and canal system, more than 6 million hectares of water bodies under reservoirs, ponds, Fish is an integral part of food and culture in India. Indian coastal region is rich in aquatic bio diversity with various identified fish species and also in aquatic resources. India with its vast and varied marine and inland resources stands third in world total fish production and second in inland fish production. India's fish production has increased from 5.6 million tonne during 2000-2001 to 6.9 million tonne during 2006-2007. During the year 2006-07, the country witnessed foreign exchange of Rs.8363.53 crore through export of sea food. Fish processing industries in India are gaining importance in recent years due to increase in demand of processed food by the developed countries. In India fishery sector provide livelihood to an estimated 10 million people and the Inland fisheries are of particular importance to the rural poor accounting for about 15% of total global employment (FAO, 2000). Sustainable utilization of available resources with appropriate farming system approaches will increase fish production and also will generate employment opportunities.

Fishery in India has been mostly traditional capture fisheries from natural resources, but with advent of aquaculture explosion, fish farming is being practised by sizeable section of population, generally in individual and community tanks or ponds. In addition comprehensive development of its natural resources call for proper scientific management to boost its fish production. In the current write up, some of the measures that should be adopted by fish growers to transform fish farming into entrepreneurial opportunity are being elaborated. Recently the culture sector, especially aquaculture, had a boom in productivity due to the availability of various technologies like, intensive and super intensive culture. The intensive and super intensive cultures, either in running waters or recirculatory systems with very high stocking densities and production potential are one of the upcoming trends in these days. The trend of production, as a result, shows a higher potential for aquaculture when compared to marine capture fisheries. In the marine sector there is a need to exploit the deep-sea beyond 50m depths, besides diversifying the activities of the fishers towards other activities like, processing and product development. It is recognized as a powerful income source and employment generator as it stimulates growth of a number of subsidiary industries and is a source of cheap and nutritious food besides being a foreign exchange earner. In India it is estimated that 12 million people are directly engaged in fishing and about 60 million are exclusively dependent on it for a living (Manasi *et al.* 2009). To meet the multiple objectives of poverty reduction, food security, competitiveness and sustainability, several researchers have suggested farming systems approach to research and development. A farming system is the result of complex interactions among number of interdependent components where an individual farmer allocates certain quantities and qualities of four factors of production, viz. land, labour, capital and management to which he has access. Farming system research (FSR) is considered as a powerful tool for the management of natural and human resource in developing countries like India. FSR is a multi-disciplinary whole-farm approach for solving the problems of small and marginal farmers. This approach aims at increasing income and employment from smallholdings by integrating various farm enterprises and recycling crop residues and by-products within the farm itself. Under the gradual shrinking of land holding, it is required to integrate the land-based enterprises like fishery, poultry, duckery, apiary, field and horticultural crops etc. within the bio-physical and socioeconomic environment of the farmers to make farming more profitable and dependable. Currently, doubling the farm incomes has become a major focus of Government of India and specifically the vision of our honorable Prime Minister. Besides, attracting and retaining rural youth to agriculture has been a major challenge for the nation. Farmers are leaving agriculture, since it is becoming unprofitable under present scenario due to numerous resource- and production–vulnerabilities besides climate change. Under such scenario, integrated farming systems (IFS) seems to be a ray of hope which has great potential in raising the farmers incomes and employment generation, minimizing the risks in farming and enhancing the resource use efficiency, thus, leading to sustainable agriculture. IFS also prove as a potential approach for retaining the rural youth in agriculture through ample farm employment under different farm enterprises and better livelihood options in small and marginal farms.

Pen culture of carps

Beels are major fishery resource contributing about 25% of the fish production. However major portions have been rendered unproductive due to excessive siltation and growth of weeds and only 33% of potential is being utilized for fishery. Currently fish yield from beels is 300kg/ha /yr the same can raised to 1,000-1,5000kg/ha/yr through applying pen culture technology.

Pen is an enclosure that can be utilized to hold fishes as captive stock with in beels and wetlands. Useful in shallow, derelict and weed choked water bodies, where aquaculture is not possible and fishing is difficult. A production of 400 kg table size Indian major carps (*Catla catla, Labeo rohita, Cirrhinus mrigala*) can be achieved from a pen unit of 0.1 ha area with stocking of 1000-1500 fingerlings of 10g/100mm size; 80kg freshwater prawn (*Macrobrachium rosenbergii*) can be produced from a pen of same size with stocking of 3000-4500 juveniles of 60-80mm size, in a grow out period of four months with a survival of 80%. The technology can also be used for raising fingerlings for stocking. The technology is eco-friendly and involves low input cost including supplementary feeding and fish health management.

The characteristics of ideal site for erecting pen are

- Shallow area having gentle slope with firm and smooth bottom having 1.5-2.5 m depth and water availability for at least 8 months
- Area with low water fluctuation
- Number of pen units to be put up depends on the size of water body and production requirement.

Management needed is

- Desiltation, Deepening of connecting channels, Deweeding, Suitable changes in lease policy and ownership arrangements

Economics worked out of pen culture in 0.1 ha pond is,

Incurring cost, Rs. 30,000 to 35, 000/; Net income, Rs. 13000/2 crops (Grow out carp), Rs. 8000/2 crops for Prawns.

Local material like Bamboo (140 no.), coir rope (35 kg) Nylon net material for screening and fingerlings of 100-150 cm size for stocking needed.

Enhancement of Seed production through pen nurseries

The increase in pace of aquaculture activity all over the country has put great demand for fish seed and pressure for pond/ tank resources. The same can be met by exploiting the natural resource of beels and reservoirs. The seed rearing business is more profitable as it needs water availability only for 2-3 months. The incurring cost and net profit although same as in grow-out culture but one can get returns from seed production within 3 months and 1-2 crops can be harvested within a season thereby making it more profitable.

Carp culture by scientific methods

The standardized package of practices for intensive carp polyculture that includes stocking of Indian major carps and exotic carps of 20-50 g size at densities of 20,000-25,000/ha; pond fertilization with application of *Azolla* at 40 t/ha/yr at weekly split doses as biofertilizers substituting for traditional organic and inorganic fertilization; provision of balanced supplementary feed; provision of aeration to keep the required dissolved oxygen level; maintenance of water depth over 2 meters and water replenishment depending on the water quality, and fish health management through prophylactic and curative measures.

Input needed are, Fish seed, organic and inorganic fertilizers, feed

Output capacity: 5-8 t/ha/yr

Unit Cost: Rs: 120,000/- per unit area of 1 ha (on lease)

Net income: Rs 73, 000/ha

The aquaculture can be made more profitable if the indigenous species preferred by local population will be cultured along with usual six species combination, or one or two species will be replaced by local variety. Following species, which have been successfully bred in captivity need to be takenup.*Labeo calbasu* (Kalbasu); *N.hexagonolepis* (Chocolate mahseer);*O.balengari* (Pengba); *S. semiplotus* (Semiplotus).

Culture and Seed production oF *Clarias Batrachus (Magur)*

Magur *(C.batrachus)* is the most popular catfish owing to its taste and good medicinal value. It can thrive well in derelict water and swamps. India having large number of swamps, the resource can be utilized effectively as the fish has ready market and fetches price of Rs 250-300/kg compared to Rs 100/kg for carps.

The technology package from breeding to grow-out culture of Magur on commercial scale has been worked out .It needs administration of hormones like carp pituitary extract (30-40 mg/kg), HCG (4000 IU/kg) and Ovaprim or Ovatide (0.6-0.8 ml/kg). Stripping after 14-17 h and artificial fertilization through sperm suspension. Incubation of the fertilized eggs in a flow-through hatchery. Water exchange and high protein diet.Larval rearing in indoor rearing tanks fed with live plankton. A two-phase, semi closed clear water technique of larval rearing for seed production. First phase rearing (10-12 days) at high density (200-400/ m^2). Second phase (20-25 days) rearing at low density (50-60/m^2). Feeding with diet of *Artemia* nauplii, and egg custard. Water exchange (50%) and monitoring of water quality parameters. Seed output ranges between 35-40/leter and cycle lasts for 35-40 days.

The post larvae are gradually acclimatized to freshwater and reared at high densities (2000-5000/m^2) for 10-15 days in hatchery, which are suitable for stocking in grow-out ponds.Juveniles are stocked in grow-out ponds (0.1-0.4 ha) and are fed daily with formulated pellet diet (2-3 mm size) @ 10% of the biomass initially and then reduced to 3% of the biomass at the end of the culture period.

The expenditure incurred in magur hatchery works out to be Rs. 160, 000/- and Net income Rs 46, 300/ for 1.5 lakh fry within 2 months. For Growout culture, it is Rs 2,43,000 and net income Rs 81,640/ha (7-8 months)

Already Magur farming is being taken up in many areas of Assam like Barpeta, Nalbari and Sonitpur and farmers are earning Rs 2- 3 lakhs per year . Same needs to be extended to other districts like Morigoan, sibsagar, Nagoan having large number of water resources in the form of wet lands and swamps

Culture of fresh water Prawn, *Machrobrachium rosenbergii*

Aqua culture production of giant fresh water prawn in India has shown phenomenal increase in recent years due to high economic returns and ready export market. However non availability of seed has been major bottleneck for commercial farming but with availability of seed in adjoining state of Tripura , the same can be taken up either as a monoculture or in polyculture replacing either common carp or mrigal from the combination.

The post larvae are gradually acclimatized to freshwater and reared at high densities (2000-5000/m^2) for 10-15 days in hatchery, which are suitable for stocking in grow-out ponds. Juveniles are stocked in grow-out ponds (0.1-0.4 ha) and are fed daily with formulated pellet diet (2-3 mm size) @ 10% of the biomass initially and then reduced to 3% of the biomass at the end of the culture period.

The incurring cost for 1 ha pond has been worked out to be Rs 1,52, 000 and net income Rs 47, 000 with a production of 800kg /ha

Ornamental Fish Culture

Ornamental fish keeping in households is becoming increasingly popular day by day throughout the world as also in India .Global ornamental fish trade currently stands at 22 billion dollars. India's overall share is only Rs 10 crores. The resources for same are mainly (80%) contributed from north-eastern states like Assam and Manipur .But the resources are mainly exploited from natural habitat which will lead to depletion of these in the long run.Dependence on natural resource is also a bottleneck for developing economically viable entrepreneurship out of it, as pre-requisite for any trade to flourish is adequate quantity and continuous supply of essentials. Therefore breeding and rearing of Ornamental fishes in captivity is essential.

At present the indigenous and exotic freshwater species, that have good demand and can be bred and reared for commercial purposes include, Guppies (*Poecillia reticulate)*, Moly (*Mollinesia* sp.), Sword tail (*Xiphophorus* sp.), Platy, Goldfish (*Carassius auratus)*, Koi carp *(Cyprinus carpio var.kao),* Zebra danio (*Brachydanio rerio*), Black window tetra (*Cymnocrocymbus* sp.), Neon tetra (*Hyphessobrycon innesi)*, Serpae tetra (*Hyphessobrycon callistus),* Angelfish, *Pterophyllum scalare,* Red-line torpedo fish (*Puntius denisoni),* Loaches (*Botia sp.),* leaf-fish like *Nandus nandus,* snakehead (*Channa orientalis)*

Enhancement of Production through Integrated Fish Farming

Successful Cases of Fish Culture in India

- Balusamy *et al.* 2003, explained that rice + Azolla + fish culture is one of the economical option in such type of area. Monoculture system rely mainly on external inputs while in integrated system, recycling of nutrients takes place that help in reducing the cost of production for economic yield. The fish in rice field utilized the untapped aquatic productivity of rice ecosystem as the rice bottom is highly fertilized on account of the production of zoo and phytoplankton and these resources are fully utilized by the fish. The data clearly advocated the beneficial effect of Azolla on rice+fish. The gross income obtained in rice + Azolla + fish was 25.7 % more over the rice crop and 6.9 % more over the rice + fish. The net income followed the same trend. Thus rice + Azolla + fish on an average gave Rs 8,817/ha more over the rice monoculture and Rs.3,219/ha over the rice + fish. This model was proposed for extensive scale adoption in Tamil Nadu.

- Fresh water is becoming scarce day by day due to increase in population and their emerging demands. To overcome this fresh water scarcity, there is growing awareness to adopt integrated farming systems to improve overall water productivity and maximize net returns. Some of the approaches of Fish farming along with other domesticated animals which have been validates are as follows

Fish cum pig farming	
Fish production: 6-7 tonnes/ha/yr	Animal Production: 4000-5000 kg/crop
Stocking rate: 8000-9000/ha	Stocking no.:30-40 pigs (60-70 kg/animal)
monthold Fish composition 6 species	Feeding rate: 1 kg/day
No fertilizer+feed	
Fish cum Duck farming	
Fish production: 3-4 tonnes/ha/yr	Animal Production: 500 kg/crop
Stocking rate: 6000/ha	Egg Production: 17000-20000
Fish composition 6 species	Stocking no.:200-400
No fertilizer+feed	Feeding rate: 100gm/bird/day
Fish cum Poultry Farming	
Fish production: 4-5 tonnes/ha/yr	Animal Production: 1500-2000 kg
Stocking rate: 6000/ha	Egg Production: 60-70000
Fish composition 6 species	Stocking no.:400-600
No fertilizer+feed	Feeding rate: 100gm/bird/day

Paddy cum Fish culture

Water is crucial for economic development and sustenance of life. Around 85% of fresh water is being used in agriculture sector which is likely to reduce to 69% in future due to other compelling demands. This calls for enhancing productivity of water to produce more food with less water. Fisheries constitute an important enterprise to raise agricultural water productivity. Indian coastal region and cold water fisheries areas offers conducive environment for such activity and the same has been in practice traditionally. This activity needs to be followed on scientific lines. The area meant for pisiculture should be netted thoroughly to weed off unwanted fishes, and fishes be provided with feed.

Some important interventions for livelihood improvement can be explored by ICAR-RCER, Patna as :

i) Rice–fish–poultry farming system-a success story from Tamil Nadu

ii) Integrated rice–fish–vegetable system for enhanced livelihood in selected backward districts of Assam

iii) Integrated fish–vegetable–poultry/pig system

- Enhanced livelihood through pig-fish-vegetable system in Assam
- Enhanced livelihood through fish-livestock-horti system in Dhalai
- Fish–duck–pig based farming system in South Garo Hills, Meghalaya
- Fish–duck–pig based farming system in Dhemaji, Assam
- Integrated poultry- fish–vegetable farming system, Meghalaya

iv) Improvement of livelihood through integration of fish with makhana (Euryale ferox Salisb.) & Water Chestnut (Trapa bispinosa Natans.) in flood prone ecosystem of Darbhanga (Bihar).

v) Utilization of upland fallows for cultivation of maize for increased income, food and nutritional security in Bastar region (Meena *et al.,* 2013).

References

FAO. The State of World Fisheries and Aquaculture. Food and Agriculture Organization of the United Nations, Rome, 2000.

Manasi S, Lathe N, Raju K V. Fisheries and livelihood in Tungabhadra Basin, India: Current status and Future possibilities. The institute for social and Economic change, Bangalore, India, 2009, 1-24.

Meena, M.S. Singh, K.M. Bhatt, B.P. and Kumar, U. (2013). Gender perspective in Integrated Farming System, Training Manual No. TM-2/PAT-2/2013, ICAR Research Complex for Eastern Region, Patna, Bihar, India.

Balusamy M, Shanmugham PM, Baskaran R. Mixed farming an ideal farming. Intensive Agric. 2003;41 (11–12):20–25.

Improved Agriculture and Allied Practices for Farmers, pp 315-330
New India Publishing Agency, New Delhi, India

30

Scientific Lac Cultivation – A Viable Option for Rural Empowerment and Enhanced Income

J.P. Singh and A.K. Jaiswal*

ICAR-Indian Agricultural Research Institute, New Delhi-110012

**ICAR-Indian Institute of Natural Resins and Gums, Ranchi-834010*

Lac insect is an insect of commercial importance like silk worm and honey bee. The silkworm produces silk, honeybee produces honey and wax whereas, lac insect produces three products namely resin, dye and wax of great commercial importance. The resin commonly called lac, is only one of animal origin and commercially available in the market as shellac or seed lac or as a button lac. The resin is in fact the hardened secretory product of lac insect secreted as a protective covering of an insect. Lac culture is one of the important sources of livelihood and rural economy for poor resource constraint people particularly inhabiting in forest and sub-forest areas. It is one of the important sources of livelihood security in many parts of the country in states like Jharkhand, Chhattisgarh, Madhya Pradesh, West Bengal, Odisha, Maharashtra, Madhya Pradesh, Uttar Pradesh, Assam, NE regions etc (Pal *et al.,* 2007). It has also high potential for generating employment for both men and women in forest and sub-forest areas of the country. It is a very remunerative crop, paying high economic returns to the farmers and also foreign exchange for the country through its export.

India is world leader in production, processing and export of lac. Other lac producing countries are Myanmar, Thailand, parts of China, Taiwan, Indonesia, Philippines, Vietnam and Cambodia etc. Being an export-oriented commodity, the country alone fulfills around 55-60% of the total global demand. The export earnings from lac and lac products are around Rs. 175 crores annually. Lac is exported to more than 70 different countries, but the major markets are Indonesia, Germany, U.S.A., A.R.E., Spain, Bangladesh, Italy, Switzerland, U.A.E and U.K. Shellac, seed lac, de-waxed shellac, aleuritic acid, bleached lac, gasket shellac, lac dye, molamma lac and shellac wax are the forms of lac exported from India. The country's production of lac was 19,577 tonnes in 2012-13 (Yogi RK, *et.al,* (2014). Lac resin is a natural biodegradable product and finds wide utility in commerce and industry. The resin finds its use in various industries as adhesive, electrical insulators, varnish, printing ink, cosmetic, leather, food, automobile and pharmaceutical etc. The dye obtained from lac is used in dyeing of wool and silk, soft drink, pill coating, confectionery and chocolate coating etc and has potential to be used as food additive. The lac wax is used in automobile and floor polish, bottle sealer, tailor's chalk, fruit coating etc. A definite demand already exists for lac-derived materials, besides a tremendous potential for much higher consumption, due to global trend for safer natural products.

Lac insect

The lac insect is a soft-bodied insect belongs to Coccids group of order Homoptera. There are Nine genera and 87 species of lac insect all over the world of which Two genera (*Kerria* and *Tachardina*) and 19 species of lac insects have been identified from India. Two species contributing nearly 99% of the national lac production are commercially important. *Kerria lacca* (Kerr) is commercially most important lac insect, widely distributed in the plains and plateau of India. Whereas, *Kerria chinensis* is a very common species in East Asian countries, but in India, confined only to the North Eastern Hills. *Rangeeni* and *Kusmi* are the two bi-voltine strains of lac insect.*Rangeeni* matures in June- July and October-November whereas *Kusmi* matures in June-August and December-February. Tri-voltine insects found in coastal areas of Odisha and West Bengal, which matures in March/April, June/July and October/ November.

Stages of Lac Insect Development

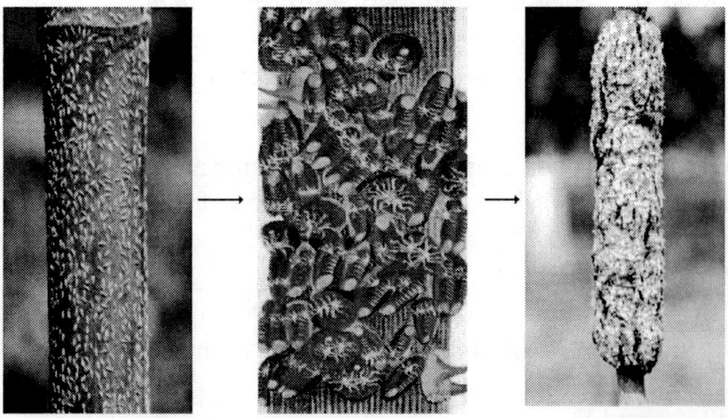

First Instar Larvae → Sexual Differentiation → Mature Lac

Life Cycle of Lac Insect

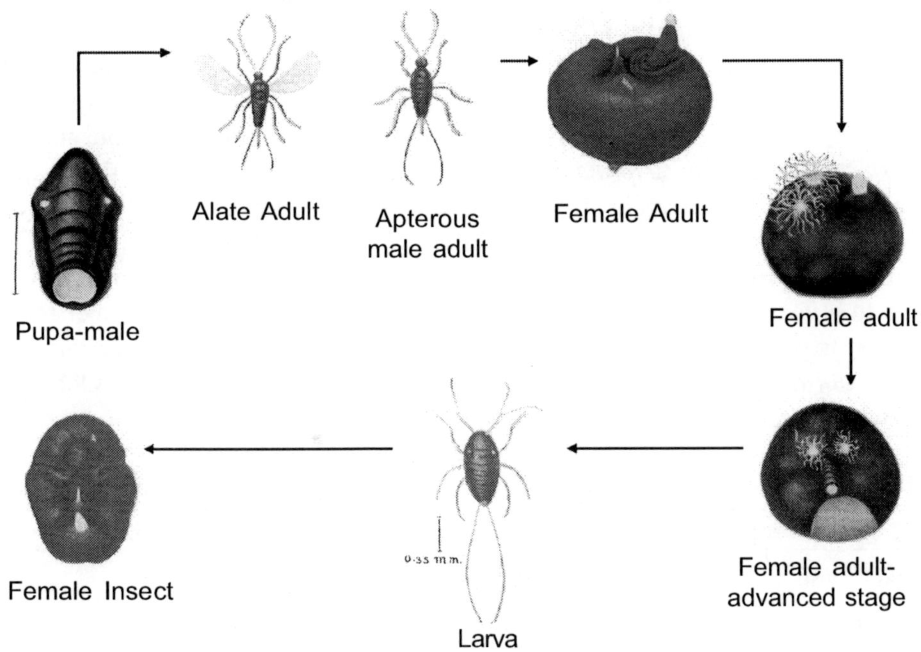

Table 1: Stages of bi-voltine lac insects and their life span in days*

Strain	Crop	Young Ones			Male		Female	
		I	II	III	Adult	Total	Adult	Total
Rangeeni	*Katki* (Rainy)	20(3)	14(2)	8(1)	2	44 (6-7)	67 (10)	109 (16)
	Baisakhi (Summer)	50(7)	40(6)	15(2)	3	108 (15-16)	145 (21)	250 (36)
Kusmi	*Aghani* (Winter)	20(3)	14(2)	10(2)	2	46 (6-7)	150 (21)	194 (28)
	Jethwi (Summer)	32(5)	25(4)	12(2)	2	71 (11-12)	90 (13)	159 (24)

Figures in parentheses indicate time in weeks

Host of lac insect

Lac insect thrives on some specific hosts called lac hosts and more than 400 lac hosts have been reported on which lac insects survive. But lac insect commercially cultivated on *Schleichera oleosa (kusum), Ziziphus mauritiana (ber), Butea monosperma (palas)* and bushy host,*Flemingia semialata (semialata)*. *Rangeeni* and *Kusmi* are two strains of lac insect, each of which produce two crops in a year (bi-voltine). *Kusmi* strain of lac insect commercially cultured well on *Schleichera oleosa* (*kusum*) trees and also on a few other lac host trees but not on *Butea mononsperma* (*palas*), whereas *rangeeni* strain grows well mainly on *palas* and also on a few other trees but not on *kusum*. Whereas on *Ziziphus mauritiana*, *kusmi* and *rangeeni* both the strains can be grown. The *rangeen*i insect matures two times in a year, once in October-November and thereafter in June-July. Whereas, *kusmi* strain matures in January–February and then in June-July. So, together there are four crops of lac in a year. In coastal regions of Odisha and West Bengal, however three crops (trivoltine lac insect) in a year are produced by *Kerria* sp on *kusum, ber* and *rain trees*. The insect matures in March/April, June/July and October/ November.

Table 2: Lac-host plants and their suitability for lac crops

S.No.	Host	Common name	Strain	Suitable Crop
Major hosts				
1.	*Butea monosperma*	Flame of forest	*Rangeeni*	Summer and Rainy crop
2.	*Schleichera oleosa*	*S. oleosa*	*Kusmi*	Summer and Winter crop
3.	*Ziziphus mauritiana*	*Plum tree*	*Rangeeni*	Premature summer crop
4.	*Z. mauritiana*	*Plum tree*	*Kusmi*	Winter crop
5.	*Flemingia semialata*	*semialata*	*Kusmi*	Winter
Other hosts				
6.	*Acacia catechu*	*Khair*	*Kusmi*	Winter crop
7.	*Albizia lebbek*	*Siris*	*Kusmi*	Winter crop
8.	*Albizia lucida*	*Galwang*	*Kusmi*	Summer crop
9.	*Albizia saman*	*Rain tree*	*Rangeeni*	Summer and Rainy crop
10.	*Croton oblongifolius*	*Putri*	*Rangeeni*	Summer crop
11.	*Ficus cunea*	*Porho*	*Rangeeni*	Summer crop
12.	*F. glabella*	*Putkal*	*Rangeeni*	Summer crop
13.	*F. glomerata*	*Dum, Z. mauritiana*	*Rangeeni*	Summer crop
14.	*F. infectoria*	*Pakur*	*Rangeeni*	Summer
15.	*F. religiosa*	*Pipal*	*Rangeeni*	Summer
16.	*Protium serratum*	*Kandeur*	*Kusmi*	Winter
17.	*Ziziphus xylopyra*	*Ghont*	*Rangeeni*	Premature summer

Lac cultivation technologies are very simple can be easily understandable and farmers friendly. Due to increasing awareness about the merits and benefits of lac cultivation, now the farmers are switching over from old traditional lac collection to scientific lac cultivation practices even by planting the lac hosts on plantation basis. Lac cultivation operation involves several major steps, which are as follows:

1. Pruning of host trees

The pruning of trees is an essential step for lac cultivation in order to have large number of new, succulent and healthy shoots at the time of broodlac (seed material) inoculation of trees with lac insect. The dead, cracked and broken branches, which are not suitable for lac culture are removed. The pruning also helps the tree to maintain its vigor, after harvesting of the crop as well as proper tree canopy can be maintained to facilitate proper and smooth cultural operations. The pruning time and techniques for various lac host trees have been standardized and should be strictly followed for sustained lac production and productivity.

The trees are pruned with axe or *dauli* / secateur or with tree-pruner. The twigs more than 2.5 cm diameter should not be cut but the twigs of less than 1.5 cm diameter should be cut from the point of origin. The twigs between 1.5 to 2.5 cm diameters should be cut, leaving 1-1.5 ft length from the point of origin. Shoots of *Flemingia* sp should be cut 6 inch above the ground. Pruning of lac hosts viz., *palas, ber, galwang, bhalia and semialata* have to be carried out 5-6 months prior to inoculation, whereas pruning of *kusum* and Australian babul should be carried out 12-18 months prior to inoculation.

Partially pruned tree of *Butea monosperma*

Pruned tree of *Schleichera oleosa*

Table 3: Pruning schedule of different lac hosts

S.No.	Lac host	Lac crop	Pruning time
1.	*Butea monosperma (palas)*	*Rangeeni* rainy season	Mid February
		Rangeeni summer season	April
2.	*Schleichera oleosa (kusum)*	*Kusmi* winter season	Jan-Feb
		Kusmi summer season	June-July
3.	*Ziziphus mauritiana (ber)*	*Kusmi* winter season	Jan-Feb
		Rangeeni summer season	April
4.	*Flemingia semialata*	*Kusmi* winter season	Jan-Feb
5.	Galwang	*Kusmi* summer season	June-July

2. Inoculation of broodlac (Infestation of host trees)

In case of lac, broodlac (lac stick with mature living mother cells ready to give birth to young ones of lac insect) is just like a functional seed of other crops. Infestation of host trees with lac insect is done by placing the bundles of broodlac on host trees at the time of hatching, for allowing young larvae to come out from mother cells and settle on shoots of lac host plants. The unwanted portion from broodlac sticks like leaves and petioles and part of twigs, where there is no settlement of lac insect, is removed. The broodlac sticks are cut into 15-20 cm length and are bundled (about 100 gm) with thread. These bundles are tied on to the branches parallel shoots on upper side. Normally 10-25 gm of broodlac for one meter of succulent shoot will be required for infestation to trees; depending on host tree as well as lac crop and quantity of broodlac/tree for inoculation should be calculated accordingly.

Inoculation of broodlac on tree species

Inoculation on bushy lac host, *Flemingia semialata*

Table 4: Inoculation time for different lac crops

S.No.	Lac insect strain	Lac crop	Inoculation time
1.	*Kusmi*	Summer season	January-February
		Winter season	June-July
2.	*Rangeeni*	Summer season	October-November
		Rainy season	June-July

3. Removal of used-up broodlac (*phunki*) sticks

After complete emergence of lac larvae from broodlac, the used-up bundles (*phunki* bundles) are to be removed from trees. This is an essential operation in order to minimize attack of predators and parasitoids harboring broodlac to new lac crop and is also necessary to avoid wastage of lac after drying of lac encrustation on *phunki*-sticks which normally fell on ground.

The *phunki* sticks should be removed as soon as emergence of lac insect larvae is over. Normally the emergence of lac insect from broodlac is over in 7-10 days during summer and within three weeks during winter months. This period should not exceed more than 3 weeks after infestation.The removed bundles are collected and lac encrustation is scraped either by simple knife or by lac scraping machine. The scraped lac from this *phunki* sticks have to be disposed to village market as early as possible.

4. Harvesting of lac crop

The lac crops are harvested either as a mature lac crop or immature (ari) lac crop from lac host trees along with host twigs. If the harvested lac is to be utilized for re-infesting (inoculation) the other trees for raising next lac crop i.e. to be used as a broodlac then only mature lac is harvested. The maturity is assessed by post-embryonic developmental stages of lac larvae inside the mother cell. Broadly speaking, it is to be cut from parent host tree when young-ones are ready to hatch from mature crop. However, the immature lac can be harvested at any time. Harvesting of immature lac should be avoided as this result into shortage of broodlac for raising next crop in coming days. The immature harvested lac is scraped, sold in market as raw lac or kept for drying. This freshly scraped lac is spread on floor at a place where there is sufficient aeration, as this is preferably dried in air. In most of the lac host plants, pruning is also carried out simultaneously while harvesting the lac crops.

5. Insect pest of lac and their management (crop protection)

Lac insect is sedentary in nature and has a complex of associated insect fauna. Predators and parasitoids constituting third trophic level cause substantial damage to the lac insect. Several species of parasitoids constituting fourth trophic are useful biological control agents keeping predators of lac insects under check. In some cases, predators and parasitoids problems are a serious bottleneck in introducing lac culture in newer areas. The major insect predators include *Eublemma amabilis, Pseudohypatopa pulverea* and *Chrysopa* spp. Besides, several species of parasitoids especially *Aprostocetus purpureus, Tachardiaephagus tachardiae and Eupelmus tachardiae* are also causes substantial loss and posing a serious threat to lac crop if not managed properly. So, lac crop protection is one of the major components in sustained lac production and quality broodlac production.

The lepidopteron enemy insect feed on lac insect in zigzag way. When the lac encrustation is thickened, the larvae conceal itself and form tunnel like structure. Sometimes a dome shape structure appears on lac encrustation which is formed from the excreta of caterpillars. The excreta of *E. amabilis* are brown and disc shaped while of *P. pulverea* is black and like a very small granules. *Chrysopa* may be identified by its eggs on top of the silken threads over lac encrustation. At the early stage of crop growth, chrysopids destroying lac culture by feeding on lac insect and gaps are visible over lac encrustation. The larvae of *Chrysopa* wrapped lac insect wax around its body and mimics lac culture to avoid its identification.

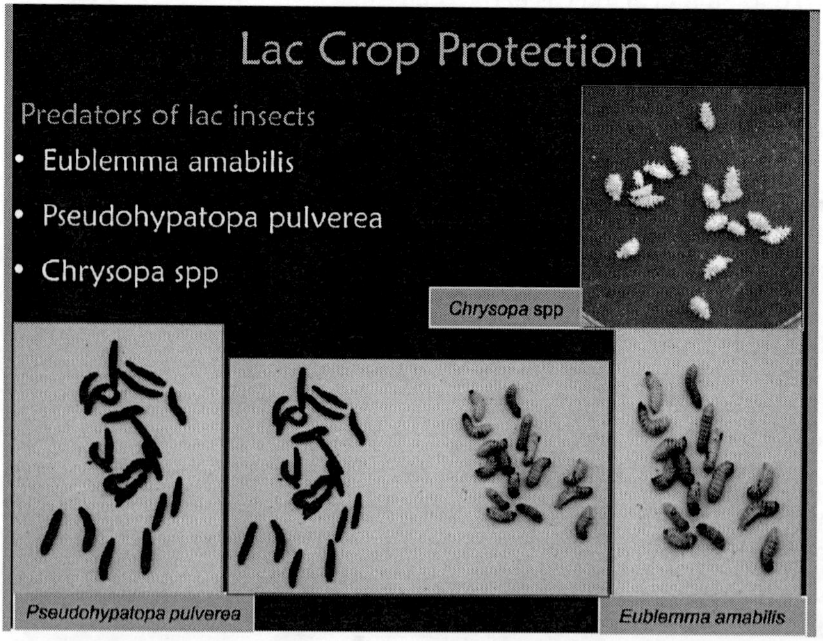

Integrated Management of insect pests associated with lac crop

- Preferably healthy broodlac free or minimum level of enemy insect infestation should be used for inoculation/raising of new lac crops.
- Inoculated broodlac bundles should be kept on tree for a minimum period only, till larval emergence continues from broodlac and ordinarily not beyond three weeks from date of lac insect larval emergence.
- It is strongly recommended to use 60-mesh nylon netting bags (27 x 12 cm) as brood container to keep broodlac (H" 50 g) for inoculation, particularly when broodlac obtained from repeatedly self-colonized culture. It helps to trap the enemy insects without interfering with lac insect emergence from broodlac. This practice is especially advisable when lac cultivation is being introduced for the first time in newer areas.
- Some trees (around 10%) should be inoculated heavily with more broodlac (twice to normal rate of broodlac) as a trap crop. Such trees should be in between normally inoculated trees. The heavily inoculated trees will attract/ trap the enemy insect which intern either can be harvested before maturity of crop or sprayed with recommended insecticides for control of enemy insects.
- Self-colonisation of lac culture should be avoided except when recommended and it should be not be more than once.
- All harvested lac crop which are not suitable/required for broodlac purpose and all *phunki* lac (used-up broodlac sticks) should be scraped as early as possible to avoid multiplication of enemy insects because lac on sticks are more suitable for breeding of enemy insect rather than scraped lac.
- Fresh scraped-lac may be offered to poultry for feeding of larvae of insect-predator of lac insect present in the lump of lac.
- The harvested scraped lac should be:

  (a) Removed from the vicinity of lac inoculated trees

  (b) Exposed to sunlight for 2-3 hours

  (c) Converted into seedlac as early as possible to kill the enemy insects and to increase shelf life of lac
- The cultivation of *kusmi* strain of lac insect in predominantly *rangeeni* areas and *vice versa* should be avoided because it provide easy access of enemy insect on each strain.

Chemical control measures

The insecticide needs to be applied to contain the menace of enemy insects. The incidence of Chrysopids starts in the beginning itself with the settlement of lac larvae. Similarly oviposition of both lepidopteran predators also begins at the same time and hatching take place within a week. After hatching, larvae of all these three predators start feeding on settled lac larvae and hence it is necessary to initiate control measures at this time. The recommended insecticides for insect pest management are selective which are not harmful to lac insect but able to control enemy insects effectively. The name of insecticides and its recommended concentrations are as follows:

Table 5: Recommended insecticides and their concentrations

S.No.	Insecticides		Concentration	Quantity per
	Common Name	Trade Name	liter of water	
1.	Fipronil 5 EC/SC	Regent	0.007%	1.5 ml
2.	Fipronil 80 WG	Jump	0.007%	0.1g
3.	Indoxacarb 15.8 EC	Dhawa Gold	0.007%	0.5 ml
4.	Indoxacarb 14.5 SC	Dhawa, Avant, Isacarb, Dhal, Lafhkar etc	0.007%	0.5 ml
5.	Spinosad 2.5 SC	Success	0.005%	2 ml
6.	Spinosad 45 EC	Tracer	0.005%	0.10 ml
7.	Ethofenprox 10 EC	Nukil, Bombard	0.02%	2 ml

Note: In case of heavy incidence by *Chrysopa* spp, indoxacarb (0.01%) may be used @ 0.7 ml per liter of water especially during rainy season as it will knock down this species quickly.

Any one of these insecticide formulation are sprayed along with 0.05% carbendazim (1 g per liter of water) on lac insect settled twigs. Rocking sprayer with Hi-jet gun is suitable for the spraying of insecticides. The spray is done up to such extent that the white filament of lac encrustation should be washed out. The spray of insecticides is normally carried out at around 30, 60 and 90 days of crop inoculation. For spraying of insecticides on *Flemingia semialata* plantation, Knapsack sprayer may be used. Insecticidal solution should be thoroughly mixed prior to spray.

The application schedules of the insecticides for different crops are given below:

Table 6: Time schedule for application of insecticides on different lac crops

S.No.	Lac	Crop	Time of application in days from date of inoculation (DOI)	Restricted time for spray application (during male emergence) in days (DOI)
1.	Kusmi	Summer season (*jethwi*)	30- 35, 60 -65, 90-95	66-88
		Winter season (*aghani*)	25-30, 38-40, 60-65	42-58
2.	Rangeeni	Summer season (*baisakhi*)	30 -35, 60 -65, 130-135	100-120
		Rainy season (*katki*)	25-30, 60-65	42-55

Normally first spray of insecticide is being carried out between 25-30 days (rainy or winter crop) or 30-35 days (summer crop) of inoculation or larval emergence whichever is later. Though second spray is normally carried out one month after first spray of insecticides but in case of *kusmi* winter crop, one more spray is required before mating period in order to save lac crop especially from *Chrysopa* spp., which is relatively more prevalent on *kusmi* lac crop. At this stage pupae of male lac insect are clearly visible.

Restricted period for spray of insecticide has been given above. This period correspond to mating time which differs from crop to crop. During this period adult male insect crawl and reaches to female lac insect for mating. Thus the adult male lac insect is now exposed to external environment and these may die if come in contact with insecticides if sprayed during this critical period. The post fertilization stage corresponds to 90-95 and 130-135 days for *kusmi* and *rangeeni* summer crops, respectively. Similarly for *kusmi* winter and *rangeeni* rainy season crop, this stage falls on 60-65 days of inoculation or larval emergence. The insecticides given above are effective against all the three major predators namely *Chrysopids* spp (Neuroptera), *E. amabilis* and *P. pulverea* (Lepidoptera) of lac insect. Ethofenprox belong to synthetic pyrethroid group of insecticide, showing quick knockdown effect on target insect. Hence its application is preferred during rainy season.

Division of lac host trees in groups (coupe)

Normally the lac growers take broodlac (lac insect seed) and immature lac (mainly resin lac) from the same tree and the trees remain infested throughout the year. They harvest lac as and when they require money without taking care for future requirement of broodlac. This practice resulted in (i) Reduction of yield potential of lac host trees (ii) Increase in incidence of enemy insect on lac culture (iii) Reduction in quality of broodlac and (iv) Overall reduction in lac production over a long period. Farmers are advised to grow lac culture in coupe system by dividing the lac hosts in different sets (specified for different lac hosts/ lac crops) for obtaining sustained lac production over a period of time without adversely affecting lac hosts.

A. *Butea monosperma (palas)*

The grouping of trees is carried out in the following way (Table1).

- The total trees of each owner may be divided in three groups (coupe) with equal number of trees in each.
- Two groups of trees treated as broodlac groups are utilized mainly for broodlac production (2/3rd trees) in alternate years. To the best possible healthy and productive trees should be kept in this group.
- Third group of trees, treated as immature summer crop is utilized every year and such trees should be in between trees meant for production of broodlac.
- For production of broodlac (lac insect seed) in both broodlac group, combined crop of one year, the summer cum rainy season is taken (inoculated in October/November and harvested in next October/November i.e. one year after inoculation).
- For immature summer crop, the trees are inoculated in October/November and harvested in April.
- Broodlac obtained from first group is utilized for inoculation of trees in second group and also in immature summer crop group of trees.
- Thus every year broodlac is obtained in October/November, from first broodlac group in 2, 4, 6 and 8 years (even years) and from second group in 3, 5, 7 and 9 years (odd years).
- From immature summer crop group, lac is obtained every year in April and utilized for resin value.
- In this way shortage of broodlac can avoided to a large extent and all the trees owned by the farmers may be utilized for this purpose.

Table 7: Lac Cultivation Schedule for *Butea monosperma* (*palas*) & or *Ziziphus mauritiana* (*ber*)

Year	Month	Broodlac set-I (*Palas*)	Immature *(ari)* set (*Palas/Ber*)	Broodlac set-II (*Palas*)
First	April	Pruning		
	Oct/ Nov	Inoculation		
Second	April		Pruning	Pruning
	Jun/ Jul	Self Inoculation		
	Oct/ Nov	Brood harvesting→	Inoculation	Inoculation
Third	April	Pruning + *ari* collection	*ari*-harvesting + Pruning	
	Jun/ Jul			Self Inoculation
	Oct/ Nov	Inoculation	Inoculation ←	Brood harvesting
Fourth	April		*ari*-harvesting + Pruning	Pruning + *ari* collection
	Jun/ July	Self Inoculation		
	Oct/ Nov	Brood harvesting →	Inoculation	Inoculation
Fifth	April	Pruning + *ari* collection	*ari*- harvesting + Pruning	

B. *Schleichera oleosa (kusum)*

- All trees are divided into 4-5 groups as per need (Table 7).
- Each group of trees are pruned at the interval of six months.
- The trees are inoculated with broodlac @ 20 g per meter of tender shoot when broodlac is available. First time inoculation can be carried out 12 months after pruning but subsequently after 18 months.
- The partial harvesting of broodlac crop is carried out, six months after inoculation, when larval emergence begins.
- Complete harvesting is carried out six months after partial harvesting or one year of inoculation and trees are simultaneously pruned.
- Broodlac is harvested only when larval emergence begins.
- Broodlac harvested partially after six months of inoculation, are utilized for inoculating trees of other group and those obtained by complete harvesting sold to market. However, it is to be sold in the market when own requirement is fulfilled.

C. *Ziziphus mauritiana (ber)*

Ber trees are utilized in combination with other host trees. For *rangeeni* it is alternated with *palas* trees and for *kusmi* normally with *kusum* trees.

Following system can be adopted (Table 3).

- Total trees of *S. oleosa* are divided into two sets, with equal number in each set and exploited for summer crop, producing broodlac in alternate years in June/ July.
- Total trees of plum are kept as one set and exploited for winter crop, producing broodlac every year in January/ February or March as crop matures.
- For the first time plum trees are pruned in January-February and later on harvesting of broodlac will serve the purpose of pruning.
- Plum trees are inoculated in July and crop harvested in January or February, when it mature as broodlac.
- The broodlac obtained from plum tree is utilized for inoculation of *S. oleosa* tree and *vice-versa.*
- Excess broodlac obtained from both hosts are sold in the market. Plum trees get six months and *S. oleosa*, one and half year rest.
- Normally immature lac from host tree is harvested in October due to financial requirement or fear of theft. This should be avoided as this will result into scarcity of broodlac and trees will be also damaged due to wrong pruning time.

Wild Indian plum tree is of little economic value to farmers although it is available in large number in forest areas. The fruit is either consumed locally or wasted due to poor market value. If these trees are utilized for cultivation of lac on its tender shoot, adequate employment and income can be generated at village level itself. Besides, there will be no adverse effect on host tree, instead harvesting of lac crop also serve the purpose of pruning and thus helps in maintaining the general healthy condition of tree in forest/rural areas. The income generated helps the poor people to manage food for their family for around 3-4 months per year. Hence action should be initiated by all R&D organisations to encourage lac cultivation on this and other tree species for livelihood of rural mass.

This tree species can be raised on barren land and also on border of paddy field without any adverse effect on rice production. The wood biomass obtained after scraping of lac is utilised as fuel by rural people

Table 8: Lac cultivation schedule on *Schleichera oleosa (kusum)*

Year	**Month**	**Set- I**	**Set- II**	**Set- III**	**Set- IV**	**Set- V**
First	Jan/Feb	Pruning				
	Jun/Jul		Pruning			
Second	Jan/Feb			Pruning		
	Jun/ Jul	Inoculation			Pruning	
Third	Jan/ Feb	Partial →Harvesting	Inoculation			Pruning
	Jun/Jul	Harvesting & Pruning	Partial →Harvesting	Inoculation		
Fourth	Jan/ Feb		Harvesting & Pruning	Partial →Harvesting	Inoculation	
	Jun/Jul			Harvesting & Pruning	Partial→ Harvesting	Inoculation
Fifth	Jan/ Feb	Inoculation ←	←	←	Harvesting & pruning	Partial Harvesting
	Jun /Jul	Partial → Harvesting	Inoculation			Harvesting & Pruning

Table 9: Cultivation schedule for *Kusmi* lac on *Ziziphus mauritiana (plum)* and *Schleichera oleosa* (*kusum*) trees

Year	Month	*S. oleosa* (Summer crop)	*Z. mauritiana* (Winter crop)	*S. oleosa* (Summer crop)
I	June-July	Pruning		
II	Jan-Feb		Pruning	
	June-July		Inoculation	Pruning
III	Jan-Feb	Inoculation	Harvesting + Pruning	
	June-July	Harvesting	Inoculation	
IV	Jan-Feb		Harvesting +Pruning	Inoculation
	June-July		Inoculation	Harvesting
V	Jan-Feb	Re-inoculation	Harvesting + Pruning	
	June-July	Harvesting	Inoculation	
VI	Jan-Feb		Harvesting +Pruning	Re- Inoculation
	June-July		Inoculation	Harvesting

D. Combination with *Schleichera oleosa* or *Albizia* spp

From this biotype, two crops are taken *viz*. Winter-crop (June-July to January-February) and summer crop (January-February to June-July). Due to absence of leaf in summer season, the lac culture is unable to survive due to direct heat of sun. Hence in most of the areas where temperature goes high in summer, it becomes very difficult to get sustained production and thus only winter crop is to be raised. In such areas, for summer crop, other *Kusmi* host tree species like *Schleichera oleosa* or *Albizia* spp are required.

References

Jaiswal AK and Bhattacharya A. 2007 Conventional and Scientific method of lac cultivation. Model training cource on 'Advanced lac production, storage and application technology for employment and income generation', Indian Lac Research Institute Namkum, Ranchi, 2007; pp. 26-31.

Jaiswal AK and Sharma KK. 2002 Lac insect its development and lifecycle. Model training cource on 'Advanced lac production, storage and application technology for employment and income generatio' Indian Lac Research Institute Namkum, Ranchi, 2002; pp. 20-24.

Pal, Govind, Jaiswal, A.K. and Bhattacharya, A. 2007. Lac Statistics at a Glance 2007, Indian Institute of Natural Resins and Gums, Ranchi, Jharkhand.

Singh BP. 2007 Management of lac cultivation on *Flemingia* sp., advanced lac production, storage and application technology for employment and income generation. Indian Lac Research Institute Namkum, Ranchi, 2007; pp. 72-80.

Yogi R.K., Bhattacharya A. and Jaiswal A.K. 2014. Lac, Plant Resins and Gums Statistics at a Glance 2013. ICAR- Indian Institute of Natural Resins and Gums, Ranchi (Jharkhand), India. Bulletin (Technical) No. 06/2014. pp. 1-38.

Improved Agriculture and Allied Practices for Farmers, pp 331-346
New India Publishing Agency, New Delhi, India

31

Improved Management Practices for Pig Farming

V. S. Solanki and J. P. S. Dabas

Centre for Agricultural Technology Assessment and Transfer
ICAR-Indian Agricultural Research Institute, New Delhi-110012

Introduction

Three hundred million poor people in Asia depend on livestock for their livelihoods and there are opportunities for improvement in livestock related livelihoods due to the rapid and dynamic changes that the livestock sector in Asia is undergoing (Njuki *et al.* 2010). Piggery farming has been recognized as one of the profitable venture among the rural masses. This venture has proved to be one of the most important livelihood option. Generally the pigs are reared for pork, considering the increasing trend of pork consumers in India; Piggery farming will certainly take a industrial form of livelihood in India in future. There are more than 12.8 million pigs in our country, out of which approximately 14.5% are graded and of exotic variety. There are 120 pig breeding farms in the country whereas pigs of exotic breeds are maintained. China ranks first in the world in pig rearing where number is around 300 million or 1/3 of the world pig population.Pig production is basically an enterprise which is able to convert inedible or inferior quality food into the food rich in animal protein. Pig is considered a machine because they convert grain, pasture and other low quality feed into Pork which is used as human food (Rahman, 2008) . Despite the fact that pig rearing is a profitable enterprise most of the progressive farmers even today hesitate to adopt it due to prejudice sentiments in our society against pig farming. For many others there is religious taboo against pig. A majority of meat eaters also and cheap source of protein. Pork as a source of animal protein is gaining popularity in India and pork products will be in greater demands in years to come. The high carcass yield attracts the farmers because of the better returns within the shortest possible time. High fertility in pigs assumes income to the rearer round the year. To bring about improvement in the productivity of pork and other piggery products extensive crossbreeding programme has been taken up to develop animals of large size, better feed conversion efficiency, high dressing percentage and better quality pork. Pig keeping also contributes to socio-cultural obligations and risk diversification and converts existing resources and low value waste products into high-value animal source food for home consumption and/or sale (NEPED, 2008).

Advantages

- Pigs grow faster than any other animals. They have higher feed conversion efficiency. That means, they have a great feed to meat converting ratio. They can convert all types of inedible feeds, forages, certain grains byproduct obtained from mills, damaged feeds, meat byproducts, garbage etc. into valuable, nutritious and delicious meat.
- Pigs can eat and consume almost all types of feed including grains, damaged food, forage, fruits, vegetables, garbage, sugarcane etc. Sometimes they even eat grasses and other green plants or roots.
- Pigs become mature earlier than other animals. A sow can be bred for first time at their age of 8-9 months. They can farrow twice a year. And in each farrowing they give birth of 8-12 piglets.
- Setting up pig farming business is easy and it requires little capital/ investment for building houses and buying equipment.
- The ratio of total consumable meat and total body weight is higher in pigs. We can get around 60 to 80 percent consumable meat from a live pig.
- Pig meat is also one of the most nutritious and tasty meat. It is higher in fat and energy and lower in water.
- Pig manure is a great and widely used fertilizer. You can use this manure for both crop production and in pond for fish farming purpose.
- Pig fat also has a huge demand in poultry feed, paints, soap and chemical industries. And this demand is continuously increasing.
- Pigs grow faster and has a good ROI (returns of investment) ratio. They reach slaughter age earlier compared to other animal. A pig become suitable for slaughtering purpose at their age of 7-9 months. Within this period they reach marketable weight of 70-100 kg.
- Pig meat has a good domestic demand. We can also earn good income by exporting pig products like bacon, ham, lard, pork, sausages etc. to the foreign countries.

- Pig farming business can be a great income opportunities for the small and landless farmers, unemployed educated or uneducated young people and for the rural women.
- Waste food, fruits and vegetables, milk and meat by-products and grains are easily digested by the pigs to convert them into nutritious pork.
- Pigs usually have a big litter size and the piglets have ability to grow faster, attains maturity early than any domestic mammal. Besides this, a sow has ability to farrow more than twice a year with a litter size of 8-12 piglets.
- Earning is made both from pork and sale of piglets. For this, piggery assures opportunity for economic growth among rural masses.
- After 8-9 months of weaning, piglets attain maturity and becomes capable of giving birth.
- Under proper care and management pig weighs 80-90 kgs at 6-8 months of age. 8. Piggery can be done in waste or uncultivable land.
- Piggery is less labour intensive than other animal farming. Dressing percentage is more in pig than other animals (65-80%) in comparison to cattle (50-60%) and in sheep and goat (45-50%).
- It also produces high quality manure. In a word, commercial pig farming can be a great business idea and income source for the people and it can contribute the national income of our country highly (Anonymous, 2009).

Related terminology

Drove or stock or herd	Group of pigs
Boar	Male adult (Uncastrated)
Sow	Female adult (Used for breeding)
Boar ling	Young male
Gilt	Young female
Piglet	Newly born one
Hog or stag or barrow	Castrated male
Suckling	Female with its offspring
Farrowing	Act of parturition
Coupling	Act of mating
Back Fatter	A fatty pig too heavy for bacon trade
Barren sow	A mature female pig incapable of producing offspring due to infertility or sterility

Carcass	The major portion of meat from body of pig after removal of viscera, head, skin and shanks.
Litter	A group of piglets born in single farrowing
Pen	A house for keeping a single pig or group of pigs
Weaning	Separation of young piglets from mother sow at 8 weeks of age
Pork	Fresh, frozen or salted meat from pig carcass.
Piggery	Where pigs are reared or pig farming

Breeds

Indian breeds of Pig have no distinct breed features and their characteristics vary region to region. They are black, brown, grey, rusty grey or admixture of these colours. Traditionally these are reared by weaker section of the community. Indian breeds of pigs are as under:

a) **Wild Pigs:** The wild boars are commonly found in jungles or forest and in Himalayan tract. The animals measure about 1.5 m in length 71 to 90 cm high at shoulder. It exceeds 136 kg in weight and males are larger than females. Color is rusty grey in young age and it becomes dark chestnut brown in adult animals. The wild pig is extremely active and when provoked may also attack human beings.

b) **Pigmy hog:** Pigmy hog is found at the base of Himalayas in Sikkim, Nepal, Bhutan and Assam. They prefer to remain in high grasses and therefore are rarely seen. They live in herds of 5 to 20. The animal measures about 32 cm over the shoulder and 66 cm from the snout to rump. They have brown and black color and weighing about 7 to 8 kg.

c) **Domesticated or Indigenous Pigs:** These pigs differ in their characteristics and color from region to region within the country depending on the typography and climatic conditions. Adult pigs weigh up to 168 kg and female possesses 6 to 12 teats. Pigs have different colours viz. black brown, rusty grey and even an admixture of any two colors. Pigs possess a long face tappering towards the nostrils. Ears are small and medium sized. Head and shoulders are heavier as compared to hind quarters.

Exotic breeds of Pigs: There are 60 well recognized breeds of pig in the world .Out of those some most common exotic breeds of pig reared in India are as under:

Selection of Breeds

Certain factors are important in selecting a breed of pigs for a prolific enterprise. The goal in pig production is to produce large litters of pigs, which can be grown out rapidly and economically, which fetch top market price to

S.No.	Breed	Place of origin	Adult body weight	Body feature
1	Large white Yorkshire	England	Boar 300-400 kgSow 250-300 kg Prolific breederBacon type breed	Entire-white, black, pigmentspots, fine skin without wrinkles & smooth
2.	Landrace	Switzerland, Denmark	Boar 300-350 kgSow 200-250 kg Bacon typeexcellent	Short legs, lean carcass, white in colour with black skin spots, longer body, skin without wrinkles
3.	Middle white Yorkshire	England	Boar 275-375 kgSow 200-290 kg Less prolific, Bacon type	Cross between large and small Yorkshire, long leveled back white in colour, fine skin, medium sized
4.	Hampshire	U.S.A.	Boar 250-330 kgSow 200-300 kg Meat type	Black hog, with a belt encircling the body includingfront legs, short legs, smaller breed
5.	Berkshire	England	Boar 275-375 kgSow 200-290 kg Good quality meat	Medium sized, legs of mediumlength, black in colour, good width and broad back
6.	Temworth	U.K.	Boar 300-350 kgSow 250-300 kg Best quality bacon	Golden brown in colour, thinshoulders, strong back
7.	Wessex saddleback	England	Boar 200-280 kgSow 180-260 kg	Robust make-up, black in colour bacon type, prolific
8.	Duroc	Jersy	Boar 350-400 kgSow 300-350 kg Good quality meat	Red in colour medium sized

gain maximum return of profit (King *et al.* 1998). Pork producers must decide which breed or breeds will fit but in their breeding programs (Theodoropoulos, *et al.* 2009).

The following factors must be given careful consideration in the selection of breed of pigs

- For commercial pig farming upgraded / crossbred or exotic stock in good health should be selected.
- While selecting a gilt or sow primary aim should be to secure a female that will produce large survivable litter and which can attain marketable weight at an age of six months or less.
- Animals those are ready to be bred should be selected.
- The animals should be active but should have good physique and be easily handled
- Breeds should be disease resistant to the maximum extent.
- Identify the newly purchased animal by giving suitable identification mark (ear notching or tattooing).
- Vaccinate the animals against diseases.
- Purchase animals in two batches at the interval of three months
- Cull the old animals after 10-12 farrowings
- Keep the animal under observations for a period of about two weeks and then mix with the other animals.

Selection of Breeding Females

- **Gilts:** Breeding pigs are selected from mother pigs having good mothering ability as shown by large litter size and weight, at weaning. Gilts should have 12-14 evenly spaced sound teats, good growth rate and feminity.
- **Weaned sows:** Sows are usually weaned at 8 weeks after farrowing. The weaned sows come into heat in 3-10 days after weaning. They may be bred on the first post-weaning heat itself since this period is most fertile. She may respond to nutritious diet and pick up fast if the condition is poor.
- **Pregnant sows and gilts:** In addition to having proper nutrition, the pregnant animals should be provided with regular exercise. Forced exercise may be brought about by feeding some distance away from the house or inducing the animal for moderate walk. Sows of the same size and condition

can be run together. The bred and gilts may be housed separately. Do not house boars along with pregnant stock. Avoid overcrowding, mixing of new and old stock, slippery ground and overexciting of pregnant animals.

Breeding Management

Age to breed gilts	8 months
Weight of breed gilts	90-120 kg
Length of heat period	2-3 days
Best time to breed in heat period	Gilts – first daySows- Second day
Number of services per sow	2 services at an interval of 12-14 hours
Period of oestrous cycle	18-24 days (Average 21 days)
Occurrence of heat after weaning	2-10 days
Gestation period	114 days

Selection of Boar

The following points should be taken into consideration while selecting a boar for breeding purpose:

- The boar should be purchased from a breeder or a farm maintaining adequate information on its performance
- The boar should be selected from a dam which has consistently farrowed and weaned high litters
- A good boar will weigh 90 kg in about 5-6 months will be of good type and will be strong on feet and legs
- The feed conversion from weaning to 90 kg weight would be the most desirable.
- The pig should have sound feet and legs
- Negative blood test for brucellosis and leptospirosis should be made during selection and the pigs should be vaccinated against swine fever
- Pigs should be free from other diseases and physical defects

Management of Piglets

It has been estimated that from 25 to 30% of piglets that are farrowed fail to reach weaning age of 8 weeks due to certain reasons like non-availability of sufficient milk and other feeds, due to some injury or disease or other unscientific managemental practices. Therefore, the management of piglets under hygienic conditions is necessary.

Care of piglets at birth

- Remove the piglets soon after they are farrowed.
- Clean all piglets and make their body dry.
- Make breathing passages of all baby pigs clear.
- Cut the naval cord with sterilized scissor leaving 3 cm from naval and disinfect by application of tincture of iodine.
- Allow piglets to suckle milk from mother sow for about 8 to 10 times in 24 hours.
- Protect the baby pigs from trampling by sow.

Feeding of piglets

Age	Daily feed	Cumulative feed in kg
Birth to 1 month	Mother milk	—
1 to 2 months	0.5 kg	15
2 to 3 months	1.0 kg	30
3 to 4 months	1.2 kg	36
4 to 5 months	2.0 kg	60
5 to 6 months	2.5 kg	75

Note: About 216 kg. Of total feed is expected to produce a pig of 70 kg body weight in 6 months.

Prevention of anemia

It is a highly fatal disease of suckling pigs caused by a marked decrease in hemoglobin and fatty degeneration of the liver.

Causes

Lack of iron and copper salts in the milk of sow kept in indoor pens, on concrete floor and limited milk diet from sow. Suckling pigs of about 3 to 6 weeks age are mostly affected.

Symptoms

Pigs are dull and inactive, fatigued, lack of vigor. Pig shows dyspnea, thumps and rough coats, depression on the slight exertion. Pig may be weak and thin, and muscles are labby. Skin over the neck and mucus membrane are pale. Pigs die suddenly and wrinkles are found over the legs.

Prevention and treatment

1. Add small amount of iron and copper in pigs diet @ 25 mg of iron and 5 mg of copper/day/pig.
2. Iron sulphate-3.6 ounces Water-5 quarts

   Note: Feed 1 dram daily.

   OR
3. Paint the udder of sow daily with the following mixture:

Iron sulphate	500 g
Copper sulphate	70 g
Sugar	500 g
Water	10 lit

   OR
4. Intra-muscular injection of iron-dextrose compound.

Removal of needle teeth

Baby pigs at birth have 4 pairs of sharp teeth on each jaw called needle teeth. These are of not any use to piglets and may cause injury to mothers udder. These should clipped by means of plier. Care be taken not to injure jaw or gum of piglet.

Weaning piglets

Wean the piglets at 8 weeks of age.For weaning separate the sow from piglets for few hours everyday Supply creep feed with 18% crude protein Deworm the weaned piglets at 10 week age

Causes of preweaning mortality

Maximum death of piglets occurs in the first week. Hence good husbandry practices in early life will be profitable. The major causes of death were found to be agalaxia, pneumonia and gastroenteritis. Strict hygienic measures may help in reducing the mortality due to infectious diseases. Piglets with birth weight were prone to agalaxia, trampling and infectious diseases. Therefore, special attention to pregnant sow nutrition may reduce preweaning mortality

Castration of piglets

Castration of Piglets is essential to achieve following goals

- To eliminate undesirable males.
- To prevent indiscriminate breeding
- To make animals more docile
- To prevent boar odour in cooked meat.
- To develop pork of superior quality.
- Castration is done in the males which are not selected for breeding purpose and this practice is performed at the age of 3 to 4 weeks.

Method

For this purpose operation method is adopted and following steps are taken:

- Secure the piglets and make it lie down on dorsal side and grasp feet under control.
- Make hands and castrating knife clean and sterilize with spirit soaked cotton.
- Wash the scrotum with an antiseptic solution of acriflavin
- Make an incision on the scrotum on each side deep enough extending the cuts well down to permit proper drainage.
- Remove the testes with its membrane by pulling them backward, and bring with them as much cord possible and break it by cutting.
- Scrotum and surrounding parts must again be disinfected thoroughly with tincture iodine.
- Apply the sulphanelamide powder mixed with iodoform (fly repellent) to prevent infection or other complication

Housing management of pigs

Adequate housing and equipment for raising pigs are necessary to provide shelter against inclement weather, prevent diseases, control parasites and save labour.

The normal requirement of floor area, water and air space in pens for various classes of pigs is given below

Class of animals	Covered floor area per animal (m^2)	Open-yard area per animal (m^2)	Water required (liters')
Boar	6.25-7.5	8.8-12.0	45.5
Farrowing	7.5-9.0	8.8-12	18-22
Weaner	0.96-1.8	8.8-12	3.5-4
Dry sow	1.8-2.7	1.4-1.8	4.5-5

Creep space

The flooring should have a rough finish and should be be of a regular masonary type made up of water proof cement mortar. Proper drains should be provided so that the effluents are disposed off. Generally under village conditions the housing can be made up of pens measuring 3 m X 2.4 m or 3 m X 3 m withan open yard of nearly the same dimension or in some cases slightly longer. Walls should be 1.2-1.5 m high from the floor. For the purposes of farrowing some of the pens could be converted into farrowing pens by providing guard rails made up of G.I pipes of 5 cm diameter, along the walls, 20-25 cm from the ground and the wall. In addtion to guard rails, creep space can be provided for the piglets along the wall by making a partition or in one of the corners with separate entrances for the piglets. This space usually of 0.75 m X 2.4 m area. In many of the farms the yard is provided with regular flooring.

Prolonged exposure of exotic breeds of pigs to bright sunshine may cause pigs to become overheated even during moderate weather. Shade helps in preventing deaths and increasing production efficiency during hot weather. While it is desirable to plant trees in the neighborhood of pens for reducing the intensity of heat. But it is not desirable to plant trees for giving regular shade because they permit rapid buildup of parasite levels.

Farrowing Pen

Pigs have very few sweat glands. In areas having warm weather mature breeding animals and fattening animals need a wallow during summer months. Instead of permitting unsanitary wallows a masonry wallow with proper drainage would be desirable. The size of the wallow will depend upon the number and size of the animals.

Feeds and Feeding of Pigs

Pig is a simple stomach animal, therefore, it can not utilize fibrous fodder and requiresmore amount of concentrate mixture in his ration. Pig is an efficient converter ofconcentrate into meat and feed cost accounts about 75% of total cost of rearing. Feedconversion ratio in pigs is most efficient (1:3) and hence needs less feed per kg gain inbody weight. Because pigs are meat producing animals, therefore, they require sufficient amount of energetic feeds like Maize, Barley, Bajra, Sorghum, Molasses, Wheat, Pulse Chuni etc. Pig suffers more from nutritional deficiency as compared to ruminant animal,therefore, their feeding programme must be efficient but economical also. For this purpose, most economical feed ingredient should be selected and small pasture providing succulent green forages is good for raising the pigs. For providing required amount of vitamins and minerals their supplements must be added to pig ration. Common salt @ 30- 40 g/pig and antibiotics 20 g/100 kg feed should be provided. Pigs need proteins to develop muscles, body tissues and offspring. They need carbohydrates and fats to provide heat and energy and to produce lard. Minerals are needed to develop bones, muscles, teeth and in blood. Vitamins and antibiotics are needed for the pigs to use feed effectively. The nutritional needs of pig vary with age and stage of production. Sows need rationcontaining 15 to 16 percent protein and fairly high in minerals and vitamins. Prestarter ration of pig should contain about 20 percent protein, 0.70 percent calcium and 0.55 percent phosphorous. Pig starter should contain 16 to 18 percent protein. The protein content of rations for pigs over 25 kg in weight is decreased gradually as the pigs grow heavier from 12 to 14 percent at 25 kg to 10 to 12 percent for 65 to 100 kg pigs.

Salient Points related to pig production

Aug.- Sep and Feb-March months are the breeding seasons of the pigs Gestation period of Sow is 112-114 days and heat period is 2-3 days In India, maximum number of pigs are found in Uttar Pradesh Age of puberty in pigs is 6-7 months and breeding age of gilts is 10-12 months.A Sow can birth 10-12 piglets at one farrowing.Piglets should be castrated at the age of 4-6 weeks.Market age of fattening pigs is about 6 months.First day of heat and second day of onset of heat in sows is proper time of mating. Farrowing interval in sow is about 7-8 monthsFarrowing interval in sow is about 7-8 months Generally sows are known to breed for 8-10 years and she can perform farrowing about 6 times during this period. Generally sows are known to breed for 8-10 years and she can perform farrowing about 6 times during this period.Average body temperature of pig is 102 F, Pulse rate 60 to 80/ minute and respiration rate 8-18/minute. Extra allowance of grains should be provided for meat, milk production and pregnancy also. Green leguminous fodder atleast 1-2 Kg/ day/ about animal should be fed.Pig feed incorporated with vegetable wastes, refuse from hostels and restaurants, damaged food grains unsuitable for human consumption and left over feed will minimize the cost of ration and fetch maximum profit Health and Diseases: It is important for the farmer/ animal keeper to know the common signs of apparently healthy pigs so that the diseased conditions of the pigs can be easily identified

Management of Pig Farm

- For the proper management of a pig farm and to earn maximum profit following points should be taken into consideration.

Age determination

- Age of pig or piglets is determined with the help of teeth in his mouth. An adult pig has a sum of 44 teeth as below:

  2(I 3/3, C 1/1, PM 4/4, M 3/3) = 44

- Eruption of all temporary teeth is completed by the age of 5 months. In case of permanent teeth-1st pair (corner) at 6 months age, 2nd pair lateral at 10 months age, 3rd pair middle at 12 months age and 4th pair central is erupted at the age of 1.5 years.

Identification

To ensure proper care in feeding and better management practices wherever needed and to maintain proper records of the animal, individual pig is identified. For this purpose,ear notching is the common method used for marking the pigs.

Pig exercising

To keep animals fit, exercise limited but it is essential each day. Exercise of pigs is performed usually by providing run – out in uncovered area and free access to pasture.

Disposal of manure

Manure produced by pigs should be removed twice daily by means of wheel barrow and shovel and disposed into a pit for decomposition. Such manure will return 75 per cent of its fertilizing value to soil. The production of manure from each adult hog is about 2kg per day. Manure pits should be about 200 meters away from the pig house.

The general signs of healthy pigs are given as below

- The pigs remain alert and active.
- The head is moderately stretched forward and upright parallel to dorsal line.

- The faeces are semi-solid.
- Ears are alert and eyes are bright.
- Tail movement is also frequent.
- Pig consumes normal amount of ration and water.
- Respiration, pulse and temperature of the body remain normal.
- There should not be any abnormal discharge from the natural opening of the body.
- There should not be any reduction in body weight.
- Milk yield should not be reduced in sows.

Some common diseases of pigs

S.No.	Name of disease	Cause	Symptoms	Prevention & Treatment
1.	Hog cholera swine fever	Virus	Fever, loss of appetite dullness, body temp. 1060F, and later it fallbelow normal.	Anti-HC-Vaccine, No effective treatment.
2.	Swine paratyphoid	*Solmonella choleraesuis* bacteria	Fever, purple spots onskin, diarrhoea.	Same as swinefever.
3.	Swine dysentery (colitis)	*Hamorrhogic, evteritis, swinetyphus* bacteria	Bloody diarrhoea, temp.103-1040F, off the feed.	Sanitation, use of sulphamezathe neor streptomy-cine
4.	Swine pox	Virus	Fever, lack of appetite, eruption on skin, ear andthighs.	Segregation andsanitation.
5.	Swine plague(H.S)	*Pasturella suiseptica* bacteria	High fever, loss of appetite, difficultrespira-tion, swelling onthroat, dysentery.	Segregation, Injof 'Dexona'
6.	Foot and mouthdisease	Virus	High fever, secretion andulterations on mucusmembrane of mouth, Footlesions are painful, hardlyto walk.	Isolation, washfoot and mouth with KMno4 solution, noeffectivetreat-ment
7.	Bruecellosis	*Brucella suis*	Abortion, sterility	Testing of pigs
8.	Swine in fluenza	*Homophilus in fluenza* suis	High fever, loss of appetite, cough and discharge from eyes and nose, difficult movement	Provide dry, clean and warm atmosphere to the affected pig.

Precautions for disease control

- Erection of physical barriers 50 metres around pig endosures to keep off visitors, cars, advisors equipments and surgeons, farm attendants etc.
- Discouraging casual callers, visitors, use of protection clothing, disinfecting feet and hands.
- Check the quality feed and water, ensure feed is steam pelleted and is packed in paper sacks.
- Liming of pasture and keeping it parasite free. Rotation of pastures. Proper sanitation in house (dry floor, well ventilated and with good lighting arrangement).

Conclusion

Pig is one of the most efficient food converting animals amongst domesticated livestock and can play an important role in improving the socio-economic status of the weaker sections of the society, particularly in the North-Eastern Region of the country. Bulk of the pig population in India is indigenous type with low growth rate and productivity. The share of pork to the total meat production has been almost static for last 15 years at about 10%. Average meat yield of pigs in India is about 35kg/ animal, which is about 55% lesser than the corresponding value of world average. Over 20% of the pigs kept in India are crossed with exotic breeds, but with a large amount of inbreeding because of non-systematic breeding and selection. In nutshell, the pig rearing is still unorganized venture that requires science and technology driven support to make it a vibrant enterprise. The various stakeholders require promotion at various levels such as technology, entrepreneurship development, and financial support to bring Indian pig farming at global level. Further, the shrinking resources in terms of land availability, water as well as threats from the changing environment being conducive for emergence of new diseases are gradually expected to limit the capacity for pork production optimization.

References

Annonymous (2009). Piggery Farming. Department Of Animal Husbandry, Livestock, Fisheries & Veterinary Services Government of Sikkim.

King, V. L., Koketsu, Y., Reeves, D., Xue, J., Dial, G., (1998). Management factors associated with swine breeding-herd productivity in the United States. *Preventative Veterinary Medicine*. 85 (255-264)

Njuki, J., Pali, P., Mburu, S., and Poole, J. (2010). Pig Production, Management and Marketing in the North East Indian State of Nagaland. *International Livestock Research Institute Nairobi.*

Rahman S, Barthakur S and Kalita G (2008). Pig production and management system in Aizawl District of Mizoram, India. Volume 20, Article #139. Retrieved June 26, 2010, from http://www.lrrd.org/lrrd20/9/rahm20139.htm

Theodoropoulos, G., Stevens, K.B., Hartsa, A., Theodoropoulou, H., Pfeiffer, D.U., (2009). Farm-level factors associated with above-average production on pig farms in Evia, Greece. *Preventative Veterinary Medicine* 89: (163-166).

Improved Agriculture and Allied Practices for Farmers, pp 347-361
New India Publishing Agency, New Delhi, India

32

Goat Farming for Enhancing Profitability of Farmers

V.S. Solanki[1] and N.V. Kumbhare[2]

[1]Centre for Agricultural Technology Assessment and Transfer
[2]Agricultural Technology Information Centre (ATIC), ICAR-Indian Agricultural Research Institute, New Delhi-110012

Goat is the 1st domesticated animal in the ancient age and has been considered to be the Poor Man's Cow Goat's meat is lean & has low fat and the demand for goat's meat is increasing, with rising prices. Goats usually give birth to twins & triplets. Goat is a multi functional animal and plays a significant role in the economy and nutrition of landless, small and marginal farmers in the country. Goat rearing is an enterprise which has been practiced by a large section of population in rural areas. It can efficiently survive on available shrubs and trees in adverse harsh environment in low fertility lands where no other crop can be grown. In pastoral and agricultural subsistence societies in India, goats are kept as a source of additional income for their meat, hair fiber, and dairy, or as a pet. Goats are also used in ceremonial feastings and for the payment of social dues. According to FAO (2008) goat milk production in India were 4 million metric tons, Bangladesh 2.16 million metric tons and the Sudan 1.47 million metric tons over much of the world. Goats are among the main meat-producing animals in India, whose meat (chevon) is one of the choicest meats and has huge domestic demand (Shalander Kumar, 2007). Goat rearing is an enterprise which has been practiced by a large section of population in rural areas. Goats can efficiently survive on available shrubs and trees in adverse harsh environment in low fertility lands where no other crop can be grown. (Oberoi *et al.,* 199I) reported that in variable cost, expenditure on labour was the major component of cost and was 81% under goat farming.

In some parts of the world, especially Europe and North America, distinct breeds of goats are kept both dairy and meat production. Rising per capita income, growing urbanization and unfolding globalization are boosting the demand for high-value commodities including meat (Birthal and Joshi, 2006). Excess male kids of dairy breeds are typically slaughtered for meat. Both does and bucks of meat breeds may be slaughtered for meat, as well as older animals of any breed. The meat of older bucks (more than one year old) is generally considered not desirable for meat for human consumption. Castration at a young age prevents the development of typical buck odor. Dairy goats are generally pastured in summer and may be stabled during the winter. As dairy does are milked daily, they are generally kept close to the milking shed. Their grazing is typically supplemented with hay and concentrates. Stabled goats may be kept in stalls similar to horses, or in larger group pens. In the US system, does are generally rebred annually. In some European commercial dairy systems, the does are bred only twice, and are milked continuously for several years after the second kidding. East Asia in the next 20 years presents an excellent opportunity for enhancing export of live goat/sheep and their meat from India (Dalgado *et al.,* 1999). The capital investment is relatively low, land requirement is small, reproductive rates are higher due to shorter breeding interval and high prolificacy. Goat Rearing can be managed by spare family labour and do not require any serious housing facilities and management skills (Rawat *et al.,* 2016).

Advantages of goat rearing

Popularity of goat husbandry is increasing day by day in tropical countries with the recognition of its significant contribution for improving the socio-economic conditions of the rural poors and landless labourers.Goat and its multi-profit has paramount importance in tropical countries. Proper feeding of goats is much important as feeding of any other farm animals. Lack of attention in feeding may lead to its poor performance. The ultimate performance of a goat depends on , how a goat is fed during its growth, pregnancy, lactation and dry periods. Therefore, a scientific feeding programme of goats must be followed.

- The initial investment needed for goat farming is low.
- Due to small body size and docile nature, housing requirements and manage mental problems with goats are less.
- Goats are friendly animals and enjoy being with the people.
- Goats are prolific breeders and achieve sexual maturity at the age of 10-12 months gestation period in goats is short and at the age of 16-17 months it starts giving milk. Twinning is very common and triplets and quadruplets are rare. In drought prone areas risk of goat farming is very much less as compared to other livestock species.
- Unlike large animals in commercial farm conditions both male and female goats have equal value. Goats are ideal for mixed species grazing.

- The animal can thrive well on wide variety of thorny bushes, weeds, crop residues, agricultural by-products unsuitable for human consumption.
- Under proper management, goats can improve and maintain grazing land and reduce bush encroachment (biological control) without causing harm to the environment.
- No religious taboo against goat slaughter and meat consumption prevalent in the country. Slaughter and dressing operation and meat disposal can be carried without much environmental problems.
- The goat meat is more lean (low cholesterol) and relatively good for people who prefer low energy diet especially in summer and sometimes goat meat (*chevon*) is preferred over mutton because of its "chewability"
- Goat milk is easy to digest than cow milk because of small fat globules and is naturally homogenized. Goat milk is said to play a role in improving appetite and digestive efficiency.
- Goat milk is non allergic as compared to cow milk and it has anti-fungal and anti bacterial properties and can be used for treating urogenital diseases of fungal origin. Goats are 2.5 times more economical than sheep on free range grazing under semi arid conditions. Male goats are referred to as a 'Buck' or' Billy' goat. A male goat when castrated or neutered is called a 'Whether'.
- A female goat is known as 'Doe' or a 'Nanny' goat and its offspring, up to the age of one, is called a 'Kid'. Goats live in a herd and are collectively referred to as 'Caprine'.
- A cross-bred goat would be called a' Brush' goat.

Indian Breeds of Goats

The popular Indian breeds of goats used for rearing are given below:

1. **Assam Hill Goat:** This breed is found in Assam and its adjoining areas. The body coat is black and sometimes white and the breed is famous for meat production. It is a poor milk yielder. Selected animals of this breed may be suggested for utilization by the poor farmers of Assam and its adjoining areas for their economic upliftment.
2. **Black Bengal:** Found in Assam and Bengal. It is famous for production of excellent quality meat of an adult male. This breed coat colour is predominantly black. An adult male average body weight is about 15 to 16 kg and female is about 12 kg. Usually, they start kidding at 9 to 10 months of age. These goat kid twice a year and twinning are very common.

3. **Kashmiri:** They are found in high hills of Kashmir and Tibet.They are also popular in Himachal Pradesh and Punjab. Body is covered with a fine silky hair, 4-5 inches long, beneath, which is a *fur* like undercoat of greater fineness, the warm fur, *Pashmina*.

4. **Jamnapari:** Found in pure form in the Etawah district of Uttar Pradesh and is a dual-purpose breed with the ability to produce good amount of milk and meat. It is the biggest and the most majestic breed of goats in India. The breed has no uniform colour but they are generally white or light yellowish tan with light brown spots on the neck and face, and occasionally patches of tan or black are found on the body. The ears are very long, flat and drooping but thetypical character of the breed is a highly convex nose line with a tuft of hair known as 'Roman nose' or parrot mouth appearance. The breed has well developed udder round in shape with large conical teats. Daily milk yield varies from 1.5 to 2.0 kg per day with a total lactation yield of about 200 kg.Usually doe kids once a year, giving birth to single in 57% while twins in 43% cases, they kid once a year.

5. **Barbari:** The breed is a promising dairy type goat, which has probably originated in the city of Barbera in British Somali land in East Africa. In India, the breed is distributed at Etawah, Agra, Mathura and Aligarh districts of U.P. and Bharatpur district of Rajasthan. It is wide variation in coat color, but white with small light brown patches is most common. Ears are short, tubular and erect. Both sexes have twisted horns, medium in length and directed upward and backward. Bucks have a large thick beard. In addition to being a good milkier it is highly prolific and generally gives birth to twin and triplets. It may kid twice in a period of 12-15 months. It is dwarf breed highly suited for stall-feeding conditions and hence generally found in the cities.

6. **Beetal:** The Beetal is found throughout the state of Punjab and Haryana. True-bred animals are however found in the districts of Gurdaspur, Amritsar and Ferozepur in Punjab. The breed is a good dairy type, second to Jamunapari in size but is superior to it in respect to proliferation and adaptability to various agro-climatic zones and also to stall feeding. The breed is large but slightly smaller than Jamunapari and good dairy type. Coat color is variable, predominantly black (about 90%) or brown (10%) having spots of different sizes.The ears are long and flat, curled and drooping. Both sexes have thick, medium-sized horns, carried horizontally with a slight twist directed backward and upward. Roman nose. Male possesses marked beard while females are beardless. The tail is small

and thin. The udder is large and developed having big conical teats. Milk yield per lactation varies from 150-190 kg, averaging daily yield 2.0 kg. The litter size single 41%, twins 53% and triplets 6%.

7. **Sirohi:** This breed is found in state of Rajasthan and surrounding region. Sirohi breed coat colour is brown, white & admixture of colours in patches and their hair coarse and short These goats have compact and medium sized body. This breed has small sized horns and pointed curved direction. A male average body weight is about 50 kg and female is about 25 to 30 kg. Usually, they start kidding at 19 to 20 months of age. This breed gives a birth once a year and twinning are very common.
8. **Surti:** The white goat is distributed in Surat and Baroda. The breed is known to be a good dairy breed and is good for maintenance under complete confinement and stall feeding conditions. Medium sized breed, white in color with highly developed udder. Ears are medium in size. Both sexes have small horns directed backward.The breed is unable to walk long distances and is stall feed.They are most economic to rear as they can live on leaves or on food waste thus brings higher feed efficiency ratio. The breed is a good milk producer, yields on an average of 2.0 kg per day. By crossing with Saanen it has resulted high potentiality of milk yield.
9. **Osmanabadi:** The goats are mainly distributed at Latur, Tuljapur and Udgir taluks of Osmanabad district of Maharashtra. The goats are large in size. Coat color varies, but mostly it is black (73%) and the rest are white, brown or spotted. Ninety per cent males are horned; females may be horned or poled. The breed is considered useful both for meat and milk. Average daily yield varies from 0.5 to 1.5 kg for a lactation length of about 4 months. The dressing percentage varies from 45 to 50. In favorable conditions the does will breed regularly twice a year and twinning is common.
10. **Jhakrana:** The breed is found in Jhakrana and few surrounding village near Behror, of Alwar district of Rajasthan.The breed is large and predominantly black with white spots on ears and muzzle.The breed is very similar to Beetal, the main contrast is that Jhakrana is comparatively longer. I has highly developed udder. Forehead is slightly bulging.These goats are used mainly for milk production. Average daily milk yield varies from 2.0-3.0 kg for a lactation length of about 180-200 days. These are prolific. Kidding is mostly single but in 40% cases twins are given birth. Triplets are not uncommon. The goats are also useful meat producers, and their skins are popular with the tanning industry.

Exotic goat breeds

The details of exotic breeds of goats are given below:

1. **Toggenberg:** A breed of Switzerland. It is famous breed for milk production with an ability to produce 5.5 kg of milk per day.
2. **Saanen:** This is also a breed of Switzerland with a milk yielding capacity of 2-5 liters per day. Coat colour is white to biscuit in color with black spots on the nose, ears, and udder. The breed is polled. The goats are large in size with straight nose and erect ears pointed forward and upward. The body has a good dairy conformation and the udder is well developed. It is known as milk queen of the goat world. There appears to be a tendency for them to be sensitive to strong sunlight and thus it needed to shade them and provide indoor management. Average live weight 55-70 kg.
3. **Alpine:** Found in France.
4. **Nubian:** Found in Africa, Egypt and Ethiopia.
5. **Anglonubian:** A cross between Nubian and Indian breed.
6. **Angora**: A Turkish breed.

Selection of breeding stock

The following features of goats are considered while selection of breeding stock:

- Animals in good health and having good physical features must be purchased in consultation with Veterinarian.
- Purchase animals, which are ready to breed and in prime stage of production.
- Identify the newly purchased animals by suitable identification mark.
- Vaccinate the newly purchased animals against the diseases.
- Keep the newly purchased animals under observation for about 15 days and then mix with the general flock.
- Unproductive animals should be culled promptly and should be replaced by the newly purchased animals or farm born one.
- Animals are to be bred at the interval of 8-9 months for maximum productivity.
- Cull the old animals at the age of 6 years and above. Avoid the kidding during peak periods of summer and winter.

Management

1. Housing management

Construct shed on dry and properly raised ground. Avoid water-logging, marshy areas. In low lying and heavy rainfall areas the floors should be preferably elevated. In temperate Himalayan region the floor may be made of wood. The shed should be 10 ft. high and should have good ventilation. Bucks should be housed in individual pens.Does can be housed in groups upto 60 per pen. Provide proper shade and cool drinking water in summer. Dispose of dung and urine properly. Give adequate space for the animals. Avoid over stocking or crowding (Jagdish Prasad, 2014).

2. Feeding management

Ensure Bushes/shrubs for browsing of animals. As an alternative to above, supply of cultivated fodder from own farm or from surrounding farms may be ensured. Offer roughages adlib. As a thumb rule 2/3rd of the energy requirements should be met through roughages. Half of the roughages should be leguminous green fodders and rest half should be grasses/tender tree leaves. In the absence of good quality green fodders, concentrates must be considered to replace them. Kids should be fed colostrums up to 5 days of age. Later on they can be put on Kid starter rations. Green leguminous fodders should be offered adlib to kids from 15 days onwards. Provide salt and water to kids at all times. Additional concentrates should be given to bucks and does during breeding season. Care should be taken to meet the nutrient requirements as recommended.

3. Breeding management

It should be planned to obtain 3 kidding in 2 years period by adopting optimal management conditions. For every 25 does one buck should be provided in one breeding season. Breed the animals 12 hours after the onset of the first symptoms of heat for maximum conception.Unreadable animals must be examined thoroughly as directed by veterinary doctor for prompt elimination of causes for anoestrus or cull them if necessary.

4. Care during pregnancy

In advanced stage of pregnancy the does must be transferred to either kidding pens or separately earmarked space for kidding within the main shed after thoroughly disinfecting it. After kidding, the does should be provided with warm bran mash for two days.

5. Care of Kids

- Cares of kid is the very important for profitable goat farming. Almost immediately after birth, the kids, if healthy and strong, are on their legs and make attempts for their mother's teats. Failure to reach the teats, however, is of no consequence, because the kids do not require nourishment for several hours after birth. If more than one kids is born, it may be necessary especially when they are very young, to ensure that the smallest of them gets its due share of milk, because it may be prevented from doing so by the stronger kids. In case the udder is too full, a proportion of the milk should be drawn from as otherwise the weight of the udder will cause discomfort to the animals. As soon as there is teats should be held by the hand and pressed into their mouths. Once they have drawn a little of the milk, it will not be long before they take to the normal methods of suckling.
- Generally, male kids are heavier than the female kids. At birth, a male kid of the Beetal breed will weight about 3 kg and a female kid about 2-3 kg. For the first three or four days after kidding, goat's milk like cows milk is considered unsuitable for human consumption. This milk, the so-called colostrums, is yellowish in appearance and is viscous' it coagulates on boiling. It is nature's first provision of food for the new born and it must be given to the kids whether they are to be reared on the goat or artificially. Colostrums acts as a laxative and, because of its large contents of vitamin A and serum globulin, it confers immunity against certain diseases.
- When about two weeks old, kids begin to nibble green food or dry fodder, and it would be well to see that small quantities of these are within their easy reach at this time. It is also important that kids are allowed plenty of open air and sunlight. In the hot weather, this can best be done by keeping them in an enclosure build round a tree so that they may also be provided with shade. The enclosure should be large enough to allow them plenty of exercise.
- At the age of 2 to 3 months, the suckling may be practically discontinued and at four months the kids should be completely weaned because by this time they will become fit like the older goats to eat solid food, although they may as well be allowed to suckle a little longer.
- Male kids, unless they are required for breeding purposes, should be castrated at the age of 2 to 3 months for it has been proved that castration improved the quality of meat. Otherwise, they should be kept separated from the female kids.

- The rearing of kids may be either natural or by hand rearing and each has its advantages and disadvantages.In India, it is the natural method that is usually practiced and this consists of in leaving the kid to take what amount of milk it can obtain from its mother. Hand rearing is resorted to when weaning is practiced or when the goat dies.There are two methods of hand rearing' one consists of feeding the kid with a bottle and the other is feeding it off the pail. Both methods are learned by them easily, but bottle feeding is to be preferred because the saliva that is produced during the process of suckling the milk aids digestion. Kids will also readily take to feeding on a foster mother when they are put on her teats.
- Male kids for breeding should be fed and handled in much the same way as doe kids, except for the fact that they require a little more milk as well as gram ration than the female kids on account of the larger size they have to attain. Kids with body size below normal should be discarded, as they seldom prove good breeders when mature.They should be fed well at all ages to keep them in good condition, but excessive feeding should be avoided, particularly when they are old because, if fat, they become sluggish and are slow breeders.Where the animal is unduly fat, its grain ration should be cut. At one year, a buck should receive 1.8 kg of grain mixture the allowance being increased by 50 per cent during the breeding season. A liberal amount of fodder should be given. An average of 7 to 8 kg of green fodder per day should be adequate for a full grown Jamunapari buck when entirely stall fed.

6. Care of new born kids by providing guard rails

- Treat / disinfect the naval cord with tincture of iodine as soon as it is cut with a sharp knife.
- Protect the kids from extreme weather conditions, particularly during the first two months.
- Dehorn the kids during first two weeks of age.
- Male kids should be castrated for better quality meat production.
- Vaccinate the kids as per the recommended schedule.
- Wean the kids at the age of 8 weeks.
- Proper selection of kids on the basis of initial body weight and weaning weight should be initiated by maintaining appropriate records for replacing the culled adult stock as breeders.
- Additional feed requirements of lactating does must be ensured for proper nursing of all the piglets born.

7. Health management

Common diseases of goats are given below

Anthrax (*Bacillus anthracis*)

Clinical symptoms: Sudden high temperatures (108°F), loss of appetite, sudden death, in less per acute form, goat may live for a day and develop bloody diarrhea. Preventive care to keep the affected animal separate. Annual vaccination of goats is recommended.

Foot and Mouth Disease

This disease has shown mild effect in adult goats. Vesicles on the tongue, lips, checks, gums, dental pad and on the skin between claws, occasional blisters on teats and under.Lameness is also seen in adult goats, Deaths have been observed in young goat kids care: Isolate the sick goats and vaccinate the herd at the face of outbreak. Wash the mouth of infected goat with mild disinfectants like Alum, Potassium permanganate. Apply boroglycerine on the ulcers.

Hemorrhagic Septicaemi

Clinical symptoms: High fever, difficulty in respiration, coughing and sudden death may occur. Preventive care: Vaccination with H.S.adjuvant vaccine is recommended before onset of rains.

Coccidiosis

Clinical symptoms: Diarrhea., anemia, weakness and death have seen in coccidiosis of goat kids, in old animal milk production is reduced with foetid smell, develops.

Bronchitis

This occurs due to lung worm infection or the feed place in drum, some time goat gets irritation and bronchitis that can be controlled by dampening the mixture with water or molasses water, this way dust in the feed will be reduced. Chronic coughing also occurs in mycoplasmosis.

Brucellosis (*Brucella* organisms)

Clinical symptoms: Abortion in late pregnancy is the feature of the disease, retention of placenta and metritis are common. In male goat, infertility, orchitis and swollen joints are seen. Preventive care: Goats should be tested for brucellosis and isolated or cull the positive animals.

Big-Head

This occurs when goat heavily infested with worm, pale of mucous membrane, along with protruding nictating membrane and scouring, swollen head with Jaundice is seen in lepto spirosis, but the mucous membrane Jaundice often seen with yellow serum exudes from facial skin.

Bloat

This bloat occurs due to use of Lucerne hay, so goat becomes distressed, stamps its feet, bleats, urinate frequently, and walk with stilted action. To prevent bloat, always use dry hay or dry pasture when ever feeding fresh legumes. Goat feed should be prepared by the combination of dry pasture with leaves of legumes. To avoid bloat, peanut oil sprinkled on feed is always helpful.

Cystitis

Symptoms: It is an inflammation of urinary bladder, which shows swelling of value, staining of perineum, frequent straining with small amount of urine, pus and blood in the urine, also show depression, anorexia and excessive thirst.

Eye Conditions

Conjunctivitis: This situation occur due to foreign body like seeds, grass, thorns touch the cornea, which become swollen and reddish. This can easily be cure by washing the eye with normal saline solution (1 level teaspoon of salt plus 600 ml of worm water).

Pink-eye

Is a contagious opthalmia, animals affected by mycoplasmosis, then ocular discharge also be seen. First eyes are washed with saline water, then treatment with 2% sodium sulphacetamide or Terramycin aerosol drops used in large herds. Individual animal is given Gentamycin drops, 2 drops twice a day.

Acetonemia

Is also called Ketosis.This metabolic imbalance caused due to goat fed large quantities of concentrated food.

Abortion

Mostly occur from 6-8 weeks of pregnancy, veterinary treatment is needed to control infertility. This could occur due to drinking water containing salmonella typhinmurium. Abortion can occur in a goat fed on rich clover or trefoil.

Anaemia

Anaemia covers groups of conditions characterized by paleness of skin and membrane. Symptoms can be seen nictitating membrane of the eyes, mucous membrane of the mouth, skin around the udder and vulva.

Arthritis

In goat kids it may be Navel infection arthritis caused by corynebacterium pyogensis, streptococci, staphylococci, while non-superlative arthritis is by birth, caused due to deficiency of vitamins and minerals. In old age, swollen joints from mycoplasmosis, occurs at 3-4 weeks of age.

Dysentery

This show presence of blood diarrhea, it may be a symptom of enterotoxaemia coccidiosis and some plant poisoning.

Enterotoxaemia

Is caused by toxin of clostridium perfringens type D.Clinical symptoms: Severe pain in the abdominal region, excitement, kids jump in the air and fell down, and sudden death is there. For Preventive care kids of 3 months are should be vaccinated and repeated after 14 days. Annual vaccination is required.

Foot rot in goat

Foot rot is a contagious disease, caused by a mix of different bacteria with the essential one being *Dichelobacter nodosus*. There are two main forms of foot rot - virulent and benign, they are caused by different strains of the bacterium *Dichelobacter nodosus*. The bacterial infection cause varying degree of damage to the horn of the foot, leading to lameness and significant loss of body weight. The foot rot diseases are two types, benign and virulent. Both have been started as inflammations seen as moisture, reddening and loss of hair between the toes.

Goat Pox

Clinical symptoms: Slight fever, congested area and vesicles; appear on ear, nose and udder. After vesicles containing pus, dry crust is formed. In goat kids, high temperature is seen, but death occurs before development of skin lesions. For preventive care isolate the sick goat and daily washing the lesions with hydrogen peroxide dilute with equal part of warm water followed by antibiotic cream. Milkers' hand scrubbed in non-irritant antiseptic solution.

Hypocalcaemia

This also called 'milk fever', this occur the day following kidding, affected goats are unable to stand, breath slowly with force, fall into comma, may die , unless treated with injection of calcium arogluconate with magnesium sulphate 60 ml has to be given to the goat.

Hoof trimming

It is necessary, otherwise hooves grow to any shape, which brings extra pressure on leg ligaments and tendons which cause pain and distortion. This cause difficulty in walking, if the portion of hooves curled, then outer horn may accumulate mud and dung, which results into purification and foot-rot. Always maintain correct proportion of toe to heel on the ground. This varies in the front and hind feet. Hooves should be trimmed every 3 months.

Johne's Disease (*M. Paratuberculosis*)

The symptoms are goats show loss of body weight, weakness and emaciation.Frequent diarrhea in some adult goats is also reported. The clinical disease observed after 1-2 years of age in goats.

Mastitis

Infection Acute infection: udder is enlarged, hot painful, milk become watery, inconsistency often with flakes of blood, udder become hard, milk secretion very much reduced. In *M.agalactiae,* the affected quarter become atrophic oranalectic.

Chronic infection

The systems are milk become sours soon, curdles, unpleasant, it need rapid - mastitis test. Preventive care: After washing with $KMnO_4$solution, remove the milk carefully from the affected quarter. Introduce antibiotic 1/mammary infusion, leave for 24 hours and repeat for 3 days. But goat should be tested regularly to detect sub-clinical mastitis.

Vaccination Table for Goats

S.No.	Name of Disease	Primary vaccination	Regular Vaccination
1.	Anthrax	At the age of 6 month for kid or lamb	Once Annually (In Affected area only
2.	Hemorrhagic Septicemia (H.S.)	At the age of 6 month for kid or lamb	Once Annually Before monsoon
3.	Enterotoxaemia	At the age of 4 month for kid or lamb (If dam is vaccinated)At the age of 1st week for kid or lamb (If dam is not vaccinated)	Before monsoon (Preferably in May). Booster vaccination after 15 days of first vaccination
4.	Black Quarter (B.Q)	At the age of 6 month for kid or lamb	Once Annually (Before monsoon)
5.	P.P.R.	At the age of 3 month for kid or lamb & above	Once in three years
6.	Foot & mouth disease (F.M.D.)	At the age of 4 month for kid or lamb & above	Twice in a year (September & March)
7.	Goat Pox	At the age of 3 month & above for Kid	Once Annually (December month)
8.	C.C.P.P	At the age of 3 month & above for Kid or lamb	Once Annually (January month)

Protection against diseases

- The farmer can check few visible symptom and, first-aid which not only prevent the further deterioration of the health of the goat, but it save the life. The farmer should understand the complication of disease, and reach to the specialist or health worker as early as possible. In case of danger of spreading contiguous diseases in the herd, immediate and time action of seeking the advice of health specialist can save the greater economic losses.
- Be on the alert for signs of illness such as reduced feed intake, fever, abnormal discharge or unusual behavior.
- Consult the nearest veterinary aid centre for help if illness is suspected.
- Protect the animals against common diseases. and de-worm the animals regularly.
- In case of outbreak of contagious diseases, immediately segregate the sick animals from healthy one and take necessary disease control measures
- Examine the faces of adult animals to detect eggs of internal parasites and treat the animals with suitable drugs.

- Provide clean and uncontaminated feed and water for minimizing the health disorders.
- Strictly use the recommended vaccine schedule as given in vaccination table.

Conclusion

Goat rearing is a good source of capital storage, income and employment generation and house hold nutrition. Goat contributes milk, meat, fiber, skins and manure to the subsistence of small holders and landless rural poor. Majority of the households in rural areas are below poverty line and most of them belong to landless agricultural labors, marginal, small farmers and rural craftsperson. The incidence of poverty and unemployment is relatively more acute in rain fed farming areas.In such areas goat rearing enterprise could be adopted and expanded by the rural poor with low land base. Initial investment is less for goat farming, thus land less labourers, marginal and small farmers can also adopt the goat farming. It is good earning source if farmers adopt the goat rearing in the form of farming system approach.Proper selections of breed with good management practices are crucial in goat farming enterprise to enhance the profitability of the farmer.

References

Birthal, P.S. and P.K. Joshi 2006. High Value Agriculture for Accelerated and Equitable Growth: Policy Brief. No. 24, National Centre for Agricultural Economics and Policy Research, New Delhi, December.

Charray, J., Humbert, J.M and Levif J. 1989. Manual of sheep production in the humid tropics of Africa. Translated by Alan Leeson. Published by C.A.B International. Technical Centre for Agricultural and Rural Cooperation.

Dalgado, C., M. Rosegrant, H. Steinfedl, S. Ehui,and C. Courbois 1999. Livestock to 2020. The Next Food Revolution. Food, Agriculture and Environment Discussion Paper 28, IFPRI, Washington.

Jagdish Prasad (2014). Goat, Sheep and Pig Production and Management; Kalyani Publishers, Ludhiana (Punjab) p 496.

Oberoi, R.C., Morti, T.V. anti Sharma, A.K. 1992. Comparative economics of sheep and goat rearing (A study of tribal farms in Western Himalaya). Livestock Adviser 17(1):25-30.

Rawat S.K., Narayan S., Awasthi M.and Dwivedi S. 2016. Socio-Economic Analysis of Goat Rearing Farmers in Mahoba District of Bundelkhand. *Agro Economist - An International Journal* (2): 2, 29-34.

Shalander Kumar 2007. Commercial Goat Farming in India: An Emerging Agri-Business Opportunity. Agricultural Economics Research Review Vol. 20 (Conference Issue) 2007 pp 503-520.